Nature and Culture in the Andes

Nature and Culture in the Andes

Daniel W. Gade

1999

The University of Wisconsin Press

The University of Wisconsin Press
2537 Daniels Street
Madison, Wisconsin 53718

3 Henrietta Street
London WC2E 8LU, England

5 4 3 2 1

Printed in the United States of America

Library of Congress Cataloging-in-Publication Data
Gade, Daniel W., 1936–
Nature and culture in the Andes / Daniel W. Gade.
302 pp. cm.
Includes bibliographical references and index.
ISBN 0-299-16120-X (cloth: alk. paper)
ISBN 0-299-16124-2 (pbk.: alk. paper)
1. Indians of South America—Ethnobotany—Andes Region. 2. Indians of South America—Ethnozoology—Andes Region. 3. Indians of South America—Andes Region—Social conditions. 4. Human ecology—Andes Region. 5. Human geography—Andes Region. 6. Human-animal relationships—Andes Region. 7. Andes Region—Social conditions. I. Title.
F2230.1.B7G33 1999
581.6′3′098—dc21 98-49023

to Christopher Pierre Gade (b. 1969)

Contents

Figures

Tables

Preface

This cultural history of nature and the ecological history of culture focus on land and life in an unusually distinctive realm. Western South America from Ecuador to Bolivia is physically dominated by the Andean cordilleras, whose highland cultures have greatly influenced the western coastal and eastern piedmont zones at lower elevations on both sides. The seven main chapters cover a range of interrelationships between people and biological phenomena in which a larger configuration emerges from an empirically derived focus. They are the result of a curiosity focused on quite different aspects of the living world that have grown out of periodic observations in South America over a span of 35 years.

Remarks in the introductory and concluding chapters suggest the perceptual screens that have filtered the content and approach to the main body of the work. But the intent is more ambitious than that, for by writing out of my psyche I have also sought to grapple with the forces that have led to the construction of my research and to larger philosophical perspectives. At another level, the book cultivates a multivocal reflection on knowledge as a boundless garden of ambiguities and encroachments. Trained as a cultural geographer, I believe that many practitioners in that specialty will recognize the contents of this volume as reflecting an identifiable approach and attention to the subject matter. They are firmly rooted in the materiality of the landscape, whose combination of elements manifest strong regional character. But the busy intellectual intersection of this book's topics has heavy epistemological traffic. Several chapters can be characterized as historical ecology and others as ethnobiology. Acceptance of that overlap and others makes it more possible to concentrate on the central idea of holistic understanding.

Field observations have inspired this volume, but the bibliography suggests the importance of retrieving foundational knowledge. I want to acknowledge the help received at the following repositories: the Bailey-Howe and Dana libraries of the University of Vermont; the Library of Congress in Washington, D.C., where an editorial association with the Hispanic Division has enabled me to take special advantage of their vast holdings; and the National Library of Medicine in Bethesda, Maryland. Two summer research fellowships in Latin

American studies enabled me to benefit from the collections at the University of Pittsburgh Library in 1985 and the Olin and Mann libraries of Cornell University in 1987.

In addition, several archives yielded relevant documents. Much insight into the colonial Andes was derived from a year's work at the Archivo general de Indias in Seville in 1988–1989 funded by the Comité conjunto España–Estados Unidos. The Archivo nacional de Bolivia in Sucre, whose director then was Dr. Gunnar Mendoza, yielded substantial documentation for two chapters. In addition I have worked on the unpublished Carl Sauer Papers at the Bancroft Library, University of California, Berkeley, and have delved into the National Anthropological Archives at the Smithsonian Institution in Washington, D.C. Two of the four year-long sabbatical leaves and two other research leaves from the University of Vermont were partly devoted to field research in South America and also to library retrieval and writing this book. I am also grateful to my home institution in Burlington for other kinds of support over the years. The Department of Geography provided teaching-load reductions and the College of Arts and Sciences and Graduate College periodically awarded funds for my travel and maintenance.

Encouragement has come from many people. I wish especially to acknowledge the help I have received from James J. Parsons (1914–1997), Frederick J. Simoons, William M. Denevan, Hildegardo Córdova Aguilar, Karl S. Zimmerer, and Erich D. Langer. Assistance on specific themes has come from the following individuals: chapter 1 from William Speth, of Ellensburg, Washington; chapter 4 from Gustavo Cárdenas P., G. and P. Meruvia, of Mizque, Nestor Prudencio, of the Centro de salud de Cochabamba, and the Franciscan fathers of Mizque Parroquia; chapter 5 from Frederick Simoons, of Spokane, Washington; chapter 7 from Gregory Deyermenjian, of Watertown, Massachusetts; chapter 8 from Elizabeth Benites Estupiñan, of the Instituto de higiene y medicina tropical "Leopoldo Izquieta Pérez," and Edison Villamar, of the Dirección provincial de salud pública de Guayaquil; and chapter 9 from Richard Palmieri (1945–1997). Robin Whitaker skillfully copyedited the whole work.

Most of all, I am indebted to Mary Scott Killgore Gade, who has been a frequent field companion, a tough but loving critic of much of what I have written, and the cartographer who prepared the final maps. Over the years she has also provided many of the conditions that have enabled me to execute this long-term project. Of course, I am responsible for any errors.

Nature and Culture in the Andes

1
Reflections and Trajectories

Strange, that having read almost everything on the City of the Sun I lacked so largely the appreciation of the convergence of nature and culture that exists there.
—Carl O. Sauer from Cusco, Peru, quoted in West, *Andean Reflections*

People, plants, animals, the land, and everything else on earth are interlaced in singular and sometimes dysfunctional ways. To understand the intricate connections of the human elements with the other living components of the earth is to understand the deployment of *Homo sapiens* over the earth. Inquiry into this vast realm takes on different styles and goals. Science, with objectivity as its cherished ideal, has informed us much about these relationships. The chapters in this volume broaden the epistemological perspective by recognizing that the subjective, whether acknowledged or not, enters into all reporting on biotic-anthropic articulations. New layers of insight can emerge when respect for accuracy combines with an empirical method, a reflexive attitude, and willingness to speculate from the evidence. Holistic understandings emerge from that to counteract the reductionism implicit in Newtonian science.

These vital connections are especially abundant in western South America with its sharp altitudinal zonation, its ways of life that reflect both the unique fusion of indigenous and Hispanic inheritance and a complicated relationship with the resource base that remains in large part paleotechnic rather than modern. Some of these confluences of people in their environments are important in agriculture, pastoralism, forestry, and public health; others are scarcely known at all. Scales vary in magnitude from the regional to the local. In each case study, the various elements interlock to create a larger configuration which can be called the culture/nature gestalt. (A fuller explanation of the special meaning and treatment of this term follows in the next section.)

Culture/Nature as an Idea

Culture and nature as ideas seem to be such discrete ontological zones that each has been put into its own little box. In this classic dichotomy, nature becomes that part of the universe without apparent human artifice, whereas culture is learned behavior which humans acquire living in particular groups. Lévi-Strauss, who built an anthropological career by dichotomizing all manner of phenomena that exist as continua, described nature as being a state of unconsciousness and culture as one of consciousness, insisting that this division is a universal structure of the human mind. But if that were so, we would not find groups such as the Achuar people of eastern Ecuador, who consider plants and animals to be living beings with souls (Descola 1994). It seems plausible that at one time all humans unconsciously identified with their enveloping milieu and its individual elements. At some point ego consciousness intervened to differentiate the realm of the human from that of the nonhuman. The Neolithic era, with its new agricultural way of life, opened a psychic gulf between people and everything else. Domesticated foodstuffs gave humans a certain power over their environment. The subsequent rise of chauvinistic sky gods and priest-kings introduced powerful intermediaries who had a vested interest in the separation between humans and all else.

The polarity later hardened into an abstraction. Human language congealed the differentiation. The ancient Greeks distinguished between *physis,* that which is not tied to human agency, and *nomos,* that which comes from the human capacity to conceptualize. The biblical foundation of Western civilization forced the separation of humans from nature by insisting on a single divinity—no competition allowed—that was beyond ontological comparison with any earthly phenomena. St. Augustine set the religious tone for Christianity when he emphasized the biblical idea of a created world that on the seventh day was followed by a created humanity. Early Christianity suppressed the *genius loci* that inhabited springs, forests, and caves in ancient Rome. By checkmating the pantheistic impulse, the divine was relegated to churches, not to the sacred forest.

Knowledge of the natural world as something existing independently from our relationship with it hardened with the rise of rationalist thought in the sixteenth century. Nature became reified as a distinct category of phenomena. Our use of language has made us falsely believe that nature is a material entity. Expansion of cities and the artificiality of urban life contributed to the separation from nature. Bateson (1979) viewed the mental separation of nature from culture as a product of Western individualism and dichotomous thinking. As the technological capacity to control nature grew, the disconnection of phenomena into two discrete realms followed. The practice of science beginning in the Enlightenment did much to objectivize nature as a separate category.

In academic circles, nature versus culture became institutionalized early in the twentieth century. Anthropologists favored the split for pragmatic reasons; if culture could be claimed as its legitimate bailiwick, anthropology could be an independent field free from intellectual poaching (Horigan 1988:9–33). If nature was that to which the laws of biology applied, culture was construed to be knowledge which biology could not explain. The false dichotomy was intellectually entrenched, and the symbiotic relationships between people and nonhuman elements were simply ignored until the partition finally became intolerable. The anthropological corrective to that exaggeration dates from the formulation of cultural ecology in 1950. The rise of geography as a discipline had in its formative history the opposite assumption from anthropology. Interlacing of humans and the earth was seen as the proper terrain for geography's autonomy as a field. As early as 1907 human ecology was proposed as the true content of and justification for geography. The approach faltered, however, when the physical environment came to be seen as determinant in explaining human culture and economy.

Configuring the two elements is a problem of language; our use of language has also made us falsely believe that nature is a separate material entity. The expression *nature and culture* conveys two distinct modalities, whereas *nature-culture,* hyphenated but without the conjunction, implies a bond of some sort. Latour (1991:140) offers a hyphenated plural, *natures-cultures,* to suggest diverse cultures adapting to different natures. I, on the other hand, am using a unifying virgule in this term—*nature/culture* or *culture/nature*—to communicate a mutually interactive skein of human and nonhuman components rather than opposing polarities or separate entities.* Conceptualized in that way, *nature/culture* is a holistic configuration that has the additional advantage of foiling the binary opposites implicit in cultural and environmental determinism. Its gestalt character comes from a unity that transcends the individual parts.

Nature/culture as interlocking can be rationally grasped by observing how organisms in landscapes and on the entire earth have undergone progressive humanization. Technological advances have accelerated that trend but are less important than most people think. My own landscape reading suggests that at least 60 percent of the world's land surface unambiguously shows human impact and that another 20 percent manifests that impact in subtler ways. Potsherds under tropical forests and stone walls deep in the Vermont woods are powerful reminders that seemingly natural vegetation or edaphic compositions may actually reflect past land uses.

Humans have transformed particular species of plants and animals through selection. They also domesticated the environment especially through the use of fire. Both kinds of interventions enormously enhanced material survival and the possibilities for survival. Landscape modification may also have had

*The use of the virgule to indicate a combination is reserved for this term alone, to convey the special sense of the inextricability of the two components. In other terms that involve a simple combination of elements, the conventional hyphen appears. The slash has also been used conventionally elsewhere to indicate interchangeability or the sense of "or."

an aesthetic component. René Dubos (1972) proposed that the carefully managed and groomed landscape of northern France is more appealing to the eye and soul than anything nature has evolved by itself. Even wilderness has become a human creation through our management of it and ultimately through the conscious societal decisions to preserve it as such. Promoters of the untrammeled wild carry suitcases full of dubious assertions and *ad hominems.* Implicit in the wilderness idea is an anti-human bias that seeks to present nature as the antithesis of culture. White (1995) has suggested that environmental theorists who disdain rural work, that of farming, ranching, fishing, and logging, promote false dichotomies between people and the nonhuman part of the world.

Humankind, to be sure, has also created disharmonies, some so serious that they have compromised the durability of regions and countries. The devastation that people have wrought dominates most North American discourse about the environment. Removal of vegetation through cutting or fire has sometimes destroyed the soil and water capabilities. Either through direct assault or habitat alteration, hundreds of plant and animal species have become extinct, endangered, or rare. Ironically today these trends are greatest in the least industrialized countries of the world, where pressures are high on rural resources. But to set up a dichotomy between humans and "wild nature" reinforces a false view of life on earth. At another level, *Homo sapiens* is a biocultural entity, a "two-in-one creature" with a brain that functions with regard to both. Landscapes, the human animal, an important group of plants, and the other animals all form living examples indicating that the dichotomy of nature and culture is a polarity only in the human mind.

Place, time, object, and process provide four organizational frameworks within which to assess the intermingling of people with other living organisms. Focus on a place or area imparts a spatial coherence to the interrelationships. Not only rural places, but also cities, those prime creations of human artifice, manifest associations of organic life. A nature/culture configuration can also draw principal attention to a chronology that emphasizes the historic or prehistoric past as a way to capture the present. Third, an object or group of objects provides a focus: a tree, but also the vegetation; a crop plant, but also a cultivated field; a microscopic parasite, but also the disease. Another part of this gestaltic kaleidoscope concentrates on domestication, introduction and diffusion, and landscape changes as processes that place the focus on deciphering the transitions in the nature/culture relationship. Each of these frameworks occurs in a rural-to-urban continuum, though the biological organisms at each end are not the same.

Grasping the full meaning of nature/culture relations requires crisscrossing diverse epistemological spheres. A composite of "science" and "nonscience" that combines the empirical and intuitive offers that richly textured perspective. Johann Wolfgang Goethe (1749–1832), known best as an exceptional literary genius, had a vision for such an alternative kind of knowing which brings different phenomena together to clarify their intrinsic relatedness. In Goethe's science, the aim is to see comprehensively. To reach that stage of enlightenment, the constituent parts are examined in their singularity (*Eigentümlichkeit*). Pure description of the parts without the infusion of deductive constructs, hypotheses, or theories offers the possibility of relatively unencumbered results. In Goethe's empiricism the observer learns to trust his or her senses and to consider morphology with special care; form is often the best way to understand function. Synthesis of these observations leads to a holistic consciousness in the form of a gestalt whose unity transcends the sum of the parts.

The quest for understanding and self-knowledge sets the Goethean vision apart from modern science (Bortoft 1996). Understanding comes through intuition when the unity is grasped. Cause-and-effect explanation of normative science offers the logic of reason but does not encourage seeing comprehensively. Self-knowledge comes from immersing oneself in a constant interplay between the phenomena under study and self-reflection. In its search for holistic consciousness, Goethe's method provides a needed corrective to Newtonian reductionism. The Goethean paradigm has special applicability to geography, especially cultural geography, as well as to natural history, but also to other fields that seek synthesis as a goal (Seamon 1978). Goethe, who introduced the gestalt concept to nineteenth-century Western thought, offered the alternative view of humans as part of the biosphere. Humboldt (1850) went even beyond that to conceptualize the natural history of the mind.

Charles Darwin rightfully removed anthropocentrism from its pedestal, but social ideologues later distorted some of his ideas about selection. Their nefarious effects were felt in the first half of the twentieth century when spurious generalizations prevailed about environmental control over human life and the justification of war and extermination of minority peoples. Reserve about holism has eased with the rise of sociobiology (Wilson 1978). It builds on Darwin's natural selection to explain many aspects of human behavior and to elaborate how biological and cultural traits coevolve. Wilson (1998) has subsequently extended this line of thinking by asserting that to gain true self-awareness we must acknowledge human life as a physical phenomenon.

Holistic revival is also contained in Lovelock's (1995) formulation of an intricate and self-regulating system of interaction on the earth between organic and inorganic elements. This so-called Gaia hypothesis does not follow all the Darwinian logic of selection, but if evidence continues to accumulate, it may offer the key breakthrough in conceptualizing human culture and physical process as part of an interlocking whole. These two Copernican perspectives have enabled people to grasp the nature/culture gestalt at an individual, intuitive level and integrate it into their being.

Steps in Understanding

Formative Influences

Awareness of the investigator vis-à-vis the investigated can also be part of the discipline of knowing, for it brings out meanings and connections that can lead to a more richly textured understanding of each. An author willing to suspend the convention of impersonality may enable readers to comprehend formative factors so as to evaluate better the meaning or legitimacy of the assertions. Readers may also note ironies missed by the author. In delving into a narrative of the self, I have had to overcome a stubbornly rooted notion about personal reference in the text. Aspiring scholars of the 1960s were admonished to write themselves out of the work under the illusion that doing otherwise compromises objectivity. Once that ethereal state of scholarship is exposed as a chimera, it opens the door to a retrospective reflexive stance to connect author, experience, and the printed text.

Personal history gives insight into the scholarly explorations in this book. Events beyond one's control, impressionable experiences, and making free-will choices are all involved as backstage elements of this montage. Growing up in North America starting in 1936 positions me as a bearer of certain cultural values. Living in three different locations and three different kinds of places—village, city, and town—spanned my childhood in segments of six years each. Experiences in those places helped to shape later perspectives of theme, topic, approach, and philosophy.

My sense of wonder about living things of all kinds had emerged already in the earliest years. The village community in western New York State into which I was born was founded by refugees from the European Enlightenment. It had two churches, schools, and cemeteries and one volunteer fire department, general store, and blacksmith shop. The premachine era of transportation was then ending as animal-pulled farm implements and delivery carts dwindled and finally disappeared. The cadenced clanging of Ostwald's forge across the village square still reverberates in my brain as a metaphor for a placid time gone by. Wonders of the countryside were close at hand in the surrounding small-farm landscape dominated by orchards and grains waving in the wind (the township is, in fact, Wheatfield). My grandfather's 6 ha (14 acres) within 2 km of the Niagara River represented a primal enclosure, even though I never actually lived there. Its polycultural array thrived only because the fertilization applied over the years had made the clay soil productive. That dirt was as much cultural as it was natural. A vineyard and large strawberry field got the major investments of labor, but it was the barnyard geese (which bit me) and wild rabbits (which I couldn't catch) that conjure up that earliest stage of seeing. Most of the buildings were nondescript, but hidden beneath the clapboards of the house was a half-timbered and clay construction whose vernacular inspiration came straight out of the Uckermark.

The powerline coming from the hydro station on the Niagara River set a boundary to that horticultural perimeter. Five kilometers distant, its waters poured over the dolomitic escarpment which had also been a maelstrom of family tragedy: less than a year after I was born, two uncles swimming together were killed by lightning, one of whom was swept into the river's swift current and over the cataract. The roar of the falling water could not be heard from the farmstead, but its presence asserted itself nevertheless in another way. Niagara's cheap power encouraged industry that opened alternative opportunities to farming.

Unspectacular but illuminating changes occurred over the years, but their larger ecological implications I understood only later. Ring-necked pheasants, which seemed so well adapted that no one imagined they were introductions from Asia, disappeared after my grandfather stopped growing maize. Bluebirds, once yearly nesters in the bird houses, also left but for different reasons. A major change had occurred in eastern North America; by the 1960s, 90 percent of the New York bluebird population had disappeared, casualties of competition from cavity-nesting starlings and English sparrows and the use of DDT.

I remember too a pond scooped out of the sticky ground at a time when farm extension agents promoted them. Within 2 years, the bare ramparts of dirt surrounding that water-filled hole were covered with ruderal plants; in 15 years, cattails had come in on the margins. Silt later completely filled in the pond. In less than half a lifetime, full ecological succession had occurred. Not until I started to read about ecology in the 1950s did I realize how classic an ecological laboratory was this transformation through successional stages.

In the 1970s the farm, by then already abandoned, was sold for potential industrial development, and the buildings, including the classic half-timbered cottage, were removed. Once used in intensive crop production that had overcome the limitations of a mediocre substrate, the soil finally supported no one. Such a total land-use transformation has been repeated several millions of times in North America in this century. The strongest memory of it remains a nostalgia for the polycultural ideal and a technology of human scale.

From that quasi-rural community in Niagara County, I moved with my family to Pittsburgh. The classic industrial city of steel mills geared up for wartime production had a multiethnic population, which presented to me a new range of cultural expressions. Our residential neighborhood between tony Shadyside and working-class East Liberty had as its main biotic elements ailanthus and horse chestnut trees and a large resident rat population. Such a stark inventory of living things raised basic questions but only much later.

What was it about exotics that enabled their seedlings to volunteer so successfully? How did the backyard rodents survive the neighborhood war on rats, which used poisoned hamburger and burrow fumigations? My combined fascination with and repulsion toward bold but untamed rats surely came from being cornered by one on a dark back porch when I was 10 years old. In 1985, four decades after having moved away, I returned to find an elegantly gentrified neighborhood. Several horse chestnut trees had survived the city's once soupy air, but one resident with whom I spoke refused to believe that his neighborhood had ever harbored rats.

The dioramas at the Carnegie Museum of Natural History endlessly fascinated a nine-year-old on wintertime Saturdays in the city. These still lifes had an impact on me that people weaned on television can scarcely appreciate. Each detail of a plant's and an animal's morphology could be endlessly scrutinized at the same time as the ecological coherence of the whole scene could be intuitively grasped. Moose munching aquatic plants of the northern muskeg and the skunk robbing a bird's nest in the eastern mesic forest conveyed the web of nature as no book or photograph could. By comparison the articulated dinosaur skeletons in adjoining halls seemed preposterous in both their appearance and their time on earth. Though visited rarely, the zoo was also of compelling interest. Watching living exotic creatures move about compensated for their rankly artificial settings. Zoos stunt our understanding of animal behavior, but they allow inspection of the essentials of external form in a way unobservable in the wild. The tapir was a special favorite to examine for its strange appearance. Much later I came to appreciate the horse analogies with European animals that sixteenth-century Spaniards devised to describe the tapir for the first time.

Between the ages of 11 and 18 I lived in a placid county seat of 16,000 people in Indiana. Situated on a lake-studded moraine at what had once been the doorlike juncture ("La Porte") between prairie and forest, it offered new landscapes to discover. High adventure at age 12 was to bike on Sunday afternoons to Rolling Prairie, a village 16 km to the east. The succession of cornfields, dairy cows, and white frame houses on that stretch of the old Lincoln Highway remains today my rural image of the American Midwest. Within a few years, auto dealerships, gasoline stations, and a drive-in cinema had taken over part of that roadside.

During those years I was decidedly not in the teenage mainstream. I enjoyed being in the woods to test my self-reliance rather than hanging out at the soda fountain. Much of the time I was disconnected and self-absorbed, and my offbeat curiosity and uncommon introspection put me at odds with my reference group. Classroom lessons did not much engage my interests, which covered an encyclopedic range. The exotic was found in books about great adventurers and descriptions of places, but the prosaic things also raised issues that to most of my friends were not even worth examining: the success of thistles in pastures, the arrangement of mole hills at the golf course that I noticed while looking for my lost ball, and the volunteer watermelon seeds that sprouted to form a tangle of vines in the beach sand only to be killed off by the first frost. Simple observations of wayward flora in human association formed the germ of reflection that decades later, but for different places, was communicated to an audience (Gade 1976, 1991b).

Toward a Certain Frame of Mind

Knowledge transmitted in universities helped put phenomena into meaningful frameworks: climate according to the Köppen classification, landforms as a residue of continental glaciation, and plants categorized into the Linnean system. Still, learning outside the classroom left the greatest impressions, especially weekend reconnaissance hikes not related to my formal courses or professorial intervention. With topographic maps in hand, especially the USGS sheets at a scale of 1:62,500, I reconnoitered a piece of land to identify terminal moraines, locate abandoned rural cemeteries, and trace the courses of obscure creeks to their sources. Occasionally I discovered a map error and thus learned to appraise all maps critically.

Three ecological sites in northern Indiana evoke for me real entities, uncommon habitats, and defining experiences in fostering inquiry. The Indiana Dunes at the south end of Lake Michigan were a special place for exploration, especially where the sand invaded the maple-basswood stands, full of sassafras in the understory. Thanks to the key in Peattie's 1930 guide to duneland flora, a model of taxonomic clarity, I delighted in discovering at a college age the dunes' botanical diversity. Years later, I learned that H. C. Cowles (1899), one of America's pioneer ecologists, had made his groundbreaking studies on ecological succession there in the dunes.

Less than 20 km south of Lake Michigan is Pinhook Bog. It lies hidden in a cold hollow once occupied by a huge block broken off the last glacier, which deposited the surrounding Valparaiso Moraine. My first undergraduate research project focused on the vegetation of this bog, whose plants are typical of wetlands only in climates far to the north: sphagnum moss, pitcher plant, tamarack, and cranberry. To understand this plant cover, it was necessary to delve into other aspects of the site. Advance and retreat of the Wisconsin phase of the glacier, plant migration, highly acid substrate, and microclimate were part of the total story. The curmudgeon who owned this magnificent little relict did not appreciate visitors, even those with scientific interest. Determined to pursue what had galvanized my imagination, I waded through water and then gingerly treaded over a dangerously thin part of the quaking mat to enter the bog unseen. Pinhook Bog, now owned by the National Park Service, is transparently coherent, but one has to study its various elements to understand the meaning of the whole. Goethe taught that lesson 150 years earlier with poetic imagination and grand allusions. Perhaps more than anyone in the history of science, Goethe also emphasized the inappropriate compulsion to theorize that unwittingly accompanies even the most fundamental act of perception.

Another wetland area was much different and less easily understood. Kankakee Marsh covered the wide floodplain of the river of the same name before the marsh was drained for agriculture. I became fascinated with this area in 1957, when I learned about Indian burial grounds on its sandy knolls. That led to my imagining how this area had looked to those Indians who had once lived there, an approach I later appreciated even more when I read Carl Sauer's (1941) sage advice to look at the cultural landscape through the eyes of its former inhabitants. To make further sense of Kankakee Marsh, observation was combined to good effect with reading. Right at hand was Alfred H. Meyer's (1936) monograph on the historical geography of this wetland. Mikesell (1978:7) has evaluated it as one of the classic geographical studies of sequent occupance. Meyer's snapshot descriptions of different periods engaged the mind, but at another level, his approach incorporated a fatal flaw. By implication, it assumed an upward cultural evolution toward continual human progress in its transition from a territory of aboriginal hunting to one of Euro-pioneering before the Civil War and modern American corn farming on the drained areas in the twentieth century. Only in North America, with its short and peculiar history, has that spiral of assumptions about progress been expressed so simplistically.

Observing the Kankakee helped me to cross a threshold of undergraduate understanding about what we see with our own eyes and how observations are abstracted in a written text to fit a particular conceptualization. But it also demonstrated how written sources raise questions about and show the gaps in what we know. Although sequent occupance lost its appeal, that experience aroused in me a concern for the past which was to become vital for my appreciation of place. Meanwhile, history courses did not excite me; those that I knew about put the nation-state or the great political leaders at center stage. Neither social history nor environmental history had yet hatched.

Graduate work reinforced my belief that observation is essential to geographical understandings, even though these exercises took place in the charmless rectangularity of east-central Illinois. Methods learned there in a summer field course improved some observational skills which were applied in the fall term to the intensive study of one square mile of countryside. Particulars of modern farming on the three properties piqued my interest less than did the remnants of Osage orange hedgerows along the property margins. Before barbed wire was invented, these little trees with inedible fruit were planted in the nineteenth century to enclose fields and properties. Thirty years later on a visit, that same midwestern rectangle was unrecognizable. Suburban housing had extended far beyond the Champaign-Urbana of 1959. Barns and hedges had disappeared, and the rich black earth that had nourished such prodigious crops was feeding trimly cut grassy swards with no redeeming food value.

The active discipline of looking and the passive one of seeing gave me confidence that I could make essential distinctions, systematize facts, and eventually derive from them other meanings. Local field studies were not part of my next graduate school program at the University of Wisconsin. With my eye on far-off places, little was learned about Wisconsin or even very much about the city of Madison. Instead, those three years near Lake Mendota unleashed an endless flow of questions about the connections between biological and cultural phenomena. Linkages were pursued into recondite realms, from the dispersal capacity of coconuts to the origins of milking as a cultural trait and the hallucinogenic value of human urine from Siberian tribal people who had ingested fly agaric. Stretching my mind in unfamiliar directions to gain a coherent and sophisticated picture of the world beyond the modern industrial society came from a new-found appreciation that the arcane was also knowledge worthy of pursuit for the larger ideas it enfolded. That group of scholars conveyed the subliminal message that nothing in the world is insignificant and that it all depends on what Goethe called the "*Anschauungsweise,*" or how one views something.

Projections of the Mind and Soul

South America called me primarily for the promise of adventure in the spirit of Alexander von Humboldt (1769–1859). No stronger exemplar for original inquiry can be found than Humboldt, whose extraordinary curiosity about so much of what he saw was matched by perseverance, intellect, and a vision of humanity sharing a co-natural world. In Humboldt's empirical and synthesizing scheme of inquiry, travel is necessary, but the facts recorded are nothing until viewed in their interconnectedness. My exploratory impulse in which the unknown is a challenge became part of a larger configuration of thriving on ambiguity and uncertainty with no fondness for structure or rules.

Dissertation research presented the opportunity to learn first-hand a piece of the world's cultural and environmental diversity beyond North America. To me the importance of exotic places was not just the scientific, but a more inclusive meditation that also integrated nonscience and a mythopoetic view.

To see things beyond my culture and time could not be accomplished by contemplating the navel. One may never satisfactorily sort out for oneself the enigma and ambiguities of life or be able to accept the fact that we are strangers to ourselves. But to confront these things is to open one's mind to a multidimensional existence and a determination not to accept life in a predecided manner. It is not a question of seeking "truths," which is a prime conceit of platonism and a theologically ordered cosmos that a revealed order exists independent of the human will and that a lack of received moral guidance leads to personal disintegration. Rather, it is the sense that life without myth is impossible, for what has happened in the past happened many times before. In that eternal return lies many clues to human life on earth.

Larger meaning comes from self-knowledge, which occurs when one becomes an attentive microcosmic participant in the macrocosmic activity of life on earth. My readings of anthropological accounts of spiritual journeys into the heart and soul involved more than simply acquiring ethnographic information. They resonated with my own sense of the hierophantic and the notion of certain self-selected individuals as active seekers of knowledge. Before gods appeared in culture history, there were shamans. Individuals of shamanic inspiration have emerged in perhaps all societies, not as the result of diffusion but as archetypes of the collective unconsciousness located in the neurological substrate of the brain as a result of several million years of human evolution.

Shamans have been people with an inner curiosity about their surroundings and who have become repositories of knowledge and provide interpretations of it. Esoteric information about biotic elements is put to practical use, for knowing more about animals and their habits makes hunting more successful. Magic is merged with that knowledge to ensure a bountiful catch, cure disease, and divine the future.

Ultimately shamanism as a feature of primitive societies has been a response of the human psyche to an uncertain and threatening environment that cannot be controlled. A shamanic epiphany is religion in its purest form, for it provides direct experience of appeasing the secret intelligent powers without being filtered through a bureaucratic overlay. Shamans have sought to unfold the cosmic imago to the universal enigmas of origins of the earth and of people. To do this, they have to grasp intuitively the interconnectedness of phenomena. Indeed, budding shamans may have been self-selected for their holistic minds, which come from the unusual capacity to balance the two hemispheres of the cerebrum. The right brain thinks in terms of symbols and images that create patterns as wholes, whereas the left brain uses scientific observation and logic in its cognitive process.

Several other aspects of the archetypal shaman fit my own sense of self-identification: wide-ranging interests, affinity for wild or remote settings, psychological dissociation, special interest in altered states of consciousness, and a larger world view than most people. Other aspects of the shamanic archetype so frequently reported in the ethnographic literature—introversion, reflection, empathy, capacity to heal oneself, and listening to voices in the psyche—fit me perfectly. In a paleolithic incarnation, I can see myself as a person who has undergone what Campbell (1991) called a hierophantic realization of seeking larger meanings in the surrounding phenomena.

Instead, that questing took a path more in accord with the values of mid-twentieth-century America. A predilection for discovery had solidified already in my days as an undergraduate student. Geography had become the rubric in which I chose to pursue this predilection, for I had appreciated quite early that geography captures the breadth of the mind in its integration of humans and the environment. As it has for some other determined convergers of seemingly disparate knowledge, geography became my intellectual center of gravity. As one topic after another piqued my interest, I made note of them and filed them away literally, a procedure that Mills (1959) asserted was itself intellectual production. Some files became dead ends of incoherent facts, but many of them eventually demanded attention because they were taking up so much space in the drawer. Free writing about the information in those folders had a therapeutic value in exorcising demons, but gradual emergence of larger ideas about them had the advantage of letting the facts speak for themselves without a blatant a priori bias.

That inductive sequence is probably what Goethe meant by letting the facts themselves speak for their theory. Intrinsic interest, not vogue, prompted these topics, and few of them corresponded to the clusters of set themes that seem to grip scholars in every generation. Every file started with pure self-direction of something I had observed or in a few cases read. In some cases direct encounter had an intellectual impact, such as my own contraction of falciparum malaria or discovery of mucocutaneous leishmaniasis in a wretched *campesino.* Often they triggered extensive reading programs well beyond the narrow confines of the original topics. On several occasions, the topic totally energized me to return to the field to get my story by meeting all the local pooh-bahs, fending off vicious guard dogs, hiking into lost canyons, or crossing a rushing stream on a hand cable. The strange power of curiosity motivated risk-taking, overcame fear, and kept at bay what Nietzsche called the "troglodytic minotaurs" which inhabit the subterranean realms of the self always ready to cast doubt.

Curiosity is a general human trait, but it has not always been encouraged. Early Western societies were not hospitable to the inquisitive mind, reasoning that the secrets of nature were divine mysteries about which questions should not be formulated. St. Augustine scorned the "passion for knowing unnecessary things" (Eamon 1994:59). Medieval negativity about such things reached its most precise definition in the thirteenth century when St. Thomas Aquinas defined *curiositas* as an unhealthy appetite for finding out the unknown. He contrasted it with *studiositas,* or the worthy devotion to knowledge already cataloged. In the seventeenth century curiosity shifted from being a contrary trait to being an admirable one. *Curiositas* belongs to civilization's progression, but some Western societies have valued it more than others have. Living in both France and Spain for extended periods made me appreciate the high value that French culture places on curiosity. But the purest model of an inquiring mind that was both free and disciplined at the same time was for me an American geographer of personal memory.

Exemplar and Framework

Carl Sauer as Inspiration

During my academic apprenticeship, Carl Ortwin Sauer (1889–1975) emerged as someone whose ideas called for sustained reflection. Professor of geography at the University of California between 1923 and 1956, his distinctive inquiries, based squarely on German geographical thought, gave rise to American cultural geography and what is frequently called the Berkeley school. Sauer himself classified his work in different ways—historical geography, cultural geography, biogeography—which reflected his unwillingness to poke knowledge into mutually exclusive subsets. Even between disciplines, subject matter boundaries were of little concern to him, which may explain why he never sought to stake out a geography empire at UC–Berkeley. His California post gave him what he most wanted: the intellectual freedom to strike out on research paths notably different from those of his midwestern roots and contemporaries. Sauer's scholarly values also had an impact on others by the subliminal messages conveyed. Pursuing one's curiosity wherever it leads is paramount. Learning is its own reward, but to be validated, a piece of writing needs to be published. As Sauer wrote once in a letter to a student, "The treasure the scholar lays up on earth is largely the printed page" (Leighly 1976:345).

Early intellectual exchange with Berkeley anthropologists, especially Alfred Kroeber, stimulated Sauer to think about the concept of culture in a way that few geographers did. To Sauer the time dimension was critical to understanding its usefulness: "Culture is an accumulation of experience that works" (SP, COS to HSS, 2-2-1939).

The time element was most critical in Sauer's thinking, whether that be the past for its own sake or use of the past to assess the present. Like Johann Herder, he had a "congenial empathy into the spirit of ages" (Dilthey 1986: 256). Also like Herder, Sauer had a sense of reflective discernment conveyed by the untranslatable word *Besonnenheit.* Questions dealing with nonreplicable events in culture history—colonization, land use, domestication—were subsumed under F. Ratzel's idea of diffusion and migration ("der Gang der Kultur über die Erde"). In his inquiry, Sauer used facts generated in other fields ranging from linguistics, ethnography, and archaeology to botany, zoology, and geology.

Neither in print nor in letters did Sauer link specific philosophers to his own world view, derived from German idealism and Romanticism. The connection is clear, for like all idealists, Sauer sought a wide perspective on the broad sweep of history. His work showed the extent to which he placed mind as central to knowledge and being. Parallels between Sauer and Friedrich Nietzsche are noteworthy, even though the latter was often critical of German idealism. Both were closer to the temper of the end of this century than they were to that of their own times. Like Nietzsche, Sauer had a detached perspective on humanity which enabled him to consider from where we humans have come. The bipedal animal made the transition to culture without ceding its natural character. Nietzsche considered man to be a little eccentric species of animal; for Sauer we are a perilous social animal. In pondering the fatal passage to a hominid form of organization, Sauer (1962) perceived the seaside, not the savanna, to be the most plausible ecological setting for human evolution. Subsequent inventions of fire-making and plant domestication vastly enhanced humans' control over their surroundings. Sauer, like Nietzsche before him, believed in the human ascent to culture but not in a linear sequence to greater progress. Yet neither of these two men wrote anything that advocated an idealized return of people to nature.

Although Nietzsche (1844–1900) died when Sauer was 11 years old, they shared in their adulthood certain traits. Neither sought salvation in a transcendent ideology. Nietzsche was emphatic that the creative will rules the self; Sauer pointed in that same direction implicitly by how he lived his life and through his incarnation of what Nietzsche called a "*vornehme Seele*"—a noble soul in quest of universal knowledge and devotion to an examined life.

Both were opposed to the idolatry of the state or dominance of the political in the lives of people. They each reflected deeply on their relationship with the countryside. "Modern" technology was not considered to be a good in itself, a view that for Sauer meant bucking a powerful American belief. Nietzsche and Sauer had a passionate desire for knowledge, yet neither tied this to rhetorical authoritarianism or an explicit philosophical system.

Sauer's life and ideas have received much attention since his death in 1975. He possessed a genius for grasping the details of diverse phenomena and synthesizing them into larger wholes. In his concern for long-term patterns and the *longue durée* of culture history, Sauer used a wide range of evidence to construct his geography. But his ideas and conclusions were frequently controversial. Sauer, for example, was excessively partisan about diffusion as a process, perhaps as a reaction to ecological theorizing, which he did not like. Diffusion surely accounts for much of the present world pattern of technology and popular culture. Within the past 500 years, the black rat, maize, and *Plasmodium* parasites have each depended on human carriage. In prehistory, diffusion was also important, as witnessed by the spread of certain biological organisms among continents. However, many ideas, practices, and beliefs are not primarily the result of transfer from one group to another. Aboriginal land burning, the shaman as an institution, idea of plant or animal domestication, preparation of fermented beverages, and the magical meaning of twins were among the many traits that arose separately in both the New and Old Worlds. Sauer was willing to cede very little to independent invention of any kind. Not surprisingly, Sauer ignored Adolf Bastian's (1826–1905) idea that the psychic unity of mankind accounts for many historic and prehistoric distributions. Bastian, who assiduously scoured the world to catalog the diversity he found, interpreted similar cultural forms (*Völkergendanken*) which he found in different geographical settings to be variants of universal patterns (*Elementargedanken*). Bastian's travels led to insights of many kinds, for example, his observation that priestly castes around the world have tried to control the sex drive of people to keep them in submission. Bastian in 1860 was the first to comprehend the cross-cultural use of hallucinogens, the phenomenon of shamanism on every continent, and how shamans and hallucinogens have been tied together. Shamanic cultures widely separated in time and space have manifested the same configuration. Only the collective unconscious makes sense of the replication at different times and places of deeply etched patterns of the human psyche. This does not mean that knowledge is innate as platonists have asserted, but that humans biologically evolved psychological predispositions to feel and behave in certain ways. Archetypes are a priori forms of experience, and archetypal images are the deposits of our ancestral experiences.

Some other aspects of culture are less matters of predisposition than matters of trial and error. Given enough time, humans are likely to discover the usefulness of the fauna and flora around them. For example, the use of different species of *Tephrosia,* a source of rotenone to stun fish, has been found in Australia, Africa, and South and North America. On each continent people independently discovered for themselves the usefulness of this plant. Yet diffusion is unquestionably a much more important process in the history of the world than independent invention is. In the twentieth century, diffusion has been awesome in its scope. The interconnected planet of the twenty-first century will enable ideas and artifacts to diffuse almost instantaneously.

Other ideas reflect a line of thinking that merits fresh reassessment. Sauer also channeled into his writings the belief that people are always the active, decisive force in establishing their workable adjustments with land and biotic resources. Nature always bends to the human will, can never be the motive force. His distaste for environmental determinism puts that one-sided perspective into the academic context of his time. For at least the past half century, geography has avoided the study of influence of the environment on culture.

Sauer also did not place his ideas into larger philosophical frameworks in spite of the seamlessness of his historicist world view (Speth 1987). He made no effort to provide an explicit philosophical justification for his work, nor did he answer critics who were not sure where Sauerian scholarship fit into the larger scheme of knowledge. Unwillingness to articulate his position in that way hindered a deeper justification for geography as a field of study. Finally, Sauer assumed the objectivity of his work. In fact, its subjectivity was patent (Speth 1993:52). At least in print he did not acknowledge that the "innocent eye" of the observer can never be unbiased.

What is unassailable about Sauer as a scholar was his thematic originality. Donkin (1997:264) described Sauer as posing distinctive questions, which is largely what attracted so many budding scholars educated in other fields to the ranks of cultural geography. In framing those questions he seems to have followed Goethe's dictum that nothing is insignificant; it all depends on "how you look at it." In the breadth of his themes, Sauer ferreted out little-known angles without regard to some supposed importance or practical applicability. Philip Wagner saw Sauer's main contribution as conveying a philosophy, one that "rests on the widest observation of mankind" (SP, PLW to COS, 6-27-1969). In many respects Sauer did not *teach* a philosophy; he *was* a philosophy. Sauer firmly stamped his mark on the most distinctive tradition in American geography. Parsons (1996:22) has written regarding Sauer that "it is the awesome breadth of his inquiring mind, his gentle wisdom, the originality of his insight, the magic in his turn of phrase, the enduring quality of his humanity that admirers continue to celebrate."

Sauer and Biotic Phenomena

Work on the interrelationships of biota and people occupied Sauer's thinking and writing during his long and productive life. To American geographers at the time his plants-and-people focus was unconventional, and to some it was even inappropriate. These topics gained secure acceptance only in the 1950s, when a like-minded group of critical mass had emerged to validate the approach. Sauer pursued two main lines of inquiry on this subject: the plant as cultural tracer, and human impact on vegetation.

Crops were central to Sauer's work on agricultural beginnings and transfers. The plants were the key components and had to be understood before agricultural origins could be discussed as a generalization. He once wrote of that interest in crops and weeds as a "special kind of plant geography." [1] He did not use *ethnobotany* to describe it, perhaps because that term then still connoted functionalism divorced from the past and a larger context. Another theme of inquiry was the human role in changing the face of the earth. Sauer's attention to it focused mainly on the changes in the vegetation cover imposed through human activities.

Early experiences influenced Sauer to think about the geography of biological phenomena. At a young age in the Missouri countryside, with his father as guide, he learned to identify trees, herbs, weeds, and crops.[2] As a graduate student at the University of Chicago, Sauer took a course from H. C. Cowles, America's first professional plant ecologist, on the spatial patterns of ecological succession. Sauer's own concept of ecology added humans to the equation. In his doctoral dissertation work, he determined that both aboriginals and white settlers had used fire to create the grasslands and open woodlands of the Missouri Ozarks (Sauer 1920). His preoccupation with cultivated plants spanned the period from 1935 to ca. 1955. Questions posed about organic life grew mainly out of his field ruminations. Stimulus and reinforcement also came from correspondence and conversations with professionaly trained plant scientists.

Of greatest significance to Sauer's education was his exchange of ideas with Edgar Anderson (1897–1969), a brilliant polymathic scientist at the Missouri Botanical Garden. Initial correspondence in 1941 between Sauer and Anderson centered on the mutual interest in Mexican maize, which they both started studying about the same time. Anderson wanted to classify corn in a more sophisticated way so as to give insight into genetic origins, whereas Sauer sought to clarify the role of maize in aboriginal life and to understand its culture history. Interested in learning what the other knew, Sauer and Anderson cross-fertilized their research.[3]

Sauer's concern for time made him see his interests as a form of natural history. In 1956 he wrote:

> The field of biogeography requires more knowledge of biology than can be demanded of most of us. It is, however, so important to us [that] we should encourage the crossing of geography with natural history wherever the student is competent. In particular, much more knowledge is required of the impact of human cultures on plant cover, of man's disturbance of soil and surface, of his relation to the spread or shrinkage of individual species, of human agency in the dispersal and modification of plants. (Sauer 1956:295)

But natural history lost visibility in American university circles, although less so in museums. Reductionism in biology marginalized natural history as a pseudoscience. The twentieth century was not ripe for an alternative science to enlarge the canon of biology. After 1935 Tansley's idea of ecosystem as energy transaction through different trophic levels recast ecology more into the mainstream of acceptable science. During the environmental movement that surfaced in the late 1960s, that abstract approach remained strong; in the 1970s, with the worldwide fossil fuel crisis, energy grew into an ecological obsession. In the 1980s thinking about plant ecology shifted sharply with the realization that functionalism lost connections with real places and times. Energy transactions totally ignored form and the idea of landscape. Abstract models prevailed.

About that same time the core assumption of ecology—that nature moves toward a normative equilibrium—was shattered. Wilderness once implied a stable climax equilibrium. By the 1980s that static view of vegetation stability was replaced by one of periodic disturbance and constant cyclical renewal. Without a base line from which to measure anthropogenic change, who can say what the human impact has been. This shift in the ecological paradigm has made environmental change more difficult to assess (Demeritt 1994; Williams

1994; Worster 1995). However, at the same time, recognition of the importance of human disturbance has greatly increased and is taken more into account now than it was formerly. Landscape ecology, a growing field concerned with edges and borders, has embraced human impact and a spatial approach. However, it does not seriously consider culture as an element of human agency in creating landscapes.

In retrospect Sauer was correct in his appraisal that human impact on vegetation is more important than was recognized at the time. Yet he viewed disturbance as more narrowly anthropogenic than it should have been. Fires start and spread from lightning strikes much more than he was willing to acknowledge. Extreme weather events working alone or in concert with human agency make ecological stability an illusion.

American Cultural Geography

Sauer's influence and that of his own students account for the content of cultural geography over the last half century. The culture/nature interplay is a focus that now is often placed under the rubric of cultural ecology to define more clearly its concern and approach (Zimmerer 1996a). Biotic elements, material culture, settlement, and adaptation have formed themes around which this inquiry has been mainly conducted. In most of this work, an unabashed empiricist perspective has prevailed. Parsons (1977:15) described it as the ". . . supremacy of observed geographical data over any pyramid of deductions or formal theories, however powerful the apparatus brought to bear." Given that thrust, the emphasis on field studies is not surprising, and within the larger discipline, fieldwork is often considered to be a special feature of cultural geography which has remained important after many other geographers abandoned it (Rundstrom and Kenzer 1989). The wave of logical positivism that dominated the discipline in the 1960s and 1970s scorned the empirical field-based approach. Positivists, who were not shy about pointing out the probative deficiencies in other's work, led the bandwagon of scientific geography under the banners of maximum precision, universality, and objectivity. But cultural geographers marched to a different epistemological drummer. They paid little heed to the scholarly revolution swirling around them, for they knew that many worthwhile ideas could not be effectively handled in the ways proposed.

Other geographers realized soon enough that the choice did not involve a correct or an incorrect answer, but instead what might be the best answer one can deliver. Positivism rapidly disintegrated as the controlling paradigm in geography in the early 1980s, for its promises, like those of Marxism, were finally recognized as unkeepable and its objectives unattainable. Its logocentric character and self-deception surely help to account for the spectacular rise of postmodernism in the social sciences. In geography this critique of modernist values asserted itself in the 1980s and even more strongly in the 1990s.

It has served as a valuable corrective to the pretensions of science in general, though many biologists have so far not wished to confront this issue.[4]

Cultural geography of the Berkeley tradition has always had an alternative vision. In many respects this branch of the field never was really modern and so cannot be postmodern either. It has been the one subfield of the larger discipline which for decades took an ethnographic gaze of the world and incorporated the "Other" into its inquiries. Cultural geography in its monographs, articles, and textbooks has imparted messages that human diversity is good, all cultures have a right to survive, and material advances are bad if they lead to the destruction of cultures and habitats from the outside. Carl Sauer, who brought these ideas into cultural geography, in turn may have indirectly assimilated them from Johann Herder, the first to articulate the need to preserve primitive cultures. Cultural relativism offered its detached perspective to cultural geography, although when it came to environmental degradation, moralism was invoked. As for cultural materialism, cultural geographers have certainly not denied that many patterns on earth are clearly connected to the resource base, but they are more likely to invoke ideas as the major force in explaining many interrelationships and adaptations.

Much work produced in cultural geography has rested on the unstated assumption that cultures were to be understood in their own terms and as the result of idiosyncratic historical experience, not as a point on an evolutionary continuum from primitive to civilized. Cultural geography also contested modern notions of progress at a time when it was social heresy to do so. Material adjustments of life in Western societies, especially the United States, have been less a focus than the ancient echoes of things as they were. Elsewhere in the world, attempts to develop nonindustrial countries or regions economically have often had negative side effects. Cultural geographers and anthropologists have been the skeptics and doubters of unilinear notions of economic development. The Green Revolution, which requires large adjuncts of pesticides and hybrid seeds, has a downside that started to assert itself several years after its introduction in 1970. Purported solutions imported from industrial countries have created as many problems as they have solved. For instance, the introduction of tube wells in foreign-aid projects dealing with the Sahelian drought of the 1970s increased desertification of grazing lands by allowing increases in herd size.

Now public policies and social scientists in droves have rejected "developmentalism" and the short-term fixes that such narrow notions of progress imply. Imported technological solutions are now viewed with a reserve unimaginable in the 1950s and 1960s. In their place we find a new-found respect for the reservoirs of folk techniques, experience, and wisdom that local communities in diverse cultures possess as part of their inherited tradition. In the Andes enormous amounts of local knowledge remain to be recorded on a community-by-community basis. Everywhere in fact, including the United States, agricultural practice has moved away from academic models to "articulating situated knowledges" (Kloppenburg 1991).

To the rebuttals to hegemonic thinking that had swept the world, one must add the obsession to deconstruct accumulated knowledge. Eagerness to rip open the seams of the old, rather than to stitch new configurations, ultimately works against the generation of new knowledge about the world. Critical reflection surely has its place in exposing the many self-delusions that can be deciphered in a text, but at the same time our quest to know the world must not stop. Cultural geography's empiricist primacy and classic realist tradition contribute in an inimitable way to that endeavor. More than most other subfields of the discipline, it is positioned to continue the birthright of geography to study people in the fullness of their environments. Stoddart (1987) has warned how far geography has fallen off the track at century's end. Cultural geography offers richly textured accounts of the nature/culture interplay that merges the diachronic and synchronic. Price and Lewis (1993) have argued that the Berkeley tradition retains a compelling focus that gives it intellectual superiority over more recent claims on the geography of culture.

But if my work is called geography, it can just as readily be cast as some kind of "ecology." Cultural ecology has often been defined in primarily adaptational terms (Denevan 1983; Vayda 1986). Many practitioners who use this label have been functionalists, although Zimmerer (1996a) describes "cultural historical ecology" as a variant corresponding to the Sauerian approach. No more powerful example of the salutary benefits of a diachronic perspective is seen than Denevan's (1992) masterful "pristine myth" study, which placed a millennium of New World time in a context of major environmental change wrought by paleotechnic humans. Historical ecology is cultural ecology of the past, but since the 1970s, it has often been conceptualized more broadly to encompass environmental history and human ecology, particularly as that fusion manifests regional change (Crumley 1994; Balée 1997). Headland (1997) surmises that such a perspective is greatly needed, partly to burst the fond claim of the general public, who surmise that native people are natural conservationists.

"Ethnobiology" is another rubric that can subsume at least part of this book, especially if it implies the broad definition rather than the narrow. What makes this multidisciplinary domain so attractive is not simply that it integrates a wide range of factual information, but also that it has sensitized humanistic scholars to the need of understanding different levels of biological organization even if they do not use scientific method. In turn, it has helped to open biologists to an awareness that context and complexity are at least as important as clarity. Some of this subject matter also falls under the purview of biohistory, which has been defined as a field that studies the changing patterns of interplay between cultural and biophysical systems as they relate to processes of adaptation and quite different time scales (Boyden 1992). All these definitions of content are valid; the borders of knowledge spheres are fuzzy and every year become more so. Only when an intellectual territory stands to be defended against academic interlopers does the drive to segregate subrealms of knowledge assert itself.

Sierran Meditations

Like a moth to a candle, I was first drawn to the Andes by mountain zonation and folk culture. I wanted to understand in particular the workable connections that Andean people have made with the biotic resources in a series of vertically telescoped environments. The Vilcanota/Urubamba study area extends from high *puna* above 4,300 m to the tropical forest below 900 m. This valley has a rich culture history, stunning landscape beauty, and a peasant population still in close touch with the land and its elements. To me, coming from a North America engulfed in constant technological change and the lost landscapes of my youth, it seemed like an anchor of timelessness. Through one complete growing season and part of another, I recorded large numbers of systematic and unsystematic observations and collected and dried several hundred plants. A photographic record was taken, and maps were made at various scales. Quechua peasants and other people with whom I conversed contributed some of their knowledge. Some assumed I was a development agent, but in truth I wanted only to learn about their agriculture and plant use.

I followed scientific procedures in gathering information but also went beyond them. As a hunter follows clues that lead to an unseen quarry, a quester of knowledge builds a series of seemingly unrelated observations into a configuration which could not have been guessed by the marginality and subtleness of the phenomena used as evidence. In what Eamon (1994) calls a "venatic paradigm," linear logic and precise measurement are only some of the possibilities.

Few readers seemed to agree on how to categorize the product of my year-and-a-half inquiry. The dissertation, and later the book that emerged from it, has been variously annotated in the bibliographies as cultural geography, cultural ecology, biogeography, or ethnobotany (Gade 1975). However classified, it was for me an autobiography as well, even though none of the deprivation, culture shock, loneliness, and gut-wrenching illness I suffered was mentioned even in the preface. Data acquisition of the most elementary sort was an everyday challenge.[5] I later realized that only while I was in my late twenties could I have carried out such a project in that same scruffy, ingenuous way. By that age, I knew enough to understand what I wanted to find out, and I also had the flexibility needed to deal with the local conditions. Years later, I came to appreciate that this project in the Sierra was more than the acquisition of field data; it was also a testing ground for the self in eliciting a Nietzschean response to "self-overcoming" and in dealing with the Dionysian pulls and Apollonian pushes of life. Living in Andean culture allowed me to anthropologize myself and to question the substance of my own thinking that had been earlier inculcated with the false notion that there is a universal, immutable human nature.

Age slows down the body, but elapsed time also allows one to see the shifts in the public-policy *Zeitgeist.* The developmentalist mentality employing foreign technology enjoyed its acme of acceptance between 1950 and 1975. Peru in the 1960s was overrun with advisors, most of them living comfortably in Lima. In Cusco an American development specialist from Stanford University, with whom I discussed my project, strongly urged me to put it into a framework of economic development. From his perspective, the best way to change subsistence-oriented peasants was for the industrialized countries to exercise their moral duty to transfer a successful technology to these poor benighted folk in order to improve their standard of living. Without hesitation I rejected his vexing assumptions about recasting my project to conform to that objective. Two decades later the roof fell in on those certitudes of international discourse. Developmentalist projects went out of business or changed their objectives.

Andean agropastoralism is seen now in quite different terms. Wendell Berry (1981), in a lyrical essay on his agricultural journey to Peru, found in my work support for his arguments about sustainability of small-scale polyculture. It was a concept not too different from my grandfather's farm in Niagara County. Rural deprivation in the Andes and migration to cities have much less to do with the quality of the resource base or farmer ineptitude than with government policies. Price controls and cheap imports have long prevented highland farmers from getting fair prices for the food they grow.

The Goethean Aesthetic

Collections of objective data of the material realm fulfilled only part of my Andean quest, for behind it was a search for the symbolic meanings. Nothing written about my initial Andean experience conveyed the sense of harmony that this region conjured up for me. The view of fields on the valley floor as seen from the solitude on the ridges high above conveyed an image of the world when mankind was at one with nature. By contrast, the Sierra landscape of snowcapped peaks was devoid of human presence and a *royaume de l'absence* for living things except the condor. A double grandeur—one humanized, the other not—elicited two reactions. One was "to be there to marvel": "Zum Erstaunen bin ich da," as Goethe put it. The other was "to see the gestalt," that is, to let the pieces fall into place to reveal an interlocking nature/culture pattern. These moments of solitary contemplation unleashed flashes of understanding, although they have also occurred through encounter and engagement (figure 1.1). The desire to know and the capacity to marvel did not seem like a contradiction, but I was not willing to suggest, as Goethe did, that the magnificence of the landscape transcended the limitations of reason. As Shweder (1991:11) has phrased it, the Romantic spirit is there "to dignify subjective

Figure 1.1. Daniel Gade in 1964 examining a preconquest stone block of sacred significance in an *oca* field on the Island of the Sun in Lake Titicaca (Bolivia)

experience, not to deny reality; to appreciate imagination, not to disregard reason; to honor our differences, not to underestimate our common humanity."

A sense of aesthetic harmony is not conveyed through just grandeur and beauty. Remoteness and isolation evoke a similar appeal, perhaps because in these locales one has the possibility of coming face to face with the original something—the *Ur* principle. Search for the *Ur* meant to me retrieval of lost origins that could open the door not just to the past but also to understanding new uses or inventions. The nooks and crannies of western South America still hold secrets that even Andean specialists scarcely know about. Highland topography has imposed an economic constraint on road construction, and there are settled areas still accessible only by long walks over rugged terrain. Off the beaten track, behind that mountain, one invariably discovers something surprising. Humboldt was right that every discovery opens up the imagination further. In subsequent research trips to the Andes, my mantra was "Hinter dem Bergen wohnen auch Leute." What intrigued me on these long hikes was the relative contentment of people in their isolation. In North America, even the tiniest village is connected to a road or by air, and even closed ethnic communities want to be an accessible node in a larger commercial network.

In the Andes, isolation has occurred either because the road has never reached a place or because it is self-imposed. During the colonial period and after, hiving off to establish a new community was a way to ensure cultural, familial, or individual survival. Centuries of isolation characterized the community of Q'ero (also Queros) in Paucartambo Province, Peru. Q'ero was part of Cusco folklore as a mysterious Quechua-speaking community left behind through isolation beyond the trackless cordillera. In 1955, Cusco journalists and scholars formed an expedition to reach this classic *Ur* place. They found an ecologically multilayered community whose territory spanned environments from 1,500 m to 4,500 m above sea level. That intense verticality, as well as their customs, dress, religion, and livelihood manifested many presumed pre-Columbian characteristics. The "discovery" of Q'ero received wide national publicity in Peru, with the result that its lands, then actually part of a huge sprawling estate, were legally returned to the community in 1962.

The presence of ruins was for me another strong aspect of a Romantic construction of the Andes. They evoked an awareness of temporality, fragmentation of a former unity, and past cultural achievement. Certain sites, notably Machu Picchu, Ollantaytambo, and Pisac, represent the ultimate Inca achievement in building in stone; others, such as Pikillakta, date from centuries before the Inca. Remains of the past before the Europeans arrived are abundant, and the remote locations of some of them made me wonder if they had all been described. It was that visible record of past civilizations, combined with the contemporary Quechua folk, who themselves have certain pre-Columbian features, which made me feel that I was experiencing the ancient echoes of things as they were.

I also realized that these approaches to the Andes were more a statement about me than a generally agreed-upon picture of the region. I wanted to believe in a preindustrial harmony of contemporary Andean life and a glorious past before the Europeans. In fact, evidence was abundant that a quite different perspective could and should legitimately be taken. Andean communities manifest frequent internal conflicts. Some of these have involved ethnic divisions, but many have been driven by the personal characteristics of greed, envy, and distrust, known everywhere. Moreover, maladjustments between people and their resource bases have been common in the Andes and part of the explanation for grinding poverty. They forced me to reflect on the gap between a Rousseauean vision that grew out of a largely preindustrial *genre de vie* and the social disorganization around me. Each person invents a world suited to his or her own being, for illusion is as important as truth is to human life.

Observations in the Field

Foreign experiences of fieldwork are stressful and time consuming, but receive remarkably little reflexive comment from those scholars who have conducted research in that manner. Some of this reticence may be a sense that only those who have done it can understand it. In retrospect, method is much the lesser part of fieldwork; what stands out is the experience of living and working in such a way as to achieve the goals set. Beyond the data collected, my fieldwork was a launching pad toward greater insight, maturation, and transformation. Analogous to the passages through which a shaman goes, that fieldwork was an initiatory experience that involved testing of the self and a sense of duty toward those who had invested time and effort in my education. Henceforth I felt authorized to become a professed quester, part of which was to consider fieldwork as the prime source of information. It was also a mystique.

Subsequent travels to the Andes opened new vistas and spawned new projects that extended to Bolivia (1968, 1980–1981, 1993) and Ecuador (1986, 1995). My fieldwork over the years and the living experiences associated with those visits represented an investment in intellectual capital whose greatest benefit has been to develop a mind attuned to the possibility of discovery. What Louis Pasteur called an *esprit préparé* has its special benefit in regional studies. J. J. Parsons (1977) discussed how accruing knowledge about a region has its payoff in formulating new lines of inquiry from the most casual of references seen, heard, or read. Parsons (1915–1997), one of the most accomplished representatives of the Berkeley tradition, was living proof of how that works (Wallach 1998). His many inquiries in different parts of Latin America and elsewhere reflect a creative imagination and a clarity of expression.

In addition to generating a cascade of new ideas, repeat visits also helped me measure changes in the larger national society. My first stay in the Andes occurred during an increasingly tumultuous period of peasant resentment and hostility toward the stranglehold that haciendas had on the land. Strikes, takeovers, roadblocks, frequent security checks, and general mayhem were common occurrences. A large estate I had hoped to use as one of my bases of operation, Hacienda Huadquiña, owned by the Romainville family, had been mysteriously abandoned weeks before my arrival. By my second visit to the Sierra in 1968, the land-tenure system instituted in the colonial period had almost no defenders. The push toward social justice in the Peruvian Sierra had passed the point of no return. In my 1970 research period, I found the military government in the process of dismantling quasi-feudal arrangements. Its aggressive rhetoric contributed to a climate of nationalistic hysteria, which seems to have been designed to gain popular support for the subsequent state takeover of most of Peru's means of production and distribution. On the 1978 trip another element, drugs, came to the fore. Cocaine manufacture and trade had become the bolster of an otherwise faltering economy. In the Huallaga Valley, Tingo María had turned into an ominous town of shady characters.

A return in 1986 indicated the negative economic consequences of land reform. Haciendas turned into state-owned cooperatives (SAIS) had even less agropastoral production than in the old days. Massive food imports, out-of-control inflation, rampant unemployment, and politically inspired terrorism were roiling the country. The day I arrived in Cusco in June of that year, the tourist train to Machu Picchu was blown up while still in the station; some people were killed and many injured. Organized terror spawned by resentments of class, race, and even gender meant that for the first time in history, large numbers of Andean peasants had armed themselves with guns in self-defense. Coca growing for the cocaine market had expanded exponentially, and even if the raw material is not observed, it rapidly becomes clear how much the money generated from it has had a corrosive effect on Peruvian society.

Upheavals over these 35 years took their toll in violence, mistrust, migration, and land abandonment. Yet the deep-rooted patterns etched in the Andean landscape and in the collective memory have survived, just as they survived the much more powerful trauma of the Spanish conquest itself. Indigenous traits still dominate the rural landscape, although Spanish colonial elements are important, either syncretistically or in juxtaposition. The peasantry is conservative, with labor still coming from their own muscles and that of their draft animals. In the process of making a living from the land, people preserve knowledge from the past. Power of tradition is the key to understanding continuity from one generation to the next. Repertoires of crops, domestic techniques, religious views, musical traditions, and patterns of sociability form an Andean configuration shared by Peru, Bolivia, and Ecuador.

Yet tradition is not the entire cultural canon of Andean society. A group's reaction to the new, whether accepted from elsewhere or locally invented, also forms part of any culture. Over the past half century especially, Western ideas and technology have relentlessly advanced through much of the Andes. Trade, roads, contact with various levels of government, ways of getting information, and knowledge of ways of life other than their own have expanded considerably. Means of travel have changed a good deal. Though the Andean countryside has not yet joined the modern world, every peasant today does some things differently from his or her grandparents, and for rural migrants to the cities, the changes have been traumatic.

Experiences and understandings derived from western South America have also transformed me. Much of what I have internalized about the nature/culture gestalt has come from this realm. At a personal level, Andean people taught me the blessed art of patience and to appreciate the fleeting delight of the present moment. Certain enchanted moments in that realm have nourished the soul for a lifetime: condors soaring above the ridge, drinking chicha to the accompaniment of panpipe music at village fiestas, and the unrestrained joy of the maize harvest. These memories are forged by a fascination with the exotic and by placing history and system together both in the environment and in society. My pursuit of the Andean realm led to a heightened sense of how holistic thinking could be plugged into the values of order and lucidity. It was also in the Andes that I began to listen to all the voices within myself. From that, occasional flashes of insight crackled through my brain.

My inquiry was also about the building block of knowledge. Thinking about the study areas and the larger region raised questions about who had followed the example set by Humboldt. The trickle of foreign researchers—Hiram Bingham, Isaiah Bowman, Carl Troll, and others—who had come to the Andes over the past century produced work that was generally retrievable. But who among the local people had engaged themselves in this pursuit? Carl Sauer sought to answer this question in every place he stopped during his 1942 journey through western South America, for he believed in Royce's concept of the provincial and the creativity that can sprout in off-center locations. Individuals arise out of a regional milieu to formulate knowledge about the place to which they have home allegiance. In Cusco scholarly and scientific inquiry into the surroundings was remarkably recent.

Andean land and life, so rich in cultural-historical content, were given little attention before the 1930s, when systematic observation of anything guaranteed a contribution to knowledge. The second half of the twentieth century saw a quantum leap in the formulation of Andeanist knowledge. The development of a focus on regional scholarship in the Cusco region fits into an unusual social and historical context. The dominant figure in initiating place-based research among local scholars was Albert(o) A. Giesecke (b. 1880), an American educated at the University of Pennsylvania. More than anyone Giesecke turned the Cusco intelligentsia from scholastic speculation toward studying its own region (Tamayo 1978:131–34).[6] That shift was part of an awakening in the Peruvian highlands to the social and economic conditions of indigenous people. Two scholars, Felix Cosio and Luis Valcárcel, helped Giesecke to link programmatically the grandeur of the past and the reality of the present.

Giesecke institutionalized history, archaeology, and ethnography in the University of Cusco (Super 1994). He started an archaeological museum and a regional archive. He also began the *Revista universitaria* as a forum for locally generated scholarship. Guidelines on archaeological excavation were established as an outgrowth of the apprehension wrought by Hiram Bingham's digs and the many boxes of valuable artifacts sent to the United States. Giesecke's part in Bingham's discovery of Machu Picchu was not revealed until 1961. Giesecke had made a mule trip down and back on that trail to discover for himself the forested realm which then seemed to have so much economic potential. On the way he learned from local peasants the existence of ruins high above the valley floor but because of heavy rains was unable to reach the site. When Bingham came to Cusco in 1911, Giesecke gave him the information, which led to Bingham's international acclaim.[7]

All these associations form the drama of the Andes as one of the world's vivid edges of a folk-inspired way of life. Together with my own memories, they have allowed this region to grow in my mind in complexity and depth.

2
Andean Definitions and the Meaning of *lo Andino*

The Andes form a fundamental region of South America, but also a mythic space whose configuration has changed through the past 500 years. A definition of the Andes and of regions within the Andes appears on the surface to be straightforward and even obvious. Four observations contravene that assertion: (1) the mountains of western South America were not always seen as composing a single range having a geographical unity; (2) the Andes cannot be delineated by simply applying a physical criteria or altitudinal threshold; (3) the region labeled "the Andes" has focused, whether articulated or not, on a core; and (4) the Andean core has taken on symbolic meaning that goes far beyond the place itself.

Unfolding of the Andes as Region

However clearly the Andes stand out on a raised relief map of South America, their physiographic coherence was not generally appreciated until the nineteenth century.[1] Climatic differences and the effects that they wrought on bodily humors were a primary concern through most of the colonial period (Orlove and Godoy 1993). The complex mountain topography that dominates western South America from Venezuela to Tierra del Fuego long had no recognized unity. On many maps the mountainous character of western South America was not even shown. In the colonial period, "the Andes" signified a different place from the one the term now describes. Most often the name referred not to the continuous highlands, but rather to the forest-covered hot valleys east of Cusco (Vázquez de Espinosa 1969:272; López de Velasco 1971:81).[2] *Anti* (pluralized by the Spaniards to *antis*), the word from which *Andes* is derived, originally meant not the mountains, but people: the forest dwellers at the eastern margin of the Inca Empire. Antisuyo was the eastern quarter of the Inca Empire, an area of wooded highlands and lowlands. The

Figure 2.1. *Heart of the Andes,* by Frederic Church, was painted in 1859 from sketches he made during his 1857 travels to Ecuador. In this landscape Church combined a display of highland grandeur with subtle indication of human presence.

serranía de los Andes referred to the easternmost mountain range that extends to the Amazon piedmont. In spite of sharp differences in relief, the continuous forest cover made it a definable unit in the colonial mind.

Shifts in uses and meanings came from outside South America beginning in the eighteenth century. The European Enlightenment brought attention to the mountainous character of western South America, and its protagonists recognized that as its most significant attribute. The mountains of western South America became generalized as the Andes (Alcedo 1967, 1:69–70). Maps gradually provided a synoptic conceptualization of its vast north-south extent and virtual unity. Two contributions more than any others inscribed the Andes as a highland region in the mental geography of the world. Alexander von Humboldt (1769–1859), who traveled there between 1799 and 1804, elaborated precise descriptions of Andean topography by means of cross-section diagrams.[3] These profiles, the first ever published, allowed easy, if exaggerated, visualization of comparative elevation and relief differences. On this physiographic schema, Humboldt superimposed the arrangement of climate and vegetation to show a biophysical verticality that communicated another layer of mountain complexity. His vision went beyond science to incorporate art as a basic part of his thinking about the Andes (Bunkse 1981:137).

Humboldt's descriptions formed a direct encouragement to Frederic Church (1825–1900), who traveled to Ecuador in 1853 and again in 1857. In 1859 he painted, from sketches and memory, the canvas he called *The Heart of the Andes,* which most captured the grandeur and diversity of the Andean landscape as it was then known (figure 2.1). In Church's panorama the rich organic life of birds, trees, and flowers in the foreground contrasts with the humanized expanse of village, fields, and peasants in the middle distance. Looming in the far distance is Chimborazo, the majestic snowcapped sentinel that Humboldt had climbed more than half a century earlier. Church's painting, although perhaps not a work in the annals of world art, received critical and popular acclaim at the time (Avery 1994). Much of its appeal came from the authoritative image that it conveyed of the Andes to viewers, who saw it as scientific illustration more than art (Manthorne 1989:55). It greatly enhanced awareness of the Andes as a place where the tropics took on a benign character and where humans made productive use of a mountain environment. This canvas also opened a broad appreciation of the Andes as notably different from the mountains of North America.

The Central Andes as Body Parts

The vast Andean mountain system, stretching 6,000 km in a huge north-south distance, cannot form a cohesive geographical unit. Relief configuration and vertical zonation, as well as climate, vary by latitude. The expression *tropical Andes* has been applied to a long swath from Venezuela to northern Argentina. But the indigenous culture of the northern Andes differs from that farther south in language and tradition. South of latitude 30°S. in Chile and Argentina, harsh climate and absence of fertile basins have always excluded dense highland settlement of a permanent nature. The midsection—Peru, Ecuador, and Bolivia—captures the Andean essence.[4]

However, the territory included in descriptions of that centrality has somewhat broadened over the decades. The expression *Central Andes* is attributed to the North American geographer Isaiah Bowman (1878–1950), who first used it to refer to the highlands between ca. Huancayo, Peru (lat. 12°S.), and Salta, Argentina (lat. 25°S.) (Bowman 1909, 1914). The researcher Carl Troll (1929) followed Bowman's interpretation. Both described what they perceived as highland adaptational responses to the environment. A deterministic discourse emerged from that focus on physical phenomena. Particularly extreme was Carlos Monge's (1948) notion of climatic aggression, which sought to explain Inca expansion and organization in terms of the environmental superiority of the highlands. So-called telluric influences of the highlands as a physical unit that evoke particular human responses became part of an Andean literary tradition (e.g., Mendoza 1925). The way out of that impasse was to conceptualize the Central Andes as a cultural-historical, not an environmental, region. Rather than its unity being set by a contour line or isotherm, the definition of the Central Andes shifted toward the geographical locus of pre-Columbian civilization in South America.

Recasting the meaning of the Andes could not, however, be done until its culture history had become better known. Early in this century Max Uhle (1856–1944) was a key figure not only in understanding the extent of the Inca Empire, but also in providing a general framework and chronological sequence for Andean civilization before the rise of the Inca (Uhle 1909). Pre-Columbian cultures were not simply the outcome of developments in one neatly defined environmental unit, but also the manifestations of the connections among the coast, the highlands, and the jungle. The archaeologist Julio Tello (1930) was the pioneer of this idea of multienvironmental ties. They are especially obvious in the prehistoric diffusion of plants and animals from one environment to another, an array of species which, considered together, make the Central Andean realm one of the richest regions of domestication in the world.

A cultural definition of the Andes goes back at least to the early 1930s, when Philip Means (1931) grouped together the coast and highlands.[5] Julian Steward's *Handbook of South American Indians* played an influential role in identifying cultural distinctions on the continent. Western South America was the region of "high culture" but one that did not extend to the whole sweep of the Andes. In the *Handbook,* Bennett (1946) grouped under the Central Andes the Peruvian highlands and coast and the Bolivian highlands as a "cultural block of considerable time depth." On the basis of what is now known, highland and coastal Ecuador and the Amazonian piedmont from Ecuador to Bolivia would also form part of that Central Andean block. Highland and coastal fusion of regional attribution became the standard approach among anthropologists (Steward and Faron 1959).

Incorporation of zones outside the highlands into the cultural definition of the Central Andes is also justified in contemporary terms. Many people of highland origin moving to lowland cities, and the countryside on the west and the east have blurred the previously sharp ethnic differences. Lima today is a city of Andean culture to an extent that it was not in 1940. Most other sizable urban centers not in the mountains—among them Guayaquil, Trujillo, Ica, Tacna, and Santa Cruz de la Sierra—have received large numbers of highland migrants. Using prehistoric, historic, or contemporary criteria, no altitudinal thresholds can define the overarching Andean cultural tradition (Hornborg 1990). The most profound meaning of the Andes thus comes not from a physical description, but from the cultural outcome of 10 millennia of knowing, using, and transforming the varied environments of western South America.

There is no agreement as yet in regard to where one places the outer limits of the Central Andes.[6] Many definitions of *Central Andes* are undergeneralized; some of them have, implicitly or explicitly, gone so far as to limit the term to the Peruvian highlands (Millones O. 1982; Arnold and Yapita 1992; Kent 1993:446). Ecuador is rather frequently excluded (Bennett and Bird 1964; Dollfus 1982; UNESCO 1975; Sick 1969:452). Nevertheless, Ecuador

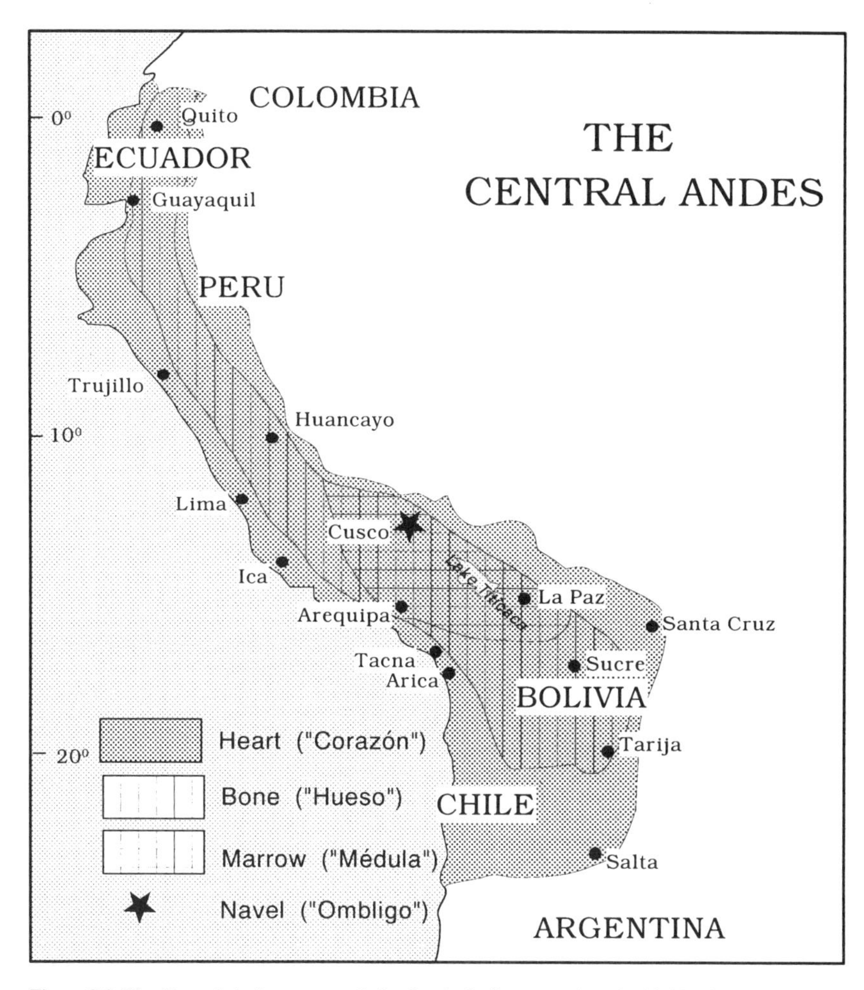

Figure 2.2. The Central Andes are now defined as including more than the highlands.

has much in common with lands to the south. Bolivia is occasionally placed within the south Andes (Wachtel 1992). The definition proposed here includes the coast, highlands, and eastern piedmont of three countries—Peru, Ecuador, and Bolivia—with projections into northern Chile and northwest Argentina. General agreement about a common definition for this geographical expression greatly facilitates communication. It also promotes a unity that a political emphasis has made unnecessarily divisive.

If the Central Andes are the heart (*corazón*) of the Andes, further subdivisions make certain useful distinctions (figure 2.2). The highlands above

Table 2.1. Proposed subdivisions of the Andes

Region	Boundaries	Analogy
Western South America	From Venezuela to Tierra del Fuego	Body
Central Andes	Highlands: Ecuador, Peru, Bolivia, NW. Argentina	Heart
	Coast: Ecuador, Peru, N. Chile	
	Eastern slope: Ecuador, Peru, Bolivia	
Highland domain	Highlands: Ecuador, Peru, Bolivia	Bone
Sierran nucleus	Ayacucho to La Paz	Marrow
	Departments above 2,500 m	
Cusco	The city	Navel

2,000 m between Ecuador and Bolivia enclose the strongest area of surviving Andean tradition from the pre-Columbian and early colonial periods. This zone can metaphorically be called the bone (*hueso*), just as the zone in southern Peru and northern Bolivia has greater specificity: it is the "marrow" (*médula*) that harbors outside its cities the classic realm of Andean values. Within that is the "navel" (*ombligo*), or Cusco. It remains the spiritual center of Andean culture, just as it was in the late fifteenth century before the Spaniards arrived, although much of that content is now symbolic (table 2.1).

Andean Essences and *lo Andino*

The highlands are the Central Andean heart, and there the Andean cultural essence is strongest. The term *lo andino* defines the culture complex that has survived the harsh acculturation imposed during the colonial period and the technological changes of the modern world. Some indigenous traits have become extinct; others have blended with Old World introductions to become Andean tradition (Gade 1992).[7] More than elsewhere in the Americas, Central Andean folk have manifested a consensual resistance to wholesale assimilation of Western ways (Pease 1981:105). Why, in fact, so many pre-Columbian traits have survived over the last five centuries in the Central Andes is a question that still awaits a careful response, for it is more complicated than it might seem.

Many autochthonous elements, practices, strategies, and symbols, both material and nonmaterial, make up the sum of lo andino. Its cultural-ecological centerpiece is an agropastoralism that takes advantage of environmental diversity in order to minimize subsistence risk. Relative self-sufficiency is an outcome of the verticality that is the basis of that diversity. The Andean essence is also captured in community levels of social organization, long-standing patterns of mobility, reciprocity of obligation, duality of spatial organization, pantheism, and weaving as a form of artistic expression.

Lo andino has material emblems of its identity that are recognized as such around the world (table 2.2; figures 2.3 and 2.4). Some of these—the llama and the poncho—are widely recognized as Andean throughout the world; others, such as the *quishuar* tree and popping bean (*nuñas*), are known less. The endurance of cultural elements of colonial Spanish inheritance (for example, the

Table 2.2. Twenty material symbols of *lo andino*

Element	Distribution
Plants and Plant Products	
Cherimoya (*Annona cherimolia*)	Ecuador, Peru, Bolivia, and beyond
Chicha (maize beer)	Ecuador, Peru, Bolivia, and beyond
Chuño (dehydrated potato)	Peru, Bolivia
Ckara (tree) (*Puya raimondii*)	Peru, Bolivia
Coca (*Erythroxylon coca*)	Peru, Bolivia, Colombia
Maca (*Lepidium meyenii*)	Peru
Nuñas (popped *Phaseolus* bean)	Ecuador, Peru, Bolivia
Oca (*Oxalis tuberosa*)	Ecuador, Peru, Bolivia
Potato (*Solanum* spp.)	Ecuador, Peru, Bolivia, and beyond
Queñuar (tree) (*Polylepis* spp.)	Ecuador, Peru, Bolivia
Quishuar (tree) (*Buddleja* spp.)	Ecuador, Peru, Bolivia
Animals	
Guinea pig (*Cavia porcellus*)	Ecuador, Peru, Bolivia, S. Colombia
Llama (*Lama glama*)	Ecuador, Peru, Bolivia, N. Chile, N. Argentina
Mountain tapir (*Tapirus pinchaque*)	Ecuador, N. Peru, Colombia, Venezuela
Spectacled bear (*Tremarctos ornatus*)	Ecuador, Peru, Bolivia, Colombia, Venezuela
Processed Objects	
Chaquitaclla (digging tool)	Peru, Bolivia
Inca ruins	Ecuador, Peru, Bolivia, N. Chile, N. Argentina
Panpipe	Ecuador, Peru, Bolivia, and beyond
Poncho	Ecuador, Peru, Bolivia, and beyond
Preconquest wooden goblets	Peru, Bolivia

Figure 2.3. Campesinos near Chiara (Cusco) carry home-made *chicha* to market in earthen jars.

scratch plow of Middle Eastern and Mediterranean origin) also merit inclusion as manifestations of lo andino. Together the indigenous and the introduced create an Andean identity of two principal strands that became blended by the late eighteenth century, when the first major reflection on Andean character crystallized in response to the oppressiveness of colonial power. The cultural particularism of the Central Andes derives from a corpus of knowledge and experience drawn from a dual past but living in the present. Andean people are represented as possessors of a noble tradition that the world outside seeks to know everything about. Their paleotechnic land use and agriculture are scrutinized for their sustainable aspects. Germplasm contained in their crops is now considered to be a treasure chest of inestimable value (National Research Council 1989; Carmen 1984).

Application of lo Andino

Lo andino offers certain low-tech solutions to land-use problems, but it also reasserts a cultural identity more profound than the nationalisms that were imposed in the nineteenth century. Through the 1960s, government ministries and international donors imposed a development model based on extraneous technology and values. Yet much Andean behavior cannot be explained by economic maximization models or those that claim energy efficiency. Partly for that reason, outcomes never matched expectations and many development projects failed. These attempts at innovations controlled from the outside allowed foreigners and national elites to gain further domination over rural people.

In the face of costly failures of schemes imported from outside the region, it was realized that the Andean tradition offers alternatives. Now 10 different communities in the Peruvian highlands corral, shear, and then release vicuña as during the Inca period. This preconquest approach to conservation of this rare animal contrasts with the preservational concept of inviolable reserves. Land uses typically overlap. Cochineal is harvested in Ayacucho from *Opuntia* scrublands (*tunales*), thus making productive use of communal grazing lands and supplementing peasant livelihood. Cochineal is a New World dyestuff which comes from the dried, pulverized bodies of the female scale insect *Dactylopius.* The polycultural ideal has favored maintenance of many minor crops. Quinoa, oca, *achira, tarwi,* and others that Western-trained agronomists once tried to coax rural folk to give up as "Indian foods" have held their own in the Andean agroecosystem. Those indigenous components in Andean agriculture now have a new lease on life. Coca continues to be grown as a plant of ritual importance to many Andean people. They have rightly rejected the transfer of culpability about cocaine use to producers of the leaf. The aggressive eradi-

Figure 2.4. *Chuño* is made in the high country above 3,800 m (Paucartambo Province, Cusco) from potato tubers by treading on them to remove the water. This manifestation of *lo andino* occurs only in Peru and Bolivia above 3,500 m.

cation campaigns orchestrated in Washington, D.C., hurt Andean livelihoods and mock its traditions.

Pre-Columbian land practices gain respect as their sustainability and productivity become known. Ridged fields make productive farming possible on marshy land (Erickson 1996). Bank terraces, one of the glories of preconquest technology, have undergone reevaluation as a useful heritage from the past (Treacy 1994). In Peru alone more than 500,000 ha of slope land were converted into bank terraces. Not just flat parcels on steep slopes, these cultivated spaces have combined irrigation, transported soil, and microclimate. Terraces have provided food surpluses at the same time as they control land degradation. Neither past achievements, nor local knowledge, nor romanticism based on communitarian nostalgia can solve intractable land-use problems. Andean people must decide for themselves what works and what does not. Strong international interest in indigenous knowledge has placed much more focus on native solutions than ever before. At the same time outside funding sources have reassessed the simplistic, linear assumptions of economic development made in the four decades following World War II.

In this milieu of dissatisfaction with Western solutions, Andean people themselves have begun to articulate their cultural affirmation. They have sought to "reidentify" the land in which they live as a place with a heavy emphasis on Andean tradition. Less and less are they willing to let outsiders determine their destinies. For Grillo Fernández (1998:194) the aim is to decolonize the Andes of Western ideas in order to have "recourse to our own knowledge, to our own capacities: to what we know and can do with certainty." Andean folk are acquiring a self-conscious awareness, which may be the main way that their traditions of working the land can remain strong and viable. No longer are Andean people isolated or unaware of the outside world impinging upon them.

Lo andino has also provided a focus for Andean studies which can be distinguished from the more encompassing topic of Latin American studies. Next to specialists on Mesoamerica, those who study the Andean region make up the most thriving international research group working in Latin America. More than 1,000 researchers trained in different disciplines focus on Andean studies, and anthropologists form the single largest group (Starn 1994). It is this group and their use of the written word that has opened Andean knowledge—still largely based on oral tradition—to a wide audience. In recent decades researchers have increasingly conceptualized their work in Andean rather than national terms of reference.[8] Since the 1980s, regional specialists have tended to define themselves as Andeanists, a term that has still, however, not entered the dictionaries.[9] The researchers come from North America, western Europe, and Japan, but principally from the Andean region itself. Though few themselves have actually participated in the rural peasant tradition, scholars born in the Andean countries bring many intuitive understandings to their research.

Andeanists from afar bring other perspectives, some of which effectively transcend nationalistic frames of reference. However, much work in Andean studies is overly representational and anthropocentric, which blocks its usefulness to the people from whom that information was derived.[10]

Assessment

The Andes, as many places, have been reconfigured as a result of changing perceptions. Over the past 400 years the definition of *the Andes* has shifted twice: from a nonregion to a region, and from a physical entity to a culture area. The Enlightenment and the *Naturphilosophen* were key forces in developing the idea of a physiographic unity for the Andes, but that coherence was later overshadowed when culture recast the definitional scope by downplaying environmental factors in explaining past and present human endeavor. Out of that has surged the notion of lo andino to capture the regional essence. Precisely how *lo andino* is to be defined or if it should be defined at all is the subject of continuing debate (Starn 1991). What people think of as the Andes has been at least as powerful as any purported objective description of their content and character.

3
Deforestation and Reforestation of the Central Andean Highlands

Any conservationist agenda must look into the past, for the preservation of species and habitats can grow only out of a sense of what has been destroyed. But what is gone is not always clearly understood, and this is particularly true with natural vegetation. For the Central Andes one strand of thinking has asserted that this region was always what it is today, that is, a treeless landscape of bunch grasses or low shrubs covering most plateaus, slopes, and valleys. Another strand acknowledges vegetation change but interprets it mainly in the context of climatic shifts. A third point of view sees the plant cover of highland Bolivia, Peru, and Ecuador to be the result of anthropogenic transformation. Long ignored because it did not fit a scientific scheme, the human-impact hypothesis has now gained many adherents.

In the first half of this century, Andean plant cover misled several notable field observers. Botanists and ecologists by training, they favored a biophysical causation for the treeless landscapes they encountered. Weberbauer (1911, 1936) and Schwalm (1927:38–39) in Peru as well as Fiebrig (1911) and Herzog (1923) in Bolivia essentially interpreted the impoverished plant cover in terms of low annual rainfall, seasonality of rainfall, and/or bedrock composition. Isaiah Bowman (1916:80) made similar assumptions, claiming that the driest and warmest slope exposures gave rise to a grass cover. Yet he was puzzled that "in some places the line between grass and forest is developed so sharply that it seems to be the artificial edge of a cut-over tract." Now these sharp boundaries between forest and grass are widely understood to be the result of burning. No thermal or moisture boundary is ever so sharp.

Erstwhile Andean Forest and Its Components

A wide range of ecological and historical evidence and my own field observations over more than 30 years convince me that the Central Andes were once mostly covered with an evergreen forest. Andean tree cover included both a closed forest canopy and an open woodland. For the most part, only in warm humid valleys did tree heights normally exceed 15–20 m. However, a general forest cover in no way suggests a continuous woodland over all the Central Andes at any point in the past. Some untreed areas would always have been there, the result of lightning fires that can spread during periods of unusual drought or landslides that locally reconfigure not just the vegetation but also the topography and hydrography. Local edaphic factors, especially porous soils, excluded or thinned tree growth in certain cases. The conventional idea of the climax community, once so enthusiastically embraced in ecological circles, conveys a stability and homogeneity that never existed in the Andes or elsewhere. Yet that does not negate the fact that the Central Andes have supported forest growth over most of their area, for rainfall is not, in spite of sharp seasonality, a major limiting factor. Dominant floristic elements had low water demands. Surviving native arboreal species, individually or in copses, offer a form of evidence of wooded vegetation and its composition (figure 3.1). Moisture and temperature differences arranged latitudinally and/or altitudinally varied the forest composition. About two dozen tree species were more common or more important among a substantially larger number (see table 3.1).

Growing mostly below 3,100 m the dominants were *molle* (*Schinus molle*), *lambran* or *aliso* (*Alnus jorullensis*), *unca* (*Eugenia* spp.), *pati* (*Ceiba* sp.), *cedro* (*Cedrela* spp.), *nogal* (a walnut) (*Juglans* spp.), *algarrobo* (*Prosopis juliflora, P. alba*), *tara* (*Caesalpinia tinctorea*). From central Bolivia southward the list broadened to include *churqui* (*Prosopis ferox*), *chañar* (*Gourliea decorticans*), *tarco* (*Jacaranda acutifolia*), *tipa* (*Tipuana tipu*), and *soto Schinopsis naenkeana*).

Above 3,100 m the forest components included *chachacomo* (*Escallonia resinosa*), *tasta* (*Escallonia myrtillioides*), *huaranhuay* or *sauco* (*Sambucus peruviana*), *intimpa* (*Podocarpus glomeratus*), *lloque* (*Kageneckia lanceolata*), quishuar (*Buddleja incana*), and *colle* (*Buddleja coriacea*). Elevations above 4,000 m harbored trees belonging to several allopatric species of *Polylepis* (known as *queñuar, quiñuar, quinuar, keñua,* or *kehuiña*). Of all the arborous genera, these were adapted to stressful conditions: nocturnal temperatures below freezing, strong winds, and intense insolation. A tree bromeliad, *ckara* or *titanca* (*Puya raimondii*), also occurred, a few specimens of which remain as witnesses to the arborous growth at these elevations.

Carl Troll (1889–1975), who did fieldwork in the Andes during 1926–1927, became the leading protagonist for climate as the determinant of the vegetation cover. He built upon Humboldt's view of plants as an ecological aggregate rather than a floristic assemblage. For Troll (1959, 1968) insufficient rainfall, a long dry season, and freezing temperatures precluded the growth of forests over most of the Andean highlands. Where trees were present, he asserted

Figure 3.1. Slopes covered with *Escallonia* woods at 3,050 m in an isolated part of Paruro Province (Cusco), Peru

that favorable microclimates permitted their growth. His ingenious three-dimensional diagrams influenced other Andeanist biogeographers to conceive of the plant cover in terms of climate (Koepcke 1961; Hueck 1966:385–87; Mann 1968; Walter 1971:183–91; Simpson 1979:15–17; Richter 1981). Troll's effective illustrations conceptualized the Andes as providing an immutable biophysical backdrop on which Andean culture history has been played out (Gade 1996). Naturalistic approaches of natural scientists favored natural causation. Many scholars trained in the social sciences assumed the correctness of that interpretation without critically assessing the evidence themselves (Ogilvie 1922:109; Winterhalder and Thomas 1978; Dollfus 1981; Gómez Molina and Little 1981; UNESCO 1981; Brush 1982; Pulgar Vidal 1987; Stone 1992:211).

As climates became better known after World War II, Troll's (1935:385) link between the physiognomy of vegetation and the regional climate could not be sustained. Tosi (1960) seriously questioned the plausibility of Carl Troll's assertions about Andean vegetation between 2,500 m and 4,000 m. Using Holdridge's now widely accepted formula that incorporates evapotranspiration, Tosi determined that all highland climates in the Andes could indeed support forests and that absence of tree growth could not be accounted for by low and highly seasonal rainfall. By way of example, in Sucre, Bolivia, potential evaporation exceeds precipitation in the five months from June to October. Seasonal moisture deficiency does not exclude forest development, even though the contemporary plant cover is now shrubs and grasses. The same five months are also deficient in moisture in the humid tropical lowland of northern Bolivia, yet rain forest is the prevailing vegetation. Rainfall receipts and temperatures of course affect biomass, which explains why highland Chuquisaca at Sucre has, with 6–7 m^3/ha, about half the potential yearly wood productivity of the tropical humid forest (10–14 m^3/ha) (Hutchinson 1967). Nevertheless, woodlands can thrive even with total rainfall of less than 100 mm/year. Parts of the arid coastal region of northern and central Peru had an open tree cover well into the twentieth century.

Troll (1952) then proposed that desiccating local winds have accounted for the sparse natural vegetation of *Trockentäler* (dry valleys); he never wavered from the idea that vegetation patterns have biophysical explanations, even when the evidence of human intervention by cutting, burning, and grazing was striking. Troll's (1931:263) naturalistic approach to Andean vegetation fit into the tradition of Humboldt, Herzog, and Weberbauer. Seemingly obvious connections were ignored. For example, he pointed out that shrub and cushion plants of the high country made good combustibles for peasant folk, but he did not grasp the larger pattern that human intervention of that sort was instrumental in creating the vegetation that he was looking that. With Troll, as with so many scientists, "nature" was treated as an entity so distinct from humans that the former could never be explained by the latter. His interpretations were also guided by functionalist assumptions of what was observed. That approach was particularly compelling in the Andes, where ecoclimatic zonations are vertically telescoped and altitudinal differences in plant growth are striking. Alternative paths of understanding emerged more slowly. Yet no ecological training was required to figure out that the Andean landscape could have undergone change. One traveler, who was struck by the Peruvian Andes as an inhabited region devoid of trees for fuel, intuited that their lack on the hillsides of Cusco was due not to altitude but to cutting over many centuries (Bruce 1913:88).

O. F. Cook (1916a:288) had strong inklings that the Andes had been deforested through human action. His interest in peasant agriculture in southern Peru and its tie with the Incas led him to observe the spontaneous tree colonization of abandoned stone-faced bank terraces. That point of view reemerged in the 1950s, when the Göttingen ecologist Ellenberg (1958, 1979) concluded that the Andes had originally been treed. Anthropogenic transformation was hastened by water stress on trees and other plants during the dry season. Other trained observers of the Andean landscape shared much the same point of view about deforestation, among them Tosi (1960:113); Budowski (1968); Gade (1975:16); Millones O. (1982:50); Donkin (1977:26–27); Seibert (1983,

Table 3.1. Reconstruction of the major components (primarily at the genus level) of the forest that once covered much of the Andean highlands.

Classification	Family	Common name(s)	Height (m)	Habitat (elev./m)	Distribution
Acacia spp. (ca. 8 spp.)	Leguminosae	jarca (*A. visco*), kini (*A. caven*)	8	up to 2,700	Peru, Bolivia
Alnus jorullensis (*acuminata*)	Betulaceae	lambran (Q.), (Sp.)	14	2,500–3,500	Ecuador, Peru, Bolivia
Anadenanthera colubrina	Leguminosae	willca, curupaú	8	up to 2,800	Peru, Bolivia
Billia colombiana	Hippocastanacea	cascarilla	6	up to 2,900	Ecuador
Buddleja spp. (ca. 10 spp.)	Buddlejaceae	quishuar (*B. incana*); kolle (*B. coriacea*)	15 8	2,500–3,600 3,600–4,300	Ecuador, Peru, Bolivia Peru, Bolivia
Caesalpinia tinctorea	Leguminosae	tara, guarango	5	2,800–3,400	Peru, Bolivia
Cedrela spp. (ca. 10 spp.)	Meliaceae	cedro	20	up to 3,100	Ecuador, Peru, Bolivia
Ceiba spp.	Bombacaceae	pati	9	below 2,500	Peru, Bolivia
Cereus giganteus	Cactaceae	caipiri	14	up to 2,700	Peru, Bolivia
Chuquiragua spp.	Compositae	chuquiragua	3	3,300–4,200	Peru, Ecuador
Cinchona spp.	Rubiaceae	cascarilla	6	up to 2,500	Ecuador, Peru, Bolivia
Clethra spp.	Clethraceae	bermejo	5	below 2,300	Ecuador
Drimys spp.	Winteraceae	cascarilla picante	10	below 2,800	Ecuador, Peru, Bolivia
Erythrina spp.	Leguminosae	pisonay (*E. falcata*)	15	below 2,800	Ecuador, Peru, Bolivia
Escallonia spp.	Grossulariaceae	chachacomo (*E. resinosa*); tasta, pauco (*E. myrtillioides*)	8	2,900–4,000	Ecuador, Peru, Bolivia
Eugenia spp.	Myrtaceae	unca, arrayán	8	2,500–2,900	Ecuador, Peru, Bolivia
Gourliea decorticans	Leguminosae	chañar	6	below 2,900	Bolivia
Gynoxys spp.	Compositae	japur	3	3,500–4,100	Ecuador, Peru, Bolivia
Inga spp.	Leguminosae	guaba	12	below 2,500	Ecuador, Peru, Bolivia

Jacaranda mimosifolia	Bignoniaceae	tarco	15	below 2,700	Peru, Bolivia
Juglans spp.	Juglandaceae	nogal	20	valleys up to 3,000	Ecuador, Peru, Bolivia
Kageneckia lanceolata	Rosaceae	lloque	4	2,400–3,300	Peru, Bolivia
Miconia spp.	Melastomataceae	colca	15	below 2,300	Ecuador, Peru, Bolivia
Myrcianthes spp.	Myrtaceae	arrayán, guapurillo, sahuinto	12	up to 2,900	Ecuador, Peru, Bolivia
Myroxylon peruiferum	Leguminosae	quina-quina	15	1,800–2,400	Peru, Bolivia
Nectandra spp.	Lauraceae	canelo	15	below 2,500	Ecuador, Peru, Bolivia
Ocotca spp.	Lauraceae	aguatillo	10	up to 2,800	Ecuador, Peru
Oreopanax spp.	Araliaceae	maqui maqui, puma maqui	5	above 3,000	Ecuador, Peru, Bolivia
Persea spp.	Lauraceae		12	above 2,600	Ecuador, Peru
Piptadenia macrocarpa	Leguminosae	willca	18	above 2,700	Peru, Bolivia
Podocarpus spp.	Podocarpaceae	intimpa, ulcumano, romerillo, pino de monte	18	2,700–3,900	Ecuador, Peru, Bolivia
Polylepis spp. (more than 20 spp.)	Rosaceae	queñuar, quiñuar, quinuar, keñua, kehuiña	8	up to 5,300	Ecuador, Peru, Bolivia
Prosopis spp.	Leguminosae	algarrobo (*P. juliflora, P. alba*); churqui (*P. ferox*)	8	up to 2,900	Peru, Bolivia
Puya raimondii	Bromeliaceae	ckara, titanca	9	3,900–4,300	Ecuador, Peru, Bolivia
Sanbucus peruviana	Caprifoliaceae	tilo, waranway, sauco	15	3,000–3,800	Ecuador, Peru, Bolivia
Schinopsis naenkeana	Anacardiaceae	soto	18	below 2,800	Bolivia
Schinus molle	Anacardiaceae	molle	10	up to 3,400	Ecuador, Peru, Bolivia
Styloceras spp.	Buxaceae	naranjillo	5	up to 3,000	Ecuador
Tecoma stans	Bignoniaceae	kellu tarku	3	up to 2,900	Peru, Bolivia
Tipuana tipu	Leguminosae	tipa	20	below 2,800	Bolivia
Vallea stipularis	Elaeocarpaceae	wantura, chicclurmay	3	up to 3,400	Ecuador
Weinmannia spp.	Cunoniaceae	waicha	5	up to 3,400	Ecuador, Peru, Bolivia

1994); Parsons (1982:257–58); Guillet (1985, 1992:16); Ansión and Van Dam (1986); Bastian (1986); Ruthsatz (1983); Schell (1987); Becker (1988:155–56); Beck and García (1991:86); White and Maldonado (1991); Young (1993); and Ulloa Ulloa and Jorgensen (1993:7–8). The perspective has now shifted, and one now starts with the assumption that the Andean highlands were covered with a montane forest, and the remaining wooded areas are understood to be the result of habitat fragmentation (CESA 1992; Young and Leon 1995:366).

The *Puna-Páramo* Problem

The thorniest question about Andean vegetation has involved the high-altitude grasslands. Many people who experience the high-country grasslands have continued to see them as "natural formations" on an assumption that their sparse population, night cold, and hypoxia have made human impact insignificant. Even researchers have frequently come to that conclusion. For example, Diels (1937:23) acknowledged anthropogenic change below 3,400 m elevation, but not above. In the 1960s and 1970s the Central Andean countries made ecological inventories based on the Holdridge model. Maps of Peru (Tosi 1960), Bolivia (MAG 1978), and Ecuador (Unzueta Q. 1975) portrayed most of the highlands as being potentially treed. However, above 4,000 m in Peru and Bolivia and 3,000 m in Ecuador, these maps show the *puna* and *páramo* as natural responses to the ambient climate. In the biogeographical scheme the vegetation was generalized as herbaceous; trees were seen to be anomalies if seen at all. Though they grew quite abundantly, trees were not part of the puna-páramo discourse (Borrero Vega 1989:37–44). Troll could not avoid mentioning the anomaly of these trees, for in his trips in western Bolivia he noted them on the slopes of some volcanoes. Nevertheless, true to his mindset, Troll (1985:307–10) interpreted this forest in terms of local climate: the *Massenerhebung* effect, whereby moisture is greater on the slopes than on the pampa below, allows trees to flourish. The sharp lower limit of *Polylepis* woodland on Mt. Sajama was assumed to correspond to a line of springs.

Aerial photography of the Andes provided new vistas of its natural vegetation (Miller 1929). Extensive forest tracts and patches of trees at high elevations became known for the first time. Air photos from Bolivia showed that more than 20 peaks in Bolivia's Cordillera Occidental are covered with *Polylepis* forests (Jordan 1983). Kessler (1995) calculated that 5,000 km^2 of *Polylepis* survive in that range, whereas only 630 km^2 of *Polylepis* remain out of what originally may have been 44,000 km^2 in the Cordillera Oriental.

In recent decades the debate has focused on how much the puna and páramo are caused by low temperatures, climatic shifts, and/or human agency. Ellenberg (1958) went further than any other observer by suggesting that the puna vegetation in southern Peru is an artifact of overgrazing. Plants found in abundance there (e.g., *Tetraglochin,* a thorn bush, *Opuntia, Stipa ichu,* and poisonous plants such as *Astragalus*) reflect intense livestock pressure. Pérez (1993) showed how cattle have destroyed the páramo plant cover farther north in the high Andes. But language barriers and a then unorthodox perspective made the anthropogenic factor languish for at least two decades. Following Troll, Marín Moreno (1961:43) saw *Polylepis* groves as relicts of a warmer geological epoch. Past climatic changes at high elevations during the Holocene cannot be denied. Confusing time scales, Cabrera (1968) saw the presence of trees there as relicts from a time when Andean elevations were lower than they are now.

Like his mentor Troll, the physical geographer Wilhelm Lauer used climate to explain Andean vegetation, but unlike Troll he later modified his view. Lauer (1952, 1976) maintained that the high-altitude puna and páramo are responses to thermal imperatives. Subsequently Lauer (1981:220; 1988:30) qualified his stand, acknowledging that human impact had lowered the upper timberline, although holding that the "present-day patches of forest on the paramo are relicts of past climatic and vegetational shifts." Lauer (1984:18) still showed skepticism about this matter by noting that Ellenberg's inclusion of human factors to account for contemporary high-altitude vegetation is a "hypothesis difficult to prove." Later he did embrace Ellenberg's hypothesis of an anthropogenically related vegetation change of the high Andes (Lauer 1993:165). Other researchers have similarly revised or modified their points of view. Quintanilla (1983) originally sought climatic variables to explain the páramo but turned to past and present anthropic action as more decisive. A suggestion of the paradigmatic shift on this question is found in Walter's survey of the earth's vegetation. Between the second (1979) and third (1985) English-language editions, Walter acknowledged the importance of human impact on the high Andean plateau.

Absence of a general vertical pattern that can be called a treeline presents an obstacle to understanding the puna and páramo as climatically determined grass or shrub lands. At Mt. Sajama, Bolivia, trees occur up to 5,300 m and in Ecuador up to 4,300 m at the Páramo de las Cajas. Even at those elevations no transition to *krummholz* occurs. The places where trees grow in the Andes are not generally defined by thermal thresholds. Nevertheless for southern Peru and northern Bolivia, Troll set the treeline at 3,800–4,000 m above sea level. That boundary was defined on a supposition of a climatic boundary that made most agriculture above that elevation a very risky proposition. In fact, however, fire destroyed most of the forests between 3,800 m and 5,000 m, and grazing has had a major effect in maintaining the grass cover. The parchment-like bark of *Polylepis* makes it highly flammable. Yet the carbonized trunk sends out sprouts, and seeds germinate well after fire. Failure to regenerate is more attributable to livestock browsing.

Figure 3.2. *Queñuar* is the highest-growing tree genus throughout the Central Andes; the species shown here is *Polylepis tarapacana* at an elevation of 4,800 m on Mt. Sajama, Bolivia.

Troll (1959) interpreted *Polylepis* to be relicts of past climatic shifts that survived cooling by occupying favorable microclimatic habitats. Lavallée (1985: 27), Rick (1980:16), and Baied (1991:49) uncritically accepted this explanation for tree growth at lofty elevations. Yet *Polylepis* has been found growing in exposed situations at exceedingly high elevations where temperatures on every night of the year fall well below freezing. On Mt. Sajama (lat. 18°S.), Bolivia's highest peak, *Polylepis tarapacana* up to 6 m tall grows between 4,300 m and 5,300 m above sea level (figure 3.2). Average annual temperature is 3.5°C. Moreover, this forest grows with an annual precipitation of only 347 mm and an intense solar radiation (Libermann 1986). Intolerance to salt in the soil best explains its absence in the pampa below (Kessler 1995:286).

Remaining groves or individuals on Sajama and other peaks are thus best explained either by fortuitous protection from fire or by sheer remoteness from human populations. Kinzl (1970:261) interpreted the continued presence of *Puya raimondii* in the Central Andes as fortuitous relicts of the fires set on the puna. A 1987 reconnaissance of the 189 stems of this spectacular plant still extant between 3,800 m and 3,900 m near Huancaray in Apurímac Department, Peru, found that all of them showed evidence of having been burned (Venero and Hostnig 1987).

Browsing also has played a major role in shaping high-altitude vegetation. Camelids once may have had foliage as a food source everywhere in the Andes. In 1995 llamas on Sajama were observed eating the leaves of *Polylepis* trees. Their wild ancestor, the guanaco, is also reported to have eaten these leaves. Today domesticated camelids are viewed as grass-eating herbivores; they are regularly conceptualized as animals of the puna, to which they are ecologically matched. But that association is the result of a human decision to put them where agriculture is too climatically risky. In the presence of domesticated animals *Polylepis* regeneration will usually not occur.

Pollen studies are too inferential to be useful in understanding whether the woody growth remaining today at high elevations represents vestiges of an incomplete anthropic transformation. Andean palynology operates on naturalistic assumptions in its study of changes in Holocene climate and vegetation after the contraction of the glaciers ca. 12,000 years ago. The same pollen profiles are open to very different interpretations (Van der Hamman and Nodus 1985). Moreover, grass pollen is extraordinarily abundant in the profiles because it, unlike that of many tree species, is wind-dispersed. Beyond that, some Andean tree genera, notably *Escallonia* and *Buddleja,* release very little pollen and sometimes none. A common inference has been that grass pollen dominance indicates a vegetation response to a climate unable to sustain the growth of trees. In some cases, tree pollen has been determined to be of "extraneous" origin if it has not fit into preconceptions of what the prevailing site vegetation should be (Kautz 1980; Graf 1981; Wright 1983).

A high index of graminaceous pollen has also been used to indicate man-caused deforestation, but the dating of it varies. Baied and Wheeler (1993) interpreted the high percentage of herbaceous taxa as signaling the onset of camelid pastoralism. Hansen, Wright, and Bradbury (1984:1462) claimed a change from trees to grass in pollen profiles from central highland Peru some time after 1000 B.C. Markgraf (1989:13) showed a sharp decline in fossil tree pollen around A.D. 1 and a simultaneous increase in weed and grass pollen. In another study Hansen, Seltzer, and Wright (1994) attributed the decrease in montane forest pollen types ca. 4000 B.C. to two quite different possibilities: a drier environment and an increased clearance for agriculture and need for wood fuels. Farther south in the Peruvian Andes a 4,000-year-old pollen and charcoal record recovered at Marcacocha, a lake above Ollantaytambo in southern Peru, conflates climatic changes, agricultural introduction, and deforestation (Chepstow-Lusty et al. 1996).

Archaeological evidence for prehistoric deforestation in the Andes is still scant. Smith (1980a, 1980b) has asserted that remains found at Guitarrero Cave, in north-central Peru, indicates a former forest cover between 3000 and 4000 B.C. Corroboration from other sites will add credibility to this study.

Little in MacNeish's work in Ayacucho leads to a similar conclusion. The historical and contemporary periods offer their own clues to Andean deforestation.

Process of Deforestation

Clearing

The near disappearance of the Andean forest involved prolonged human activities that began ca. 11,000 years ago (Lynch 1990). The Paleo-Indian hunters who first filtered into the Andean highlands from North and Central America were few in number but still capable of effecting ecological transformation. As aboriginal people have done all over the world, they burned woodland to expand the game-rich ecotone between forest and open land and also to marshal game animals in certain directions. Above 3,900 m slow regeneration due to low temperatures facilitated the transformation from woodland to grassland. At some point before 3000 B.C. agriculture and pastoralism appeared and provided the incentive for a definitive landscape conversion. Tubers and seeds domesticated in the Andes (potato, oca, *añu, ullucu, maca,* tarwi, chenopods) and maize, which was later introduced, formed the basis of agriculture between 2,600 m and 4,200 m. Remains of man-made terraces, irrigation canals, and ancient field boundaries indicate that highland agriculture near its upper limits was once more widespread and intensive than it is today (Cardich 1979; Bonnier 1986; Morlon 1992). Opening of land for cultivation was the main reason for deforestation, and it remains a major motive today (Wunder 1996).

Above 3,800 m bunch grasses and wet meadows succeeded crop fields. Little succession back to forest occurred, for domesticated camelids could flourish on this grass and sedge cover. Llama and alpaca herding played a role in creating the open treeless puna. Llamas are now largely restricted to the puna, but that was not their exclusive niche in the past. Providing herbage for them in a range of thermal environments motivated tree removal and grassland expansion.

Burning

Burning was the key means for clearing, and it had calculated outcomes, past and present. A sixteenth-century lexicon includes indigenous words for the action of burning forest for sowing (*cañani*), and nouns for the person who burns (*cañac*) and for the burned woodland (*cañasca*) (Domingo de Santo Tomás 1995). Observation yields insights into the transformation of Andean vegetation in all ecological zones. Vargas (1946:27) reasoned that the xerophytic nature of the flora between 1,000 m and 2,000 m in the Vilcanota/Urubamba Valley is a manifestation of the land impoverishment brought on by forest destruction. At elevations above 3,000 m, White (1985) observed

Figure 3.3. Burning the *páramo* on the slopes of Antisana above 3,500 m in Pichincha Province, Ecuador (Photograph by Eugene W. Martin, reprinted with permission)

contemporary peasant farmers over a three-year period in the Vilcabamba region of southern Peru. He found that all three basic vegetation categories—bunch grass (*ichu*), carpet meadow (*ch'ampa*), and shrubland (*matorral*)—were created by the use of fire. They are not biophysical responses to differences in soil, climate, or exposure, as has so often been supposed. Andean people have long had economic motivations to maintain these three types. Bunch grass and carpet meadow provide thatch and herbage; shrubs are a valuable source of firewood, which can be collected without an ax and by young children. Bowman's (1916:70) remark that the shrubby vegetation of the Andes "is of little value" is quite untrue. These unassuming ligneous plants—*Eupatorium* spp., *Dodonaea viscosa, Viguera australis,* and especially *Baccharis* spp.—have been exceptionally useful as firewood, kindling, dye sources, and in folk medicine. Species in the genus *Baccharis* collectively form the woody dominants over large areas of the Andes. Their seed production is abundant, and these plants rapidly regenerate after fires.

Without burning, the three categories would revert to the same forest that exists only in relict form. The Altiplano above La Paz (Bolivia) is now covered mostly with bunch grasses; Beck and García (1991:86) have interpreted it as originally being covered with trees, at least up to 4,200 m, before it was burned. In northern Ecuador farmers themselves have understood that the bunch grasses that prevail everywhere above 3,000 m are the result of burning (Knapp 1991:54) (figure 3.3).

Fire in the páramo-puna largely explains the form and composition of these treeless vegetation types. The Andean dry season makes the vegetation vulnerable to fire. At the same time, however, plants growing there have strategies to survive. Seeds, suckers, apical bud protection, and subterranean buds are adaptations to dry conditions that also help them through fire events (Laegaard 1992:155–56). Burning may be caused by lightning, which is exceptionally common during the rainy season, but burning is more frequently started by intentional human actions. Any one grassy tract in the Andean high country is likely to be burned every four or five years.

Wood Fuels

Need for wood formed another motive for deforestation. By A.D. 1500 an estimated 7.5 million people in the Central Andes had placed pressure on remaining forested tracts for construction materials and firewood. During the Inca period firewood was scarce in densely populated areas. As with food supply, collection of combustibles involved bureaucratic organization. The Inca placed firewood in storage facilities for redistribution, just as they did crops (D'Altroy and Hastorf 1984:340; Johannessen and Hastorf 1990). Firewood cutting, stacking, and storage were activities performed after the main crop harvest to ensure dry fuel during the rainy season, which normally began late in the year (Guaman Poma de Ayala 1980, 3:1055). Illegal removal was punished. By the time of the Incas much wood came from groves (*moyas*), whose cutting was strictly controlled (Polo de Ondegardo 1990:78).

Clearly most Andean deforestation cannot be attributed to the Spaniards. When the Europeans first arrived in the 1530s, the plateaus, slopes, and summits were described as treeless and without firewood (Jerez 1947:512; Diez de San Miguel 1964:159). Late in the sixteenth century, Ocaña (1987:218; my translation) asserted that "there is not one tree on the whole Altiplano," whereas other observers qualified their remarks on general treelessness. Otavalo was described in the *Relaciones geográficas* in 1582 as a bare open land but with many areas of forest (Jiménez de la Espada 1965, 2:235). In some places a distinction was made between the bare mountainsides and the *quebrada* bottoms, which had some tree growth (Cobo 1956, 2:235). Elsewhere the forest was on the mountainsides, not the valley bottom. Many wooded remnants had survived the Inca period in all topographies and altitudes. Individual species may have been protected for their medicinal value. Most native trees are included in the Andean pharmacopoeia as being good for something; for example, Lizarraga (1986:214) mentions that the flowers of tarco are used against syphilis.

After the conquest wood continued to be a critical material for making agricultural tools. Handles and shares were made of strong wood to prolong the life of a tool. To make the *chaquitaclla,* used for digging and turning the clod, peasants favored quishuar, colle, queñuar, tasta, unca, and lloque (Rivero Luque 1990). In the absence of these species, light woods, notably *huarango* and *waranway* were used, and in this century, eucalyptus. If no suitable wood was locally available, tools were acquired by trade at regional markets, or trips were taken to collect the raw material from distant forested areas.

Landscape Change in the Postconquest Andes

The Spanish conquest accelerated the demise of much woodland in the Andes. Controls on tree cutting were much more lax than in the Inca period, and decimation proceeded even with population decline. Postconquest clustering of rural settlements into *reducciones* placed intense pressure on trees within a walkable radius around the village (Gade 1991a). In addition to their own wood needs, some communities also met tribute obligations by supplying firewood to the authorities. In many Andean communities tree cutting for building materials, charcoal, or firewood is no longer part of historical memory (Treacy 1994:59). The new towns established for Spaniards used proportionately much more wood than those involving native people. In building construction, wood was used for roof beams, doors, and window sashes. Large beams and sometimes whole trunks still supporting roofs in buildings of colonial origin show the availability of large trees during that period. Such evidence is particularly useful in contrasting the present situation.

The introduction of seven animals, of which the sheep, cow, and goat were the most important, had a profound effect on the surviving Andean woods. These familiar domesticates re-created in the Andes a form of pastoralism that Spaniards had known in Iberia. Native people too found their own subsistence security could be enhanced by adopting European livestock that converted plant matter into useful products. Moreover, herding was as an activity that required far fewer people than farming did, and it expanded as human population densities and numbers plummeted with the high mortality caused by European-introduced diseases. Sheep, which provided wool for clothing, became the most numerous of the introduced animals. Their sharp hooves and voracious appetites took a marked toll on woodland. Browsing destroyed tree seedlings, and trampling compacted the soil. When old trees died or were cut, no saplings were available to fill those breaches. In that way forest patches eventually disappeared without assault from the ax.

Colonial authorities viewed the paucity of wood primarily as a fuel problem (Matienzo 1967:272). The presence of trees to meet firewood needs was, along with water, the critical factor in the location of Spanish settlements in the Andes. In Bolivia, La Paz's canyon had trees below 3,300 m (López de

Velasco 1971:254). As the city grew in the seventeenth century, that wood supply diminished, and the surrounding puna above had to be scoured for anything combustible (Mendoza 1976:33, 37). Wood beams for building were in short supply almost everywhere. A viceregal decree attempting to deal with timber shortage permitted removal of tree branches but not the main trunk. That idea was probably borrowed from the forestry laws of Castile, also a region with a long tree-sparse history. However, the consumption patterns of Europeans in the Andes undermined a credible tree policy. Cobo (1956, 1:236) calculated that more wood was burned in the house of a Spaniard in one day than in an Indian dwelling over a month. Whereas space heating was not part of native tradition, Spaniards introduced charcoal braziers for warmth. For that use and for cooking, charcoal became the fuel of choice in larger towns throughout the Andes. Garcilaso de la Vega (1960:374) remembered that in his youth innumerable molle trees in the valley of Cusco were cut down to be made into charcoal. The Inca knew of charcoal and called it *quillimca* or *sansa,* but the Spaniards greatly expanded its production in the Andes. The Spanish *carbón* replaced the native names for it, suggesting that its colonial associations were European, not indigenous (Jiménez de la Espada 1965, 2:22).

Although Spaniards introduced many new crops, they did not bring tree species to be used primarily as wood supplies. Cutting trees for charcoal had devastating effects where rainfall was sparse. Woodlands once had covered huge areas in the arid coastal valleys. In 1678 the archbishop of Lima wrote the king to complain that charcoal making had destroyed the woodlands (*yermos de monte*) around the city. Wood for this purpose had to be cut more than 60 leagues (300 km) from Lima. One year's charcoal needs of Lima required 35,000 ha of woodland. Wood not used in the city was shipped out via Callao to supply other places (AGI, Lima 78, 1674). By the nineteenth century constant high demand for wood affected the landscape beyond the Rimac Valley. The Paramonga Valley, north of Lima, had become a treeless desert. Many people wrongly assume that the Paramonga and all the coastal valleys have always been so barren.

European technology created new fuel demands on the meager wood resource in the highlands. Bread ovens, tile works, brick factories, and sugar cane boiling and distillation consumed large amounts of combustible material. But it was mining that had the most devastating effect on tree vegetation. Wood and its charcoal derivative were the prime fuel for smelting certain ores. At Huancavelica, in central highland Peru, for example, the trees growing on the puna there were cut down in the sixteenth century by Spanish miners to feed the mercury ovens (Favre 1975:420). Huge numbers of tree trunks were also needed as underground mine girders (*callapos*). Thousands of abandoned mines dot the Central Andes, and all of them used wood as props.

In this century railroads that penetrated the highlands contributed to forest destruction. Tree-poor Bolivia was especially vulnerable in this regard. In Cochabamba Department, much wood, especially soto, was taken out for railway ties and for the locomotive firebox (Schlaifer 1993:596). *Polylepis* forests surviving on the Bolivian Altiplano were cut to provide the fuel for wood-burning locomotives. Rail transport made profitable cutting trees to make charcoal for sale in the cities. Between 1935 and 1945 most charcoal consumed in Oruro and La Paz came from the *Polylepis* woods in the Cordillera Occidental on Mt. Sajama and elsewhere (Cárdenas 1969). A 1939 decree that prohibited further cutting of the world's highest forest had little effect. The Sajama forest was saved by the introduction of kerosene and bottled gas in cities of the Altiplano. These modern fuels do not reach rural areas of the Altiplano, which still depend on shrubs, *yareta,* and/or animal dung for combustibles (West 1987). Llama dung, which has a heat value close to that of wood, is important (Winterhalder, Larson, and Thomas 1974:98). Many inhabitants are too poor to afford alternatives, even though, despite the cold, combustibles are used primarily for cooking. The practice of Altiplano dwellers bringing, as a trade item, precooked maize from the temperate valleys suggests that fuel is held at a premium.

Elsewhere in the rural Andes firewood continues to be the major fuel. Córdova Aguilar (1993) interviewed peasants in the Sierra de Piura on their wood preferences, modes of collection, and commercialization channels. Brandbyge and Holm-Nielsen (1986:107) found that in highland Ecuador a family must gather firewood for two days to meet its needs for a week. In the Lake Titicaca region of Peru and Bolivia, one family is said to use seven tons of firewood yearly (Chevalier et al. 1990).

European control of the Andes reinforced the indigenous and Spanish custom of burning to manage land. Increased livestock populations, especially of sheep and cattle, needed more herbage. On the heights above the valleys, pastures were expanded on communally held land, whose use was open to all members. Pasture renewal became a communal activity and was placed in a quasi-religious context. The Feast of St. John the Baptist, a traditional time of bonfires in its European context, was an occasion for burning the grass cover in the Andes of Peru and Bolivia. The puna is ablaze on this night. Elimination of the dry inedible culms promotes more green shoots months later when the rains arrive. However, burning thwarts the reestablishment of tree seedlings (Hess 1990:339).

Length of the dry season has played a part in the differential deforestation of the Central Andes. Cold and dry areas had an earlier and more thorough loss of tree cover than warm and moist regions had. The west side of the Andes, an area made dry by the Pacific anticyclone and cold Peru current, lost most of its cover long before the conquest. Conversely, on the eastern Andean slope, intense rainfall makes agriculture very difficult, and for that reason little farming of the *ceja de la montaña* (the montane cloud forest) occurs. Farmers may be encouraged, however, if a month during the year is nearly dry. Under those conditions a cut forest patch will usually burn. Less interest in farming this zone is coupled with high rates of regeneration. Abandoned fields revert to forest within a decade.

When the dry season is prolonged, fires can spread over wide areas of normally rainy climate even without cutting. In August-September 1988 between 3,000 ha and 4,000 ha of forest burned over a 16-km distance up-valley from Machu Picchu (Instituto nacional de cultura 1988). Because in the previous May no rain had fallen in this sector, a huge area of forest was engulfed in fire when a peasant practiced slash-and-burn in his field. In September 1987, more fires started by swidden farmers got out of hand and again burned a large portion of the cloud forest vegetation of Machu Picchu Sanctuary.

The density of the human population also affects deforestation. Impacts on the forest are greatest where agricultural activities have been the most intense. Alluvial valleys of temperate climate have offered superb farming conditions, and as the population increased, the pressure on removing trees there became intense. Conversely, the montane cloud forest has had few inhabitants, and much of it remains intact.

Outcomes of Deforestation: Species and Communities as Relicts

One result of the wholesale vegetation removal from the Andes has been that many native tree species of the forest are now rare. Some species possibly have become extinct without any scientific record of what they were.

Extreme rarity of some species of cedro (*Cedrela* spp.) reflects their heavy human exploitation. This genus, with about 10 species once found in the Andean highlands, is seldom encountered above 2,500 m today. Its large size, straight main trunk, resistance to insect damage, and durability made cedro the preferred construction wood in the colonial period. Easily worked with tools, cedro was the prime material from which pulpits, *retablos,* moldings, and furniture were carved. Some sparsely populated temperate valleys still harbored stands of this tree at the time of the conquest. A 1590 document mentioned cedro growing in the upper Apurímac Valley at 2,700 m (Aldrete Maldonaldo 1989:6). In the late eighteenth century cedro was mentioned first on a list of 11 useful trees found in the Doctrina de Huanipaca (Apurímac Department) ("Descripción de la Provincia de Abancay" 1795:123). In his mid-nineteenth-century travels in Apurímac Department, Raimondi still found it growing wild in certain quebradas (Santillana Cantella 1989:28, 142, 144). The few cedro trees seen today in the highlands are planted as ornamentals (Herrera 1933), and almost all cedro wood has for centuries come from the still-forested eastern slope. The tree is now typically viewed as a species native only to the upper selva; most foreign botanists and ecologists seem unaware that this genus once had an ecological niche in the highlands up to ca. 3,000 m.

Nogal (*Juglans neotropica*), which has a beautiful and strong wood, was almost as valuable as cedro. It underwent rapid extraction after the Spanish conquest. In the sixteenth century Zárate (1947:468) remarked on the presence of "*nogales silvestres*" in the highlands. Today they are rare. Much more common is *Alnus jorullensis* (syn. *acuminata*). This species has been used as firewood and formerly also in making gunpowder. Its trunks have been hollowed out to make pipes for moving irrigation water, but it is not a good construction material ("Diccionario de algunas voces . . ." 1791:87).

Some zones of the Central Andes have retained large tracts of forest into the twentieth century. In Ecuador above 1,200 m there are still more than 3 million ha of primary or secondary forest. Its survival owes much to sparse human population, difficulty of access, and/or special protection (Brandbyge and Holm-Nielsen 1986:41). Not until the 1930s was the *Podocarpus* forest on Mt. Ampay, in Tamburco District, Apurímac Department, botanically identified (Vargas C. 1938). This coniferous species, ca. 12 m tall, occurs there between 2,900 m and 3,600 m above sea level. Ampay, declared in 1967 a national sanctuary of 3,640 ha, is nevertheless undergoing wood removal and animal grazing (Venero González and Tupayachi Herrera 1989). Vargas C. (1946), in his botanical explorations of southern Peru in the 1930s and 1940s, found substantial woodland off the beaten track. *Eugenia* wood near Huanca Huanca (Paruro, Cusco), though used by local people, was still flourishing into the 1970s (Gade and Escobar 1972). Only in the 1980s did these and other patches of woodland in the highland start to be cataloged (*Inventario y evaluación* . . . 1985:318–19).

On the western slope, forests were under intense pressure, but a few remnants survived into the twentieth century. Most notable has been the "bosque de Zárate," in Huarochiri District, at an elevation between 2,700 m and 3,000 m (Hondermann 1988). Known to science only since the 1950s, Zárate forest mostly disappeared over the next 30 years (Rostworowski de Diez Canseco 1981). This relict may have been part of a pre-Hispanic design to protect the upper watersheds of coastal valleys so dependent on irrigation.

Strict local control on cutting has been one reason for survivals. Communities which derive sources of livelihood from the living trees are the ones most likely to impose constraints on cutting. At Surcubamba (Huancavelica Department, Peru), wax was extracted from the fruit of *Myrica* sp. and used to make candles (Ráez 1898). In Ecuador the 20-ha *arrayán* woods (*Myrcianthes* sp.) between 2,700 m and 2,800 m in Carchi Province has survived because of its harvestable products. Its poor-quality wood is of little commercial value, but its leaves and fruit have several medicinal uses in personal hygiene and folk medicine (Cerón and Pozo 1994). In southern Peru chachacomo is the main host plant for *huaytampu,* an edible lepidopteran chrysalis (*Metardaris cosinga*), which provides protein to some peasant diets. But this advantage did not necessarily prevent cutting, for chachacomo wood has been valued for tools and mine props.

High elevations retain more forest cover than is generally recognized. Cloud forest survives on the slopes of the Pasochoa volcano, 35 km from Quito. Air photography revealed that the rate of deforestation between 1956 and 1982 was 2 percent but that, with protection, the forested area increased between 1982 and 1987 (Stern 1995). In contrast, in Loja Province in southern Ecuador a podocarp forest placed in a national park has not been protected from anthropogenic fire, which has consumed portions of the timberline woodland (Keating 1997).

Persistence of *Polylepis* forests on the slopes of Mt. Sajama and neighboring peaks owes much to the area's sparse population, relative inaccessibility, and lack of agriculture. In his visit of 1926 Carl Troll (1985:307–10) noted the use of *Polylepis* wood for construction and for firewood in the village of Sajama. Troll deplored the practice of villagers digging out the roots of *Polylepis* for fuel; he called for a law to ensure that the trees' roots remain in the ground. Although this community of fewer than 500 people has depended on this wood source, their numbers have not been large enough for the forest to be destroyed. Even the commercial exploitation of *Polylepis* for charcoal in the 1930s did not exhaust it.

In some cases superior capacities of regeneration have enabled a tree species to withstand intense human use. Of all Andean tree species, molle seems best able to maintain itself. It thrives below 3,200 m along roadsides and on dry hills. Its abundant production of seeds, which are spread hither and yon by birds, is matched by its ability to volunteer under high light conditions and on poor soils. Molle firewood and charcoal make satisfactory fuels, and the species is heavily exploited for that purpose. Its resin has use in folk medicine. Its abundant pink berries are still fermented in places to make chicha de molle. The crotches of standing trees serve as places to store harvested crops out of the reach of hungry livestock.

Given the nature of rural life in the Andes, the future of remaining tracts, islands, and patches of forests is problematical. Under the lawlessness associated with the 1980s in Peru the three-hectare *Buddleja* woods in Lampa District (Puno) were destroyed. Besides societal disorganization, earthquakes are ever-present threats in the Andes; they can create immediate demands for huge amounts of construction wood from any source.

Tree Cultivation in the Inca and Colonial Andes

The pre-Hispanic practice of planting trees to meet the demand for wood continued into the colonial period. Cultivating trees as perennial crops occurred in response to forest removal that left many communities with no other local source of wood. Prime sites for tree cultivation were the alluvial, irrigated soils of temperate valleys, even though slope locations would have competed less with field crops and would have checked soil erosion.

Planted trees survived the conquest as part of a valley landscape known as *campiña,* which includes crop fields and pastures outlined on their margins by trees (Morlon 1992:228–33). Early colonial documents record an Inca practice of distinguishing these cultivated trees (*mallko*) from those not planted (*sacha*). In the *Relaciones geográficas* of the sixteenth century, in reference to Cusco the point was made that "hay otros árboles que son puestos a mano" (there are other trees that are put in by hand) (Jiménez de la Espada 1965, 2:22). In fact, the Spaniards did not at first understand that these planted trees were private property, and cut them down. Ocaña (1987:227) wrote that "el valle de la ciudad no tiene arboleda" (the city's valley has no trees), which suggests an observation made after massive removal had occurred. Around 1590 the colonial authorities made a major effort to reforest the Cusco Valley (Sherbondy 1988) by resurrecting the Inca tree-planting tradition. Most trees were put there by ayllu members, who planted the cuttings, watered them, and kept livestock away. Each year until 1646 a "mayor of the forest" (*alcalde de las arboledas*) was elected to enforce compliance. Fines, whippings, and jail were punishments for people caught felling cultivated trees. By contrast, wild-growing trees were free sources of firewood.

Planted trees belonged to individuals, and their ownership was recorded on the *quipus* (Rostworowski 1990). One person may have had from 2 to 280 trees. Spaniards recorded them as "*pies de arboleda.*" Trees were listed along with children, crops, and livestock as part of one's estate. In southern Peru aliso and quishuar were the two major planted species (Aldrete Maldonado 1989:5–14; "Reparto de tierras en 1595" 1957; Jiménez de la Espada 1965, 2:22).[1] Even at very high elevations trees were planted, and possibly underwent genetic selection. Cardich (1984–1985:77) considered the *Polylepis* trees he found growing near settlements in the central highlands of Peru to be domesticated. Their straight trunks contrasted with the twisted trunks of wild-growing trees.

In the colonial period greater quantities of wood were needed than before the conquest. Beams, lintels, window frames, doors, and floors were new uses of wood introduced by the Spaniards (Sutter 1985). Coffins were another brand new need for wood, although their general use did not emerge until the eighteenth century, when earth burials became common. Foliage from both planted and wild trees partly sustained llamas, cattle, goats, and sheep. The Spaniards knew this feeding practice (*ramoneo*) in Iberia, where tree leaves often provided the only green fodder during the dry season.

The Case of the Buddlejas

Quishuar has had a particularly important role in Andean domestic economy. Guaman Poma de Ayala (1980, 2:760) suggested that it was the most important tree in the late Inca period. The main species is quishuar proper (*Buddleja incana*), which grows between 7 m and 15 m tall at elevations between 2,500 m and 3,600 m. Colle or puna quishuar (*B. coriacea*) is a higher-altitude version with a twisted trunk, dense branching, compact form, and small size (4–10 m tall).

Their useful qualities have elevated these two congenerics as objects of cultivation. Reproduction of both has been by cuttings (*estacas*), not seed (Matienzo 1967:274). Today cuttings are planted in containers before the rainy season and, once roots form, are transplanted to a space near the dwelling. Saplings are fertilized, irrigated, and shaded, as well as protected from hungry livestock and strong cold winds. Often enclosures of earth or stones, 50–80 cm high, are built to protect the saplings. Although wild-growing trees can be found, most *Buddleja* trees in the Andes have been planted, and O. F. Cook (1925) even included quishuar on his list of the cultivated plants of Peru.

Seed of *B. incana* and *B. coriacea* has a 75–90 percent germination rate (Brandbyge 1992:270). Nevertheless, reproduction by cuttings is preferred. These grow into trees much more rapidly than do seeds, which are tiny and do not form if frost destroys the blooming inflorescences. The suckering habit of both species produces poles that can be coppiced every three years from the same stem. When harvested repeatedly for their canes, *Buddleja* trees live about 20 years. Used as roof poles, these canes have been an item of trade to areas where they are in short supply.

The heavy wood of *Buddleja* trees (quishuar and colle) has long been prized for agricultural implements. The chaquitaclla, a tool still used for breaking up heavy sod, has traditionally been made from it. Yellowish-white *Buddleja* wood is easily worked and is a prime raw material for carving. Today spoons and ladles are made from it, but at an earlier time statues and goblets (*queros*) were also fashioned from the wood. During major festivals the Inca ritually threw wooden cups and figurines into bonfires. That custom alone must have required large quantities of wood and provided a motivation for planting this tree before the Spanish conquest. Goblets to hold chicha continued to be made from quishuar wood well into the colonial period. *Buddleja* also has been much valued as a cooking fuel, one advantage being that it burns especially well under the humid conditions of the Andean rainy season ("Memoria de subprefecto . . ." 1874:78). Charcoal made from it has also been much appreciated (Matienzo 1967:274). However, in most communities in the Andes it is now too rare to be used as a combustible.

Both *Buddleja* species have had several uses that do not involve their wood.

The leaves, said to be rich in phosphorus, have been used as fertilizer and sheep fodder. In folk medicine the leaves serve as a poultice for toothache and in infusion as a diuretic. The orange-red flowers that bloom during February–March have served as a substitute for Old World saffron in tinting food yellow (Choquehuanca 1833; Saignes 1980:12).

Naturally growing stands of *Buddleja* of both species are rare. Remote areas between Ecuador and Bolivia contain small clumps (Tovar 1973:17; Balslev and DeVries 1982:20; Venero and Macedo 1983:22–23; *Inventario, evaluación . . .* 1984:215; *Inventario y evaluación . . .* 1985). Given the peculiarities of their distribution and their presence in areas where human impact is less, it seems clear that they are the remainders of a much more widely distributed vegetation cover. A former broader distribution of *Buddleja* can also be inferred from the use of the word *quishuar* and its orthographic variants in the place names of mountains, rivers, settlements, and even agricultural parcels where these species are no longer found. In the Peruvian Andes many places have prefixes in their names that refer to this tree where none survive today. Among them are Quishuarchayoc (place); Quishuarpata (terrace); Quishuarani (compound); Quishuarcancha (enclosure), Queshuaraqui (locale), Quishuarcalla (stony area) and Queshuarhuayoc (canyon).

Diffusion of Trees

The Inca purposely took seeds and cuttings of several trees to cultivate them in other regions. A willow species (*Salix humboldtiana*), native to central Chile, was growing in the Central Andes when the Spaniards arrived. Tolerant of wet soils, willow was planted to stabilize riverbanks on poorly drained floodplains. Cobo (1956:238) mentioned willow as a fast-growing tree suitable for firewood, but its pliant branches also have been woven into baskets. Reproduction is by cuttings. Molle may have been carried from the heart of the Inca Empire in southern Peru to the northern frontier in Ecuador. By describing molle as unfamiliar in Ecuador, Velasco (1977:817, 820) suggested that it was a recent arrival there. In southern Peru and elsewhere another wild tree, *pisonay* (*Erythrina falcata*), was carried as an ornamental from its wild-growing ecological habit between 1,800 m and 2,400 m to elevations as high as 3,100 m.

Several tree species with edible fruits and useful wood were taken to valleys below 2,800 m. *Lucuma* (*Pouteria lucuma*) has a mealy, but appealing, fruit and a beautifully grained wood. *Cherimoya* (*Annona cherimola*) yields an extraordinarily delicious fruit and a good firewood. Avocado (*Persea americana*), valuable for its fruit of high oil content, is a tree of Mesoamerican origin which diffused southward in western South America. *Pacay* (*Inga* spp.) has an edible filament contained within a large leguminous seed pod, but in some places it was valued more for its excellent firewood. Nogal (*Juglans* spp.) first occurred as a wild member of the highland flora below 3,000 m but was later cultivated.

Its beautiful wood for carpentry, edible nuts, and leaves yielding a rich brown dyestuff together have enhanced its value. *Capuli* cherry (*Prunus serotina*), a tree of Mexican origin, has tasty fruit, good wood, and supple branches for making baskets. Capuli was most probably introduced to the Central Andes by the Spaniards in the early colonial period from Mexico. This tree thrives in temperate valleys where moisture is available and is propagated from cuttings rather than seed.

Other Postdeforestation Strategies

Since before the Spanish conquest, wood was a trade item in many highland communities. *Chonta,* an extremely hard palm wood (mainly *Bactris* spp.) from the upper selva, was obtained for tools. After the conquest, with the disappearance of highland wood sources, more attention was directed toward the montaña forests of the eastern piedmont. Although still partly forested today, this zone now has many fewer trees of the most preferred species because of selective removal. In this century, road access greatly expanded the wood trade from the ceja and the upper selva below it.

Another strategy has focused on alternative combustibles. Woody shrubs form a successional stage on land formerly in trees and are valued by many peasant communities as a major source of cooking fuel. They are abundant and are readily collected by small children without axes, who in that way can contribute to family chores. If ligneous material is unavailable, dried animal dung is burned. Grass is collected as kindling.

Soil erosion has been the most far-reaching outcome of highland deforestation. It is today the most serious land-use problem of the Central Andes. My own estimate, based on Andean travel, places more than 75 percent of the soils in serious erosion. Perhaps 30 percent of the area is ruined beyond easy repair. Tree removal exposed sloping land to the heavy rains, and subsequent unwise agricultural practices exacerbated erosion. Overgrazing by sheep, which have long been kept at the biological limit of their carrying capacity, is a major culprit (LeBaron et al. 1979). Some zones, such as Tarija Department in southern Bolivia, manifest serious degradation over vast areas (Preston, Macklin, and Warburton 1997). Terrace systems that safeguarded soils covered no more than 10 percent of the land cultivated during the pre-Hispanic period, indicating that some Andean soil erosion undoubtedly began before the Spanish conquest.

For more than a century reforestation with exotics has been the main Andean response to concern about the removal of the native tree cover. Though less than 1 percent of the Central Andes has been reforested, plantations are by far the most important source of wood. *Eucalyptus globulus,* introduced at various places in Ecuador, Peru, and Bolivia between 1860 and 1890, is the dominant species (Dickinson 1969:296). Today eucalyptus meets much of the demand in those three countries for board lumber, pit props, and house beams. Its exceptional rate of growth in indifferent soils is considered providential in this wood-starved region.

Three decades of eucalyptus projects in peasant communities have revealed this genus's negative effects. Eucalyptus plantations suck enormous quantities of water out of the ground, which in the Andes has resulted in lowered groundwater tables and dried-up springs. Moreover, eucalyptus emits a toxicity that precludes native biota on these plantations. Without an herbaceous strata, soil erosion is not effectively checked. Monoculture is another drawback. *E. globulus* alone represents more than 90 percent of planted trees in the Central Andes. Without natural enemies to keep them in check, pathogens and pests could rapidly spread and potentially destroy all Andean plantations (Zalles Flossbach 1984). Species diversification lowers the possibility of such an occurrence. Monterey pine and several other eucalypt species are planted to a small extent in better-watered areas.

About 200,000 ha of plantations dot the Central Andes, but more than 10 million ha of land should be forested for long-term environmental rehabilitation. Plantations are not necessarily the only way to reforest. If land were freed up for that purpose without competing uses, native species would spontaneously, though very slowly, reestablish themselves over the Andes. Outmigration and land abandonment create the opportunity for the reemergence of the Andean forest. Research has revealed more about the forestry potential of Andean species (Brandbyge 1992; Casanova Abarca et al. 1985).

Meanwhile, destruction of remaining woods continues. Even those officially protected are vulnerable. For example, the podocarp forest at the Ampay sanctuary near Abancay, Peru, shrank by 400 ha between 1970 and 1992. Inhabitants around Abancay regard it as a local lumber, cabinet, and firewood (*leña*) source. A new road always hastens removal and is often a death warrant for a relict wooded area. Extensive woodlands of chachacomo, tasta, queñuar, and quishuar around Chinchaypuquio (Cusco) on a ridge above the Apurímac canyon have now largely disappeared (Olarte Estrada 1985). Completion in 1965 of the road between Chinchaypuquio and Cusco opened the zone to wood sellers, who loaded large trucks full of firewood. If at some point forests receive real protection, some remnants may be saved.

Much more insidious is the effect of livestock, because they play the key role in preventing regeneration. The Central Andes have a domesticated herbivore population that can be estimated at 43 million animals. Sheep, which number at least 25 million in Ecuador, Peru, and Bolivia, have by far the greatest effect. They are ubiquitous, and they eat close to the ground. Nine million head of cattle have an important negative effect because of their heavy weight and trampling tendencies. Goats, which may total only about 2 million animals, are locally a problem, especially in northern Peru and the drier Bolivian valleys. The three countries also have more than 6 million domesticated camelids, which are concentrated at elevations above 3,800 m in zones of sparse vegetation where plant growth is slow. Effects of livestock even at high elevations are devastating. Without that impact *Polylepis* would quickly regenerate on grasslands (Kessler 1995:287).

Case Study: Northern Chuquisaca, Bolivia

The deforestation process that started more than five millennia ago in the Andes can be analyzed at a larger scale for the historic period. An examination of the northern part of Chuquisaca Department in central Bolivia provides a perspective on Andean landscape change in which human activity is the key. This highland region of more than 12,000 km^2 lies between the lofty Altiplano to the west and the tropical montane forest to the east. The upland surface into which valleys have been cut has elevational differences of ca. 1,000 m; the average temperature at 3,500 m is 13°C. and at 2,500 m is 17°C. Yearly rainfall amounts vary from about 500 mm in the west to 900 mm in the east.

Extrapolating from the larger Andean framework, it is likely that human occupation in northern Chuquisaca goes back more than five millennia. The emergence of agropastoralism probably predated the Inca by 3,000 years. The Yampara were the earliest defined ethnic group. They were joined in the fifteenth century by several other groups brought in by the conquering Inca (Presta and Del Rio 1993). Chuquisaca was much less densely populated in the pre-Hispanic period than lands closer to the Inca core around Cusco were. Eastern Chuquisaca was a no-man's-land because the Chiriguano people's incursions onto the plateau made settlement risky. Thus, when the Spaniards came here in the late 1530s, much more tree vegetation was intact than in the core.

Spaniards founded La Plata (now Sucre) in a temperate basin at 2,800 m, and it became a major Andean focus of settlement. Silver mining at Potosí, 130 km to the southwest, generated much of La Plata's wealth. Both cities required large amounts of construction materials and fuel. East of La Plata/Sucre five main towns were founded in the late sixteenth century: Yamparaez, Tarabuco, and Zudañez were founded as Indian reducciones, and Tomina and Padilla were established as fortifications against intrusions of Indian foot warriors from the tropical realm to the east.

Postconquest impact on the highland forests of Chuquisaca was especially heavy during three periods. Between 1545 and 1650 mining activities in Potosí required huge quantities of wood from a large radius around the city. From ca. 1780 to 1830 the anarchy and mayhem of independence suspended colonial rules and led to forest destruction. In the 1950s agrarian reform and the demise

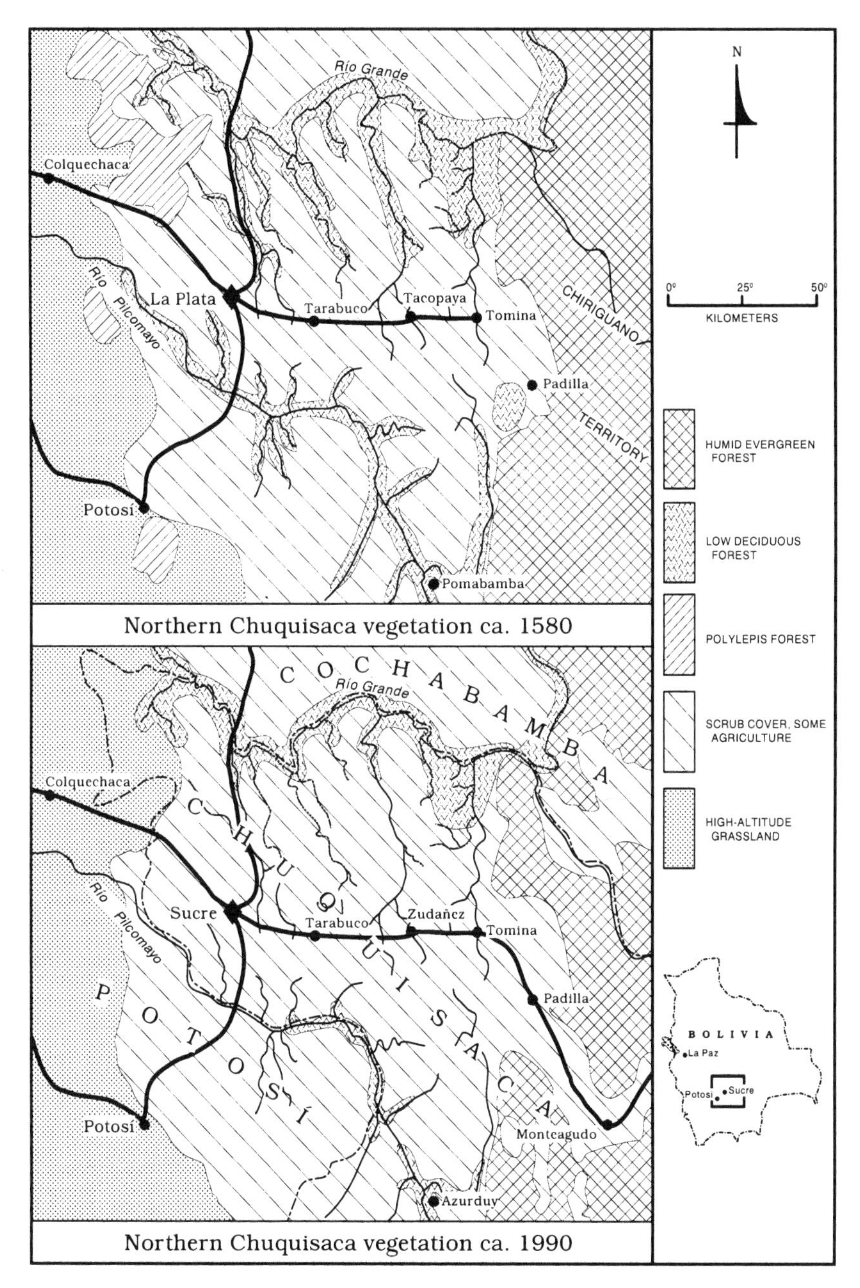

Figure 3.4. Vegetation change in northern Chuquisaca, Bolivia, between the late sixteenth and twentieth centuries as derived from travel accounts, archival documents, and field observation

of the hacienda system gave peasants free reign to cut down for fuel trees that had formerly been reserved for higher uses (Preston 1969:12–13).

The wooded tracts that still covered a substantial part of northern Chuquisaca in the Inca period contracted rapidly in the sixteenth century (figure 3.4). Woodlands on the interfluves disappeared first, whereas the warmer and more humid valley bottoms and sides maintained tree growth longer. This upland-valley topography occurred in both western and eastern Chuquisaca. Those two zones and the area within a 30-km radius around the city of La Plata/Sucre suffered from both internal and external pressures on the wood resource, which sealed the fate of the vegetation cover.

Northwestern Rural Chuquisaca

The northwestern zone of rural Chuquisaca has long been within the fuel-supply orbit for mines located to the west, in Potosí Department. Over the centuries the intensity of fuel collection has had a major effect on the vegetation. Lizarraga (1987:224) stated that the Potosí area, in spite of its high elevation, had before the silver discovery a thick growth of twisted queñuar trees good only for firewood and charcoal. The only wood used in early house construction was ligneous stems of agave for roof beams (Bakewell 1984:11). With the depletion of fuel around the city of Potosí, people had to turn to valleys 7–20 leagues (35–100 km) away to acquire firewood and charcoal (Murua 1987:567).

Chuquisaca became the source of fuel wood for Potosí because vegetation on the Altiplano west of the mines was sparse even at the beginning of the colonial period. Llamas, donkeys, and mules transported firewood and charcoal from as far as 150 km away. Charcoal, given its higher energy value per unit weight, could sustain the cost of transport beyond a 100-km radius. The high country of western Chuquisaca above 3,600 m had extensive *Polylepis* forests (Schulze and Casanovas 1988). Traveling in the final years of the eighteenth century, Helms (1807:21) noted that brushwood for Potosí City was brought from 100 km distant. As alternative sources of energy became cheaper in the 1950s, Potosí's use of firewood from this area diminished.

Two other mining centers required fuel from western Chuquisaca. In the nineteenth century Colquechaca in northern Potosí developed into a large silver mining center, where calcination and smelting activities required the use of exogenous fuel (Godoy 1990). Colquechaca became a major trade center for fuel of all kinds: queñuar, *thola* (*Lepidophyllum*), ichu grass (*Stipa*), yareta (*Azorella* sp.), peat (*turba*), and llama dung (*taquia*). Silver mining at Colquechaca faded away by the twentieth century. Llagualla in southwestern Oruro, which was a major tin mining center for more than 60 years, replaced Colquechaca as an avid consumer of all vegetal sources of energy.

In northwestern Chuquisaca construction wood came from the Pilcomayo

Valley through the colonial period. Trees with large trunks grew in the bottom of this subtropical valley, part of which was difficult to reach. Most wood came from the parts of the valley that were about 10 km north and south of the royal road between Potosí and La Plata/Sucre. Teams of Indians, not beasts of burden, carried the tree trunks out. It took four days to walk from Potosí to the Pilcomayo depression, two days to cut down a large tree, and eight days to carry it back to Potosí (Sarabia Viejo 1989:393–94).

Northeastern Rural Chuquisaca

When the Europeans arrived, eastern Chuquisaca had more forest cover in both its interfluves and its valleys than the western part of the department did. Ramírez del Aguila (1978:60) wrote of the "arboleda" of palms, cedros, and tipas around Presto, six leagues east of La Plata. Significantly the area was described as full of springs, which characteristically flowed until the forest vegetation was removed.

That regional difference persisted into the nineteenth century. In Tomina Province, forests of 10 species of large trees covered the uplands; other large specimens grew in the valleys (Torres de Mendoza 1868:320; Giménez-Carrazona 1978:25). In the Taconaya Valley (Zudañez) at 2,500-m elevation, algarrobo trees were cut in 1750 to construct the large mint (Casa de la Moneda) in Potosí (García Quintanilla 1978:57). Additional wood for this building came from other surrounding communities: Mojotoro, Sopachuy, Pomabamba, and Tomina (Abecía 1939:127).

This northeastern part of Chuquisaca was an area of active wood cutting by the mid-seventeenth century, when large trees were no longer available in the Pilcomayo Valley to the west. Higher rainfall and temperatures favored faster woodland regeneration in this zone than to the west. Moreover, settlement was less dense, in part because of real and imagined threats posed by raids by Chiriguano people from the sub-Andean forest and malaria, which was endemic in the valleys. Nevertheless, such was the demand for wood in Potosí that much timber was taken, even at this distance from the city. Teams of men hand dragged large tall trunks across the mountains from as far as 200 km away. Only their extraordinary value in the "Villa Imperial" made that economically feasible. Cedro was the prime wood from this zone for general construction. Trees of harder wood than cedro were also cut to build the refining machinery for the Potosí ore mills (Ramírez 1968:53). Drive shafts for the stamp mill required single beams 5.5–7 m long (Bakewell 1984:24). Algarrobo and to a lesser extent soto (especially for the wheel axles), tipa, tarco, and *quina quina* (an introduced species) were used for this purpose. Molle wood was used for the cogs in mill wheels.

Need for furniture and coffins in Potosí and La Plata/Sucre gave rise to a carpentry tradition in this zone. Artisans (*ebanistas*) emerged in small communities that were accessible to stands of trees of cabinet-making quality. Charcoal, though not firewood, normally could stand the high cost of transport to Potosí from eastern Chuquisaca. In addition, much wood was needed locally for domestic cooking fuel, bakeries, chicha elaboration, and sugar mills; together these demands placed additional pressures on the wood resource. Furthermore, regeneration of woodland was hampered by a grazing economy that needed pasture, for, until the 1920s, northern Chuquisaca was an important source of livestock sent on the hoof to Potosí (Langer 1989).

Such large amounts of cedro wood from this zone went to Potosí that none remained on the uplands. In some warm valleys cedro trees persisted, especially in the provinces of Zudañez and Boeto, which border the Río Grande. There, dense forests of cedro, podocarps, and nogal (*Juglans australis*) persisted into the mid-twentieth century (Ressini 1958).

Urban Radius of La Plata/Sucre

La Plata had more than 20,000 people through most of the colonial period and was a major consumer of wood. By 1950 its population had reached 40,000, and it is more than 100,000 today. Heavy wood depredation has characterized a 30-km radius around the city. When La Plata was being built, large timbers were needed for churches and other monumental buildings. As elsewhere, cedro and nogal were the most desirable woods. Their large trunks were used for beams and for carved balconies, choir stalls, altars, statuary, and moldings.

Cedro trees were still part of the natural vegetation around La Plata/Sucre in the early colonial period (López de Velasco 1971:253). Viceroy Francisco de Toledo, who visited the town in 1574, noted the destructive effects of excessive woodcutting around the city (Levillier 1925b; Abecía 1939:127–28). He was most concerned with cedro and decreed that without a permit from the municipal authorities it could not be cut. Those who acquired the appropriate permit were required to cut the cedro trunk into boards with a large saw. Wood from the branches was designated for window casements. Despite the restrictions cedro disappeared as a wild-growing plant in the seventeenth century, though folk memory of it survived into the early twentieth century (Vaca Guzman 1915:571). Several ornamental cedros are found in plazas and patios and in the monastery garden of La Recoleta.

Queñuar (*Polylepis*) cutting was also to be controlled. This "leña de queña" could not be cut again for six years after first collected, in order to allow for regeneration. A Spaniard who violated the rule would pay a fine of 50 pesos and lose the right to make charcoal for a period of six years. An Indian who did it got 100 lashes. The *ordenanza* was read at Sunday mass in Potosí to put the fear of God into the parishioners.

Toledo issued several other decrees about wood supply. He banned dry-season fires in woodland and grassland, pointing out that on windy days these can burn over five leagues in one night, destroying crops, dwellings, and even villages. No person could start a fire in forest or grassland until the Day of Our Lady of Agosto (August 15). The penalty for Indians and Negroes was set at 200 lashes; Spaniards were to pay for the damage a fire caused as well as a fine of 500 pesos.

Toledo also restricted charcoal making to the area beyond a radius of three leagues (15 km) from the city. When cutting wood for charcoal, the trunk of the tree was to be left intact to grow. Likewise bark for tanning hides could not be taken from trees within three leagues of the city.

After more than 400 years of depending on wood for fuel, La Plata/Sucre turned to other sources of energy. In the 1940s spirit stoves appeared, which used kerosene and later bottled gas. Since the 1950s the oil pipeline from Camiri greatly diminished the need for firewood and charcoal. However, firewood still comes into Sucre and is sold at a special market. Some Sucre industries, such as brickworks and bakeries, continue to depend upon firewood. Chicha made from maize requires enormous amounts of firewood; one small peasant-owned *chicheria* uses 5.5 metric tons of firewood a month (Schlaifer 1993:597).

Past practices account for the present look of the Chuquisacan landscape around Sucre. Except where eucalyptus has been planted, slopes are typically devoid of trees. Trees along paths and roadsides serve as boundary markers, and sometimes standing trees are seen in fields where they are used to store cornstalks (*chala*). Most of these trees are pollarded molle, 10–15 m tall.

Regional Outcome of Chuquisaca Deforestation

Historically in highland Chuquisaca, river terraces and valley bottoms provided tracts of flattish, good soils and warmer microclimates, in which crops such as maize, chili peppers, and cherimoyas could be successfully grown. Fire was used to clear the forest in that ecological zone as well as on the heights far above. On the heights the forest was converted into pastureland for llamas and alpacas; also crops such as the potato were grown there as a way of spreading the subsistence risk. The midslope forests were the last to be removed. Eventually, with an increasing population, those remaining trees also disappeared because the land was needed for crop growing and firewood was needed for the hearth.

Forest removal from northern Chuquisaca over the centuries forced several adjustments. Wood shortage gave an early impetus to reforestation. Already in the sixteenth century, Juan de Matienzo (1875:151–52) sent a proposal to Viceroy Toledo to plant peach trees in a three-tiered radius of four, six, and eight leagues (20, 30, and 40 km) around La Plata. Matienzo proposed entrusting this plan to the *cacique,* who would mobilize the work force. The first task was to germinate 20,000 peach pits in seedbeds, transplant them in October and

Figure 3.5. Free-ranging goats, here near Padilla (Chuquisaca), have been a major factor in deforestation in the drier areas of the Central Andes.

November when the rains come, and hire paid guards to ensure that livestock did not eat the young saplings. Matienzo envisioned these peach trees as a very good source of firewood for the city as well as a source of charcoal for the mercury mills in Potosí. The plan was that an oxcart of wood would be obtained by pollarding each tree every three years. To assure a sustained yield, only the branches would be removed. Nothing was stated about the fruit—usually the main reason for planting peach trees—but it may have been assumed that periodic harvests of fresh peaches would encourage the populace to carry out the project. This imaginative idea was not implemented, but it indicates the nature of colonial responses to deforestation.

Forest removal in Chuquisaca left the soil surface exposed to devastating erosion. Some erosion may have predated the conquest, but most of it has occurred over the last two centuries. The combination of torrential rain and unvegetated surfaces has opened gullies and washed topsoil away. Pathetically low crop yields have caused innumerable fields to be left in permanent fallow, for only grazing animals can make use of them. That has led to further deterioration. Herds of sheep, which provide wool and meat, have been kept in numbers that exceed the available forage. As a result they have compacted the soil, which lessens its ability to hold water or to grow crops at some later time. Goats are much less numerous but individually more destructive (figure

3.5). Large areas are treeless, eroded expanses which are in the process of desertification (Libermann Cruz and Qayum 1994:71–72). Two results of this resource debacle are impoverishment and migration. Chuquisaca is the poorest of Bolivian departments, and two of its provinces, Azurday and Zudañez, are the poorest in Bolivia.

People of Chuquisaca leave it for the cities of Sucre, Cochabamba, and Santa Cruz and to colonize lands in other regions. A major destination for the highland Chuquisacan peasantry has been the warm, wet sub-Andean zone centering on Monteagudo (Buksmann 1980:119). This forested piedmont, once called *la frontera,* defines a biogeographical and cultural discontinuity with the highlands. During the Inca period the piedmont became the homeland of the fearsome Chiriguano people, whose relentless westward movements and bellicosity inhibited settlement from highland Chuquisaca. After independence the Bolivian government awarded land grants there to highlanders to push back the Chiriguano. But the migrant flood dates only from the 1950s, when agrarian reform and land deterioration in the highlands propelled people to move into the area. The region continues to be a destination for impoverished highlanders from Chuquisaca.

Land-use practices in the warmer, moister piedmont region repeat the traditional ones which damaged the highlands. The forest (*chaqueo*) on the more fertile river terraces has been cut and burned. The upper slopes were cleared early for pasture, leaving forested tracts sandwiched between the two vertical production zones. Forest clearings are made for agriculture, and land no longer usable for crops is turned into pasture. Cedro and nogal are selectively logged. Charcoal operations consume wood. Thus far, vegetation has regenerated, and the forest will not be destroyed without a substantially higher population than now exists. But no conservation ethic has emerged, and with more settlers the landscape change could parallel the prehistoric and historic deforestation sequence of the highlands.

Gestaltic Reflections

My personal assessment, based on travel and satellite imagery, is that less than 3 percent of the Central Andean highlands is in native forest. I contend that biophysical causation does not account for the currently treeless highlands of Peru, Ecuador, and Bolivia. Studies at different scales and the perspective of time make clear that human agency has caused massive vegetation change, destroying over 90 percent of the Andean forest, perhaps 65 percent of it before the conquest. It took a century and a half of observation after Alexander von Humboldt for researchers to arrive at a general agreement that the plant cover of the Andes is a "*Kunstproblem des Menschen.*" The slowness of that realization can be attributed to the scientific dogma of an active group of European ecologists whose naturalistic approach blinded them to the pervasiveness of human impact.[2] However, naturalism, which considers most valid observations as coming from the natural sciences, could not continue to provide credible answers to those who reflected on what they observed. The human imprint in the Andes is ancient, and motivations for landscape transformation are compelling. After the conquest, deforestation proceeded through a new set of demands and pressures. At the same time, communal organization was often not strong enough to prevent forest removal. Not only have woodlands disappeared, but also the survival of several species hangs in the balance. For other species of the Andean forest, close human association in the past has been their salvation as living species.

The "tragedy of the commons" took its toll with the complicity of all and the responsibility of none. The biggest hindrance to a conservation ethic in the Andes was that wood was narrowly conceived as a necessity for fuel and toolmaking. If wood constructions had been important, would communities have implemented stronger measures to prevent wholesale removal?

Reforestation efforts before and after the Spanish conquest were a response to massive tree removal and the need for wood. But only with the introduction of eucalyptus within the last 150 years has reforestation occurred on a major scale. These plantings have alleviated wood shortages but at the same time are maladapted. Native species are now recognized as most suitable in slope rehabilitation. Although trees of Andean origin grow more slowly, they provide higher-quality wood for most uses and demand less water than eucalypts do. The deeper roots of native trees make them less competitive with crops for water and nutrients.

The key to understanding this story is that environment and culture operate together as long-term processes. The temporal perspective reveals this convergence of biophysical patterns and human impacts. For decades, human impact was omitted from ecological studies of vegetation, perhaps mostly because the anthropogenic was seen as peripheral to a plant ecology focused on "natural processes." Moreover, the frequent interventions of humans contradicted the metalanguage of plant ecology that centered on the notion of stability and a system in which equilibrium was the normal state of affairs.

Deforestation through prehistory carries an innate contradiction because civilization required tree removal to open the land for agriculture. Without that transformation, settled life could not have emerged in most places. However, as population expanded, excessive cutting on slopelands led to environmental deterioration that had serious consequences for crop productivity, water supplies, and fuel sources. Acceptance that the Andes were once largely forested strengthens the possibilities for rehabilitation. It is now generally appreciated that only plantations of native trees can reverse the negative effects of barren slopes.

4
Malaria and Settlement Retrogression in Mizque, Bolivia

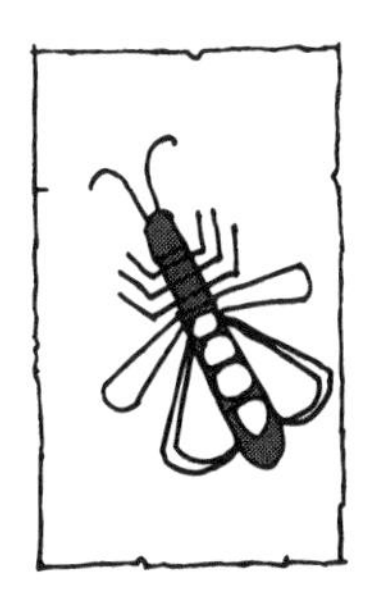

In the Andes of South America, climatic contrasts within the skein of high mountains and deep valleys have educed sharp vertical discontinuities in the incidence of several diseases. The high snowy peaks above 4,500 m and the hot valleys at 2,000 m only 50 km below show enormous epidemiological contrasts. Some of the spatial disease patterns are pre-Columbian; others appeared after the Spanish conquest because of pathogens introduced by the Europeans (Gade 1979). Malaria, a vector-transmitted malady with clear-cut environmental parameters, stands in this latter category. It has had a major impact on land and life in this part of Latin America. A circumscribed valley setting offers a scale in which to understand the many strands of this zoonotic disease in its relationship to people and landscapes. A notable case is in Mizque, Bolivia, where malaria had an enormous impact on its demography and settlement landscape. Mizque provides insights into the tight web of disease-people-environment that winds itself into a tight skein of parasite, vector, human as host and as victim, the presence of water for breeding, and the temperatures that enable the parasite and vector to survive. Many parts of the Andes have had these linkages, which marked them as unhealthy until science and technology intervened in the mid-twentieth century.

Coordinates of the Mizque Area

East of the western high plateau (Altiplano) of Bolivia, a number of valleys were choice spots for colonization both before and after the Spanish conquest. One such depression, drained by the Río Mizque, lies midway between the two cities of Cochabamba and Sucre in Cochabamba Department (figure 4.1). An upper portion of this valley at about 2,000 m above sea level widens into a 200 km^2 basin of lacustrine origin (figure 4.2). Surrounding hills rise 300–

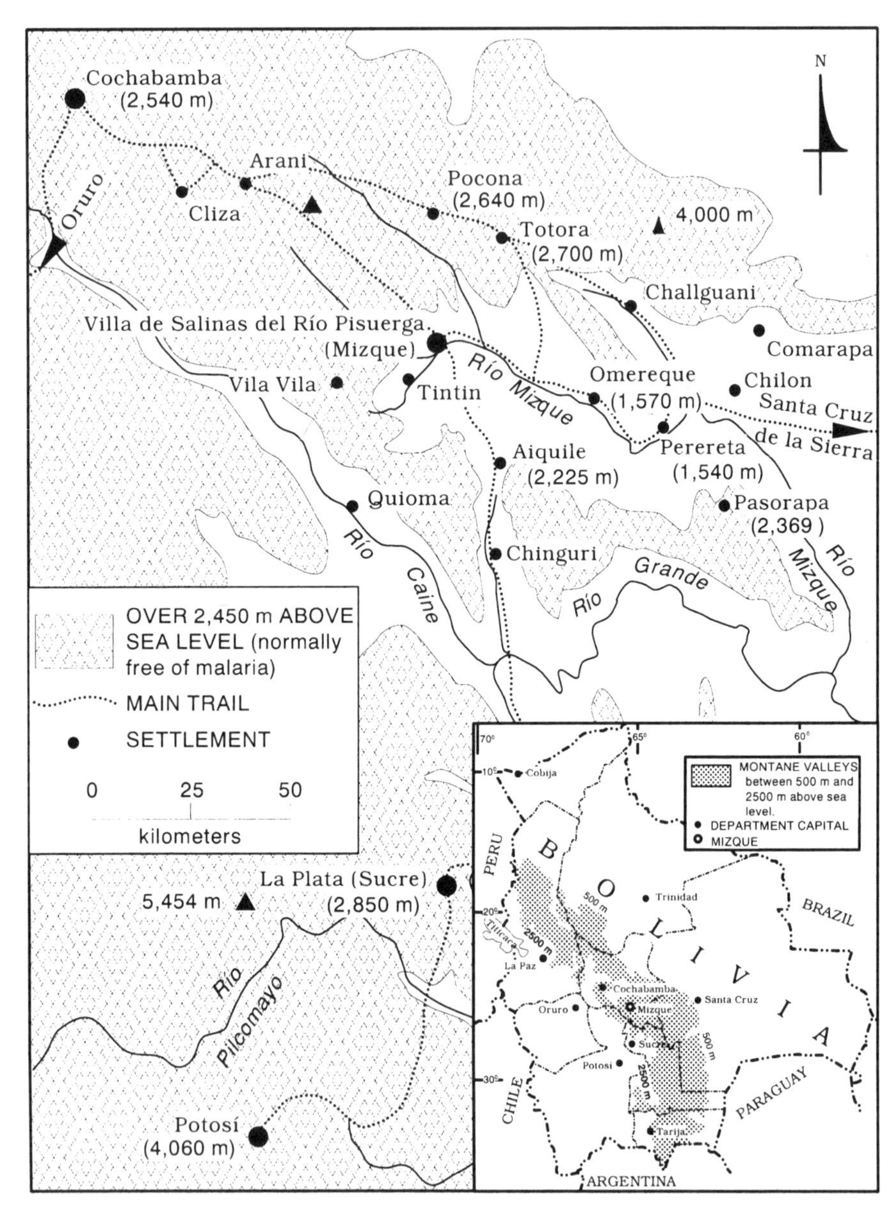

Figure 4.1. Colonial settlements and routes in the larger Mizque region and zone historically subjected to malaria

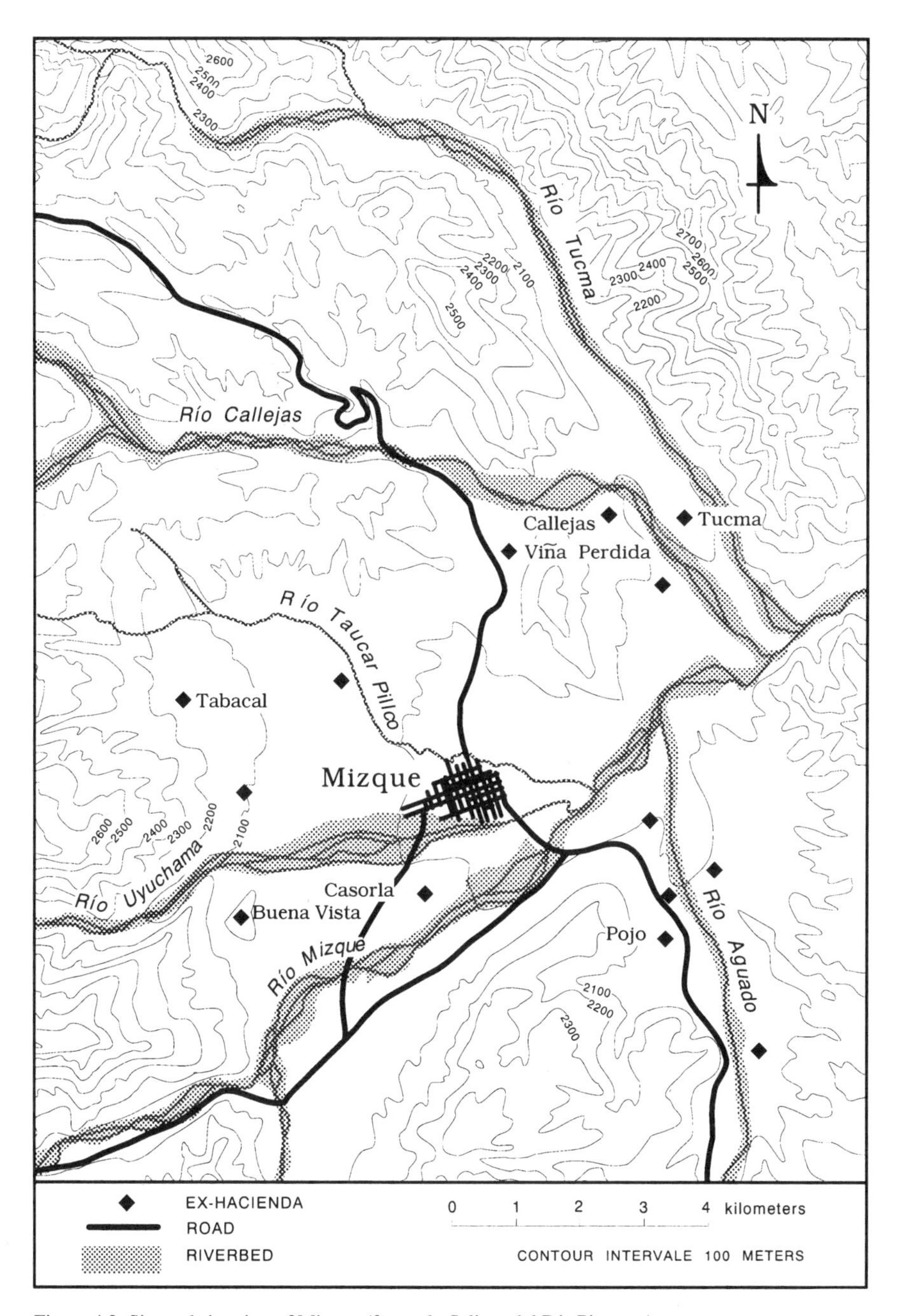

Figure 4.2. Site and situation of Mizque (formerly Salinas del Río Pisuerga)

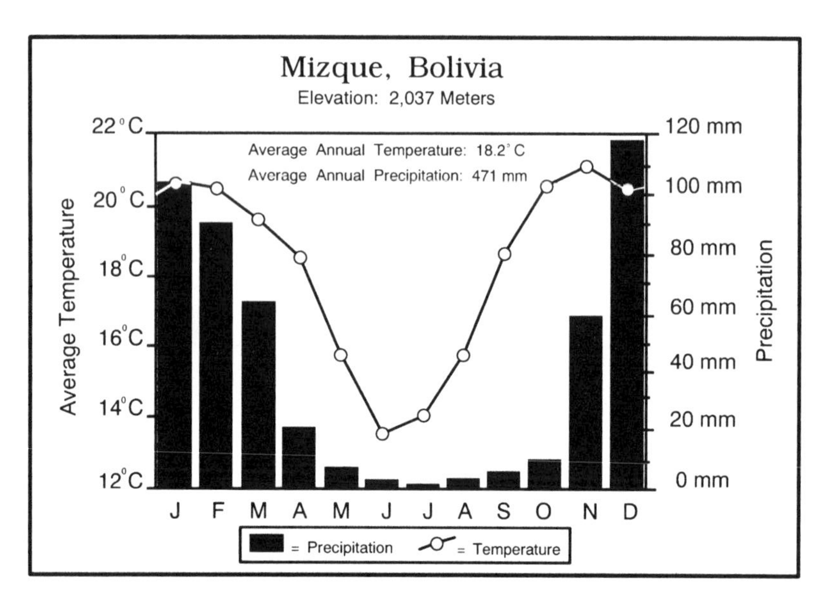

Figure 4.3. Average monthly rainfall and temperature for the town of Mizque (Data source: Servicio meteorológico de Bolivia)

900 m above the floor, and, to the north and west, mountains reach up to 4,100 m. In the basin, river terraces and alluvial fans have been formed by several tributaries of the upper Río Mizque, and in several of them stream channels often overflow. About 40 km^2 of the main valley floor is exceptionally flat. At the end of the rainy season, receding floodwaters frequently leave swamps and sometimes even small lakes. Upstream the Mizque Valley narrows and deepens to form the Tintin sector above and the Omereque Valley below. The main basin and interconnecting depressions between 1,500 m and 2,200 m above sea level form a zone with broad environmental and land-use similarities, here called the valleys of Mizque.[1]

Temperatures vary with elevation in this highly vertical environment. On the valley floor the town of Mizque (2,037-m elevation) has an average annual temperature of 18.2° C (figure 4.3). June and July are the coolest months, and five nights of freezing temperatures can normally be expected there in a year. Below 1,700-m elevation, freezes are unknown. The average annual rainfall is 471 mm, about 65 percent of which falls in the three warmest months, December, January, and February (SNM 1982). The high ranges to the northwest help to shelter the valleys from strong winds. Temperature, rainfall, and edaphic factors accounted for local variations in vegetation before human disturbance. As discussed in chapter 3, forests once covered a large part of this region. In moist bottoms, broadleaf trees have persisted or recolonized, whereas cactus scrublands dominate other sites.

Early Colonial Settlement

Mizque was the name given to the river and its main valley when the Inca established their hegemony in this zone around A.D. 1480, during the rule of Topa Inca Yupanqui. Aymara-speaking Chuy people from the Cochabamba Valley were brought as *mitimaes* to settle the complex of valleys around the Río Mizque (Wachtel 1982:201). Their warrior tradition was useful to the Inca as a bulwark against the aggressive Chiriguano, who began early in the sixteenth century to push westward toward the Andean foothills. In their incursions, these Guarani-speaking people often pillaged and burned the settlements of other polities and raped, killed, or enslaved the inhabitants. Cannibalism added to the Chiriguano's fearsome reputation. Near this perilous frontier zone, the Inca constructed fortifications to protect recently gained territory.

Four decades after the Spanish conquest of Peru, the people of six semi-dispersed Indian settlements of these valleys were resettled in a nucleated village given the name San Sebastián de los Chuyes, but more commonly called Mizque after its main valley. Viceroy Francisco de Toledo expressly founded this reducción to create a clustered population that would help to thwart the Chiriguano advance westward into the highlands (AGI, Charcas 31, 1561; AGI, Charcas 32, 1633). Of its 1,403 people, 227 were tribute payers (AGI, Charcas 464, 1583).

Spaniards themselves came to settle in the valleys of Mizque, attracted by the abundant water supply, fish in the rivers, flattish land and good soils, timber trees, and benign temperatures (AGI, Charcas 32, 1633). Initially the whole Mizque area of valleys and uplands was one vast *encomienda* granted to Pedro Fernandez Paniagua, which passed on to his widow Izabel Paniagua (AGI, Contaduria 1786, 1582). By the late sixteenth century, private properties owned by other Spaniards superseded the encomienda, which had reverted to the crown. The labor supply for running the estates was largely drawn from Indians of highland origin who had left their home communities to avoid *mita* obligations (Ramírez Valverde 1970:298). Indians were also brought from the plains of Santa Cruz (García Recio 1988:178–79). The abundant supply of landless Indians (called *yanaconas*) from the outside contributed much to the early productivity of Spanish haciendas (Levillier 1925a:136). For most of the next 300 years they were three to four times more numerous than the *originarios,* who had communal rights to land in the reducción.

The favorable reputation of the area grew as the sixteenth century advanced. Early colonial visitors extolled the *tierra templada* of the valleys of Mizque with their benign temperatures that were rarely cold or too hot (Cór-

Figure 4.4. Guaman Poma de Ayala's (1980, 2:239) conceptualization of the "villa de Mizque" around 1610. Although this hispanicized Indian never actually traveled there, his sketch captures the wooded character of the surroundings, the several monastic establishments (each indicated by a cross), and the quasi-tropical environment suggested by the parrots in the trees.

dova Salinas 1957:986; Guaman Poma de Ayala 1980, 3:985; Ramírez del Aguila 1978:60; Torres 1974, 2:305; AGI, Charcas 139, 1651). In the early seventeenth century, Vázquez de Espinosa (1948:596) called it a "piece of paradise." Mizque commonly elicited such comments from the Spaniards in Charcas/Alto Perú who preferred warm sheltered valleys to the Altiplano with its frigid nights and hypoxia. Augmenting the area's population, the climatic amenities and healthfulness of the surroundings helped to entice prospective settlers from the early core of Spanish settlement at La Plata (Sucre), 100 km to the south and five days of travel away (AGI, Charcas 140, 1609).[2]

The temperate climate also permitted the kind of agriculture that could maintain a familiar diet based on bread made from wheat flour ground in gristmills and wine fermented from grapes (AGI, Charcas 153, 1654). These advantages made Mizque more attractive to Spanish settlers than the hot, unhealthy country of Santa Cruz de la Sierra, some 300 km to the east (AGI, Charcas 140, 1609; Charcas 31, 1611; Charcas 388, 1690).[3] Moreover, Mizque's relative location opened profitable trade opportunities with the highlands to the west. In particular the needs of 160,000 people living in the sterile heights of Potosí at 4,060 m above sea level made it necessary to reach out long distances for food and materials. Guaman Poma de Ayala (1980, 3:984) accentuated the contrast with Potosí by embellishing his iconograph of Mizque with tall trees, birds, and water courses (figure 4.4).

In 1603 Francisco de Alfaro, high official in the colonial government of Charcas, established a full-fledged town given the name of Villa de Salinas del Río Pisuerga del Valle de Mizque (AGI, Charcas 17, 1603; ANB, AM, 1603).[4] Designation as a villa suggests the prestige afforded it, for few Andean towns were so honored. Its construction on a site appropriated from the reducción of Mizque contravened an earlier royal decree that Spanish towns (*pueblos de españoles*) could not be established within four leagues (20 km) of Indian villages (*pueblos indígenas*) (Gandia 1939:413). In fact, Salinas fused Spanish and indigenous sectors into one town of two different communities. Two square blocks on the settled periphery were assigned to the 500 Indians from the Mizque reducción, which had been leveled to accommodate the Spanish town superimposed on its space. This Indian barrio, composed of Aymara-speaking Chuy Indians and later Quechua-speaking arrivals, had its own parish church served by its own Franciscan priest and given the same advocation that San Sebastián had bestowed three decades earlier on the reducción (Espinosa Soriano 1980:175).

These on-site native people provided a convenient source of labor for Spaniards, but their numbers were insufficient, and yanaconas were subsequently brought from the highlands to work in constructing the city. Its regular grid-pattern arrangement with a large central plaza was carefully laid out. House lots were often quite large to accommodate subsistence gardens. Though stone buildings were few, many houses were two stories with thick walls, wide doorways, and balconies in the Iberian style.

The colonial authorities established Villa de Salinas for strategic and economic reasons. Periodic inroads by the Chiriguano raised apprehension that, without a more important and fortified settlement in that area, even La Plata or Cochabamba would be vulnerable to their barbarian advances (RAH, Colección Muñoz; Levillier 1922:316). The town agglomerated most of the Spaniards who had property in these valleys (Caravantes 1986:158). These people were also referred to as *cristianos* in clear allusion to the reconquest of Spain, where frontier nuclei had likewise been established as a military strategy to ensure Spanish occupation. Unlike the reducción village of Mizque, which had been a simple nucleus of peasant farmers, Salinas functioned as a city. As the seat of a *corregimiento,* it controlled certain political-administrative functions of the 18 villages under its jurisdiction. The institutional character was particularly impressive for such a young, isolated community. Between 1605 and 1767, Mizque/Salinas, rather than the canonical designation of Santa Cruz de la Sierra, was the de facto seat of the diocese. A succession of bishops preferred to live in Mizque rather than in the hot, isolated, and unhealthy frontier outpost farther east, at least during the rainy season. The parish church served as the unofficial cathedral for this "Obispado de Mizque." Several religious orders established themselves in the seventeenth century. Of the four monasteries—Santo Domingo, San Francisco, San Juan de Dios, and San Agustín—the last was the most sumptuous. The church of San Augustín, finished in 1619, had a separate tower with four bells; its interior boasted sculptures, 14 painted canvases, and a main altar adorned with gold leaf (Torres 1974:330–31). Three convents of nuns—Carmen, Santa Teresa, and Santa Clara—were also founded in that century, but their institutional longevity was much shorter.

In 1608 the viceroy established a hospital in Mizque. Only one of seven hospitals in all of colonial Alto Perú (Lofstrom 1983:268), its establishment was fortuitous, given the health problems that were later to overwhelm the region. Through the municipal government (*cabildo*), organized on the Spanish model, noble families controlled the town's civic culture and organized public celebrations known throughout Alto Perú (Viscarra 1967:262; Torres 1974:305). Of a total of 22,000 inhabitants in Mizque/Salinas in 1700, some 6,000 were Spaniards (Castillo 1930b). That growth represented a doubling in about half a century (AGI, Charcas 153, 1654). Such a substantial population of Spaniards gave it the critical mass of citizens needed to turn the town into a self-conscious outpost of Hispanic civilization. A large demographic nucleus required a substantial economic base. Mining was a minor activity for the town and its region during part of the colonial period. For about a century silver was extracted at Quioma near the Río Grande, and gold was taken from two small mines elsewhere. More persistent and productive than mining were agriculture and livestock raising (Ramírez del Aguila 1978:60; Córdova Salinas 1957:986; Vázquez de Espinosa 1948:596). Haciendas, some so large they encompassed several ecological zones, had a vertically staggered land use (ANB, 1709). Grazing of cattle, sheep, and horses dominated the highest zones. Below that, unirrigated slopes were used for wheat, barley, pulses, and potato cultivation. The irrigable valley floors had the most agricultural diversity. Heat-loving root crops—sweet potato, *ajipa, yacon,* and manioc—were cultivated, and the fruit trees ranged from avocado, cherimoya, pacay, banana, and orange to fig, apple, quince, and peach. Maize, peanuts, cotton, sugar cane (for sugar, later for alcohol), tobacco, and grapes occupied much valley land.

Viniculture had become commercially important there even before the Spanish town was founded (Lizarraga 1986:201; AGI, Charcas 140, 1612). Large increases in the Spanish population during the seventeenth century reinforced the grapevine as Mizque's source of greatest economic return. The Spaniards brought African slaves to do the main vinicultural tasks (Peñalosa 1629). Stone-wall enclosures around some of the 80 vineyards in the corregimiento of Mizque gave evidence of the prominence attached to this land use. Estates in the Omereque zone between 1,500-m and 1,800-m elevation especially emphasized grapes and wine. In 1620 more than 100,000 *botijas* a year came from wineries there (Vázquez de Espinosa 1948:596). One estate alone, Hacienda Perereta, annually produced 12,000 of these large terra cotta containers of wine. In 1661 Perereta had more than 60,000 vine stocks and 25 black slaves to care for them (ANB, E, 1714). In the nearby tributary valley of Chalguani, the large hacienda owned by the Jesuits made and shipped each year up to 13,000 botijas of wine to missions in the hot, wet, and vineyardless Llanos de Mojos region (ANB, AC, Mojos 1768).

In the late seventeenth century viniculture sharply contracted in the region. An often repeated but undocumented assertion is that the *corregidor* ordered the burning of the vineyards on the pretext that they had been planted without the necessary license from the king (Dalence 1851:118). Alternatively, three straight years of locust infestation may have destroyed the vineyards (Viscarra 1967:250). By the nineteenth century, commercial wine production had essentially ceased in the valleys of Mizque (ANB, EC, 1805; ANB, EC, 1802), a function of hacienda organization but also a response to low demand from Potosí and elsewhere. As the grapevine faded from the scene, capsicum pepper (*ají*) emerged as an important commercial crop in the valleys of Mizque and remained so into the twentieth century (Miller 1918:236). Mizque long specialized in *ají criollo,* also known as *ají mizqueño,* a mild-flavored variety different from the stronger *ají de indio* or *cumbaru,* traditionally grown elsewhere by Indians (Salazar 1918). Though productive, chili pepper cultivation was labor- and water-intensive, requiring irrigation for much of its 11-month growing cycle. The young plants were started in sunken seedbeds in July, transplanted in September, and harvested the following June. Dried and baled, capsicum from Mizque was an essential condiment and was sold in markets all over Alto Perú/Bolivia in the late colonial and republican periods.

In spite of high transport costs, trade items from the valleys of Mizque flowed to Potosí in response to its burgeoning demands. Until the eighteenth century, wine was the best-known commodity, famous throughout Alto Perú (Lizarraga 1986:201). The area also supplied wheat, horses, skins, wax, honey, resins, and cotton (Vázquez de Espinosa 1948:595). Much timber was sent to Potosí for drive shafts in stamp mills and for house and church interiors, the latter especially using cedro (*Cedrela* spp.), a choice cabinet wood (Bakewell 1984:24). Pack trains of either mules or llamas carried these goods the 300 km from Mizque to Potosí via La Plata. Until bridged in the nineteenth century, the Río Grande and the Río Pilcomayo were the two major natural obstacles on this road (Matienzo 1967:312; ANB, E, 1709). The Altiplano centers of Oruro and La Paz also received food from Mizque for at least part of the colonial period, and even distant Cusco got honey and wax from this area. Wine and wheat were also sent eastward to Santa Cruz de la Sierra, where neither of these commodities could be produced on the hot plains, although the transport involved 20 days by mule train and a considerable risk of Chiriguano ambush.

Decline of Salinas and the Mizque Valleys

Early in the eighteenth century the agriculture, trade, settlement, and population of the Mizque area dramatically declined (figure 4.5). In 1724 no surpluses of maize or wheat were produced, and small amounts of low-quality wine came from a few surviving vineyards (AGI, Charcas 388, 1724). Mizque could no longer be placed among the 10 productive valleys in the viceroyalty of Peru (Cobo 1956, 1:79). Lacking laborers and managers, agricultural land use shrank. Some haciendas were deserted; several others even disappeared from the registers (Viedma 1969:93). In the lateral Chalguani quebrada, the once prosperous estate la Viña Perdida was but a shell when Orbigny (1839:501) passed by there in 1830. Few traces of that former Jesuit vineyard remained, and only a handful of lethargic peasants growing a bit of maize were still living there. Through the nineteenth century much property was abandoned. Between 1838 and 1877 the number of haciendas in Mizque Province declined from 129 to 78, 45 of which were left vacant (Grieshaber 1980:245). Abandoned fields turned into weed patches, some of which later reverted to woodland.

The Villa de Salinas del Río Pisuerga lost its urban character while still in the colonial period. The geographer Cosme Bueno (1768), who passed through the area in the 1760s, noted the town's desolation, though he could still detect the vestiges of its former opulence. Municipal functions were paralyzed

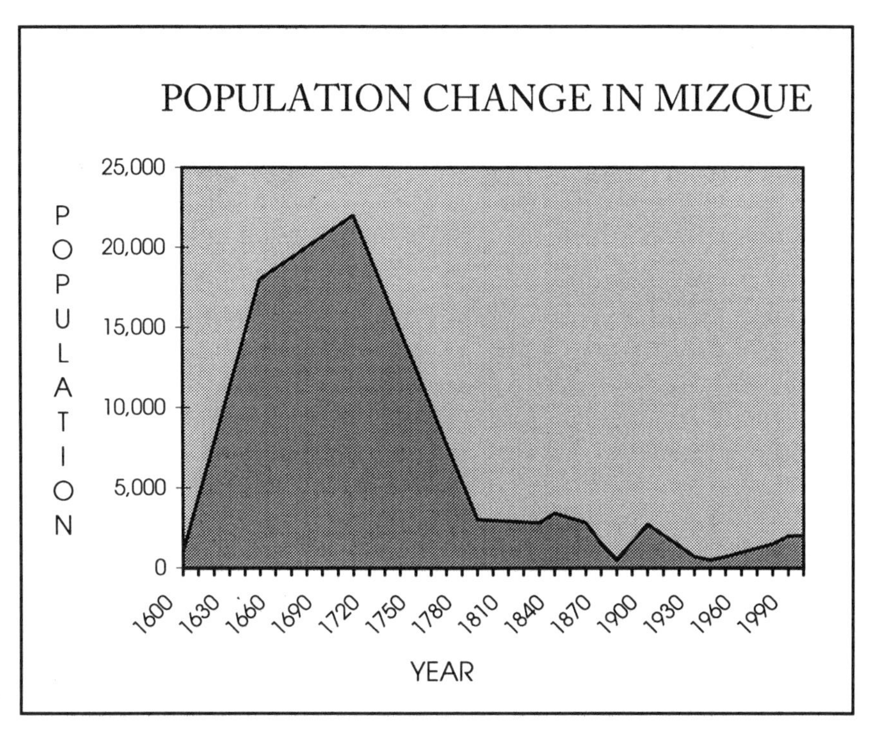

Figure 4.5. Estimated population change in Mizque, Bolivia, extrapolated from archival documents and Bolivian censuses

for lack of personnel, and the religious brotherhoods were moribund. By 1770, market trade was reduced to one day a week (AGI, Charcas 410, 1770). The Spaniards Juan and Ulloa (1978:219), traveling in the Andes in the 1740s, noted the decadence of Mizque in spite of the apparent wealth of its resource base. Viedma (1969:85), writing in 1787, clearly made the link between depopulation, disease mortality, and reduced economic activity. Even the proud Spanish honorific of "Villa de Salinas" fell into disuse around this time, and the town reverted to its indigenous appellation. The once-proud town declined further through the nineteenth century. Orbigny (1839:499) entered a settlement in 1830 and found only a few hundred residents, ruined buildings, and abandoned institutions: "Tout y annonce l'approche d'une déchéance complète." By the early twentieth century, any remaining physical evidence of the seven religious houses had collapsed into heaps of rubble. In his 1904 visit to Mizque, Viscarra (1967:193) noted that vandalism of treasure hunters and the combined effects of human neglect, time, and weathering had sealed the destruction of the monuments of this once-flourishing urban center.

Indians more than any other group underwent the sharpest population decline. Natives with legal ties to the original reducción and subject to tribute dwindled from 227 in 1582 (AGI, Contaduria 1786, 1582) to 55 in 1669 (ANB, AM, 1679), 40 in 1692 (ANB, 1692), 13 in both 1783 (ANB, 1783) and 1788 (Viedma 1969:181). San Sebastián, the old Franciscan parish founded for Indians even before the 1572 reducción, was dissolved in 1790. Massive Indian mortality simplified the encroachment of lands that had belonged to the original reducción community (Viedma 1969:182). Other Indians, those who were yanaconas because they had no lands, also languished. Since neither of the "originario" or "yanacona" categories of native people in Mizque was subject to the Potosí mita, their gradual disappearance cannot be attributed to labor conscription in the mines. After independence the remaining native population in Mizque Province as a whole dropped even more sharply, from 3,608 in 1838 to 625 in 1877 (Grieshaber 1980:263). In this province in 1846, the *cantones* of the hot valleys had only about one-fifth the number of Indians that the cantones of cool climate had (Macera 1978:13).

The Spanish population in the valleys had collapsed as early as 1724, when only four Spaniards (*vecinos*) still lived in the town (AGI, Charcas 388, 1724). In contrast, people of African ancestry increased during the colonial period. They had been brought to the zone in small numbers as slaves in the late sixteenth century. By 1787 they formed 12 percent of Mizque's *partido* (which included much land at high elevations), 21 percent of the town's population, and more than 30 percent of the population in the valley bottoms, called the *valles fuertes* (Viedma 1969:90).

Presence of Malaria: Diffusion, Perception, Adjustment

Malaria accounted for much of the decay and desolation of the valleys of Mizque. Other factors—the sharp drop of viniculture, the fall in trade as Potosí shrank in size, and the turmoil surrounding independence—each contributed to the decline. But no other region of Alto Perú disaggregated as fast and as far as the valleys of Mizque did. The presence of endemic disease and the unpredictable occurrence of epidemics kept the local economy and population from rebounding as they did elsewhere in Bolivia.

The presence of malaria in the valleys of Mizque probably dates from the early decades of Spanish settlement. Already by the 1550s the valley of the Río Chinguri, south of Mizque, had a recurrent malady suggestive of malaria (Ramírez Valverde 1970:298). In 1611 "*peste*" raged through Mizque (Torres 1974, 2:306), a term which, at that time, referred to any rapidly spreading disease and, in this case, probably referred to malaria (Basto Girón 1957:30). Archival documents suggest high mortality among Indian workers; for example, in 1709 at Hacienda Buena Vista, 22 of 66 yanaconas died, as 8 of 22 did at Hacienda Tucma (ANB, 1709). Then in 1710 Mizque was struck down by a devastating epidemic of typhus and fevers (*tabardillo y calenturas*) that reputedly wiped out half of the population (Frontaura 1935:86). *Calenturas* referred to the cyclical paroxysms that afflict malaria sufferers. In this part of the Andes, typhus (tabardillo) frequently occurred in concert with malaria (Viedma 1969:83–84; Dalence 1851:118).

In the eighteenth century the disease problem became the quintessential reality of Mizque (Bueno 1768; Alcedo 1967, 2:467; Viedma 1969:83–85; Haenke 1809:493). By the 1720s Santa Cruz de la Sierra was described as much healthier than Mizque, quite the reverse of the situation in the seventeenth century (AGI, Charcas 388, 1724). Mizque's early colonial reputation for salubrity places in sharp relief its later notoriety as a valley of death, which persisted for more than two centuries (AGI, Charcas 140, 1609). The pestiferousness of the valleys of Mizque seems to have influenced perception of the climate as torrid (*ardiente* in sharp contrast with its earlier description as temperate (*templado*).

By the late eighteenth century, the Chuy Indians at Mizque had become finished—a "*comunidad consumida y acabada.*" In 1771 the 20 tribute-paying Chuy individuals were swiftly reduced to just 7; of the others, 6 had fled the valley, 6 had died, and 1 had become too sick to pay the tax. According to the records of all five *repartimientos* of the Mizque jurisdiction, only 121 people were classified as originarios; 1,764 as other tribute payers; and 9,111 as outsiders without land (*forasteros sin tierra*) (AGN, 1772). The authorities, who had a mercenary interest in the health of Indians, were alarmed by this decline, which matched the fall in tribute revenues. A special category of *reservados interinos* was set up to cover Indians who, as a result of disease acquired during the rainy season (*enfermedad temporal*), could not pay tribute (AGN, 1772). Most of their communal lands, which previously had yielded rich harvests of maize, were uncultivated. Other parcels were rented, and in many cases haciendas were able to usurp these lands without much resistance (AGN, 1771). By this time the documents are quite specific: "Because of the spread of tertian fevers, this city is totally devoid of people and without residents and that is the truth . . ." (AGN, 1772; my translation).

Year after year the disease took its terrible toll. In the five years between 1780 and 1786, 574 burials were tallied in Mizque but only 374 baptisms. In 1792, there were 110 burials but only 93 baptisms. A major effect of epidemic and endemic malaria was the calamity of hunger that Mizque suffered (AGN, 1804).

Mizque's lurid reputation did not improve with Bolivian independence in 1825 (Moreno 1879:198; Blanco 1901:164). A U.S. government handbook promoting foreign investment in Bolivia cautioned that the "malaria common in some mountain valleys is found at its worst in the valley of Mizque"

(Schurz 1921). A similar caveat about Mizque's insalubrity was included in a government guide published to attracted prospective settlers (*Guía nacional* 1938:224). With labor perennially in short supply, a local board early in the century recommended the immigration of 200 foreign families to farm the vacant lands of Mizque (Quiroga 1906). Nothing came of that plan, for outsiders were notably reluctant simply to visit the valleys, let alone settle there, even with the offer of free land.

The long trajectory of malaria incidence in this region can be quantified only for the present century. In 1905 malaria was found in 72 of 100 houses surveyed in the town of Mizque (out of 140 houses in all). Of the 385 people living in those 72 malarial dwellings, 132 were ill with the disease (Melean and García Mesa 1905:24). If *tercianas* (tertian fevers) and *fiebres* (fevers, unspecified) are etiologically grouped together, malaria accounted for 40 percent of the 2,471 deaths in the 33-year period between 1908 and 1940.[5] For just the period 1922–1929, malaria accounted for 70 percent of the deaths in Mizque (Flores 1929:189). Children died in disproportionate numbers from the disease, which effectively eliminated whole generations. Those who did not perish could be identified as malarial by a bulging spleen and a yellowish cast to their face.

The effects of malaria were apparent in the scarce labor supply that plagued the valleys of Mizque starting in the eighteenth century. Without an abundant work force, estates could not function well and sometimes not at all. Peons still present were much less able to do sustained physical labor, because when malaria does not kill, it reduces the capacity for work. They could not meet the compulsory obligations of yanaconas in this region, who were to work at agricultural tasks 160 days a year, annually clean irrigation ditches, and keep the manor (*casa hacienda*) in repair. Wives and children of peons were required to perform domestic service and other tasks, such as preparing salivated maize quids (*muco*) for making chicha (ASNRA, 1957, 10191).

Malaria transmission was understood dimly by some, more clearly by others. Folk explanations included bathing in rivers, eating too much, consuming certain kinds of fruit, and imbibing newly fermented wine (ANB, E, 1709) or water. Mizque's preference for corn beer (chicha) stemmed partly from the perceived unhealthiness of drinking water (Corrales Badani 1930:42). Even as late as 1945 the departmental prefect anachronistically believed that the use of unhealthy water caused malaria (Araúz 1945:16). In the early part of the twentieth century, the two biggest activities in the town of Mizque were said to have been chicha brewing and coffin making (Castillo 1930b:347). Inexact as to the true reason, this idea nevertheless had the merit of correctly locating an epidemiological focus on the land. During the early months of Bolivian independence in 1825, the first president of the republic, Antonio de Sucre, came to Mizque and ordered the municipality to drain a large swamp near the town

(ANB, MI, 1825, 8). Sucre's personal directive, though backed by money to assist in the drainage, was not carried out (ANB, MI, 1826, 11). A year later, the prefect of Cochabamba Department ordered the local people to "dry up the mud," but this directive was also ignored (ANB, MI, 1827, 16). Such inaction suggests the lack of organization and will among *mizqueños* to take charge of improving their collective lot, but it may also reflect the disagreement at that time over what causes malaria. With time, stagnant water in any form appeared increasingly to be responsible. Lieutenants Juan Mariano Mujía and Juan Ondarza (1901:164), commissioned in 1845 to collect geographical data on Cochabamba Department, opined that malarial fevers in Mizque could be traced not only to floodplain swamps but also to backed-up water in irrigation ditches, caused by their faulty design.

An impoverished list of malaria remedies suggests the absence of an aboriginal experience with the disease. Folk cures ranged from frightening the disease out of a sleeping person to eating skunk meat (Aguiló 1982). Remedies of actual therapeutic value in treating those ill with intermittent fevers in Mizque came late in the eighteenth century. The roots of *tamatani,* an Andean species of gentian, were used as a febrifuge, as was cinchona bark, from which quinine was later isolated in Europe (Haenke 1809:493). Though cinchona is a native tree of the humid eastern Andean forests, it was not part of indigenous materia medica in the region.

Historically the people of Mizque were largely left to their own devices in dealing with malaria. In 1826 the Bolivian minister of the interior sent a representative to study the disease that so afflicted the people of Mizque (ANB, MI, 1826, 19), but nothing seems to have come of those good intentions. Minimal patient care was provided at the Mizque hospital, which continued to function through most of the nineteenth century. In 1835 this establishment had 32 adobe beds (*covachas para enfermos*), an administrator, a physician, and a chaplain (*Guía de forasteros* . . . 1835:154–55). Another institutional response to endemic malaria at that time was Mizque's "Junta de Sanidad," the only public health board in a provincial capital. Fed by a rank fatalism that was aggravated by ignorance of what causes malaria, high mortality continued unabated and touched a diverse population, whether it was farm workers on estates, priests in their rectories, or prisoners in their jail cells.

Since none of the medicines then used prevented malaria and mosquito bites were unavoidable, departure offered the only sure way to avoid contraction. Many mizqueños moved temporarily or permanently to areas above the zone of endemicity. It became understood that one normally would not contract malaria at elevations above the strong heat of the valles fuertes. Some *hacendados* who lived in the town of Mizque moved definitively to their landed properties, the higher portions (*cabeceras del valle*) of which were above the malaria zone. In other cases the owners left the region entirely, renting their properties, entrusting them to peons, or simply abandoning them. If free to do so, valley peasants worked above 3,000 m as agricultural laborers from December to April. Much land formerly in woodland was cut down by migrants to make crop fields above the malaria zone (Orbigny 1839:500). The inhabitants of Tintin, which was judged to be unhealthy in the rainy season, largely abandoned it during that period, transferring to the higher and colder but malaria-free cantón of Vila Vila, some 15 km distant (Blanco 1901:167). Some movement out of Tintin was permanent; the 41 haciendas in 1838 had declined to 25 in 1872 (Grieshaber 1980:245). Pasorapa (elevation, 2,360 m) became the seat of the *curato* that had been in Omereque (elevation, 1,600 m), so that the priest could remain beyond the reach of at least the most baneful form of malaria (Viedma 1969:97).

Of the seven main settlements in the then-delimited Mizque Province, only Pocona (at 2,640 m) and Totora (at 2,700 m) were at elevations high enough to be free of any form of malaria. Malarial individuals in those two locales had acquired their affliction at lower elevations. Outside the province, the town closest to Mizque above the malaria zone was Arani (2,740 m), located two days of travel away. Clerics from Mizque built an oratory in Arani, which enabled some of them to escape the risk of disease while carrying out religious observances (AGI, Charcas 388, 1759; Charcas 388, 1724). Dread of the valley intensified to an unaccustomed degree when 19 monks from Cochabamba refused their superior's orders to serve the population of Mizque (Lofstrom 1983).

These multiple dislocations led to the emergence of a regional contrast between empty valleys and well-populated uplands. At the turn of the nineteenth century, rural densities in the cantones of Totora and Pocona were two to three times higher than in Mizque cantón (Blanco 1901). Many of these high-country inhabitants were migrants from Mizque who sought malaria-free locales (Juan and Ulloa 1978:219). Most Indians who came to work at estates in the valles fuertes came from Cliza, Arani, and Tarata, at higher elevations (AGN, 1804). Totora, a collection of Indian huts in the colonial period, turned into a center of substantial economic activity and population, much of it from Mizque. In 1876, Totora became the capital of its own new Carrasco Province, after more than 300 years under the yoke of Mizque. In 1899 the Omereque sector of the Mizque Valley was removed from Mizque Province to join with a new province, Campero, with its capital at Aiquile. Once the center of a 23,000-km^2 area, Mizque, prostrate and languishing without leadership, saw its territorial integrity pared down to one-ninth its original extent. Earlier in the century Mizque's leaders had written President Sucre that their province languished with a declining population, not because of malaria, but because it was not the seat of its own department and diocese (ANB, MI, 1826, 14).

The city of Cochabamba, at an elevation of 2,540 m and malaria-free through most of its history, benefited from Mizque's decline. In 1782 Cochabamba was declared the seat of the *intendencia,* part of a new administrative geography in whose jurisdiction Mizque was subsumed as a partido. Mizque's powerful corregidor was replaced at that time by a minor bureaucrat. After independence, Cochabamba became the departmental capital. It grew from 22,305 people in 1793 to 30,396 in 1846, 35,800 in 1854, 81,071 in 1950, 209,676 in 1976, and over 330,000 in 1996. For two centuries Cochabamba has received a stream of migrating mizqueños, and today several thousand people in the capital can trace their family histories to the valleys of Mizque.

Mortality and out-migration brought population loss to Mizque, and those who remained in the valleys were unhealthy. The ravages of malaria left a residue of sallow, listless, and malnourished individuals, who were called in Quechua *chucchu puchu,* the "leftovers of tertian fever." Poor nutrition and alcohol dependency were common sequelae. Travelers commented on the unremitting squalor of these parts, the result of an unhealthy population, often intoxicated from drinking chicha, and too fatigued and disconsolate to care about maintaining elemental hygiene. More than disease and crapulence characterized the remaining folk in the valleys: Mizque was said to have included a disproportionate number of handicapped individuals, among them deaf mutes, cretins, dwarfs, and people with birth defects (Viedma 1969:84; AGI, Charcas 410, 1770). Yet long after it had degenerated into an unhealthy place of a few hundred people, Mizque remained on the title list of 11 "cities" of Bolivia. Visitors to Mizque also made adjustments to keep themselves free of malaria. The valley floors were avoided from November to April, when malaria worked its fury (Haenke 1809:493). Hacendados and traders could not readily get Indians to accompany pack trains to Potosí during that period (ANB, 1709), for the routes passed through several paludal zones besides Mizque itself. Movement along the narrow trails was concentrated between June and October. The unpredictability of the onset of the rainy season was a source of apprehension to travelers seeking to minimize the disease risk (Hoek and Steinmann 1906:10).

Indian llama drivers from Oruro Department on the Altiplano learned to deal with the disease factor when they traded with the valleys of Mizque. In a traditional exchange pattern, remnants of which survived at least into the 1960s, *llameros* from the puna spent several days in May or June of each year in Mizque acquiring maize, alcohol, peanuts, chili peppers, and honey in exchange for salt, folk medicine, and dried meat. But the presence of malaria known to generations of these itinerant traders taught them to avoid spending the night on the valley floors.

The responsibility for malaria's long and deleterious impact in Bolivia can be argued from several different perspectives: forcible imposition of the hacienda system on indigenous peoples, poverty, and/or lack of genetic resistance. Until transmission was correctly understood early in the twentieth century, explanation followed two main lines of causation. Europeans who came to Bolivia promoted the idea of bad air in low-lying areas to account for contraction of the disease. For example, the French scientist and traveler Alcides d'Orbigny (1835, 2:500) linked Mizque's endemic malaria to exhalations from nearby floodplain swamps. Since mosquitoes require standing water in which to deposit their eggs, that reasoning offered a more plausible chain of causation than the indigenous folk tradition, which related malaria to drinking bad water or eating certain kinds of foods.

Malaria and "Scientific" Understanding

Breakthroughs in understanding disease transmission punctuate the historical disease ecology of Mizque and other inter-Andean valleys. In 1880 Laveran demonstrated that malaria is caused by a blood parasite, and in 1897 Ross found plasmodia in the gut of a mosquito. Manson proved in 1900 that the bites of infected anopheline mosquitoes transmit malaria to people. Two different parasites, *Plasmodium vivax* and *Plasmodium falciparum,* were identified as causing tertian fevers (tercianas). In both, the classic triad of symptoms—shaking chills, fever, and sweating—wracks the body every third day as parasites seed the bloodstream from their hatching place in the liver. Except in children, *P. vivax* has been more debilitating than life-threatening, whereas contraction of *P. falciparum* commonly led to death before the introduction of chemotherapy. Falciparum malaria caused the epidemics of high mortality. A third parasite, *Plasmodium malariae,* causes quartan fever (*cuartonas*), in which the characteristic periodicity falls every fourth day. This form was less common than either of the two tertian fevers in the valleys of Mizque (AGI, Charcas 388, 1724). A scientific determination in 1931 found that 75 percent of the malaria was determined to be tertian, and 25 percent was quartan (Moscoso Carrasco 1931–1932:909). Two mosquitoes that carry the above three parasites were identified in the valleys of Mizque (Borda 1965). One species, *Anopheles albitarsis,* is believed to have played only a minor role in malaria transmission. The principal vector, *Anopheles pseudopunctipennis,* occurs between 500 m and 2,650 m above sea level in much of western South America. A weak flyer, it tolerates sharp diurnal temperature changes and a high evaporation rate, which may explain its adaptation to zones with a long dry season (Hackett 1945:244). All three plasmodia can be transmitted by *A. pseudopunctipennis,* a species that seeks out human victims in their habitations, but also is known to feast on animal blood.

An abundance of sunny breeding locales opened by deforestation within a five-kilometer radius of the town center favored the proliferation of the one main anopheline that cannot breed in the shade. Swamps formed on floodplains by stream overflow were permanent enough to receive local names. Especially in the main valley, water seepage through the deep gravel collected to form pools for oviposition. Mosquito breeding sites were also created by land-use decisions and practices. The seventeenth-century shift from unirrigated vineyards to heavily watered chili pepper plots may have unwittingly spread malaria. Pits banked with earth, so that water slowly percolated to plant roots, formed an unwitting microhabitat for oviparous anophelines. While male mosquitoes fed on the juices of the soft stem tissues of pepper plants, the female counterparts foraged their meals of blood from people and animals. Other man-made sites were excavation holes made to remove honey and wax from the hives of wild ground-dwelling bees, and borrow-pits formed where soil was removed to make adobe bricks. All became egg-laying loci when water accumulated in them after heavy rains.

The town of Mizque also provided breeding sites, shelter, and food for mosquitoes. Dooryard gardens had standing water in wells and ditches. Abandoned roofless buildings, pig wallows, and duck ponds were other places where anophelines could find pools and puddles in which to lay their eggs. Even footprints and hoofprints served if filled with rainwater. Insects found shelter in the weeds and brush that invaded plazas and vacant lots. Without mosquito nets or window screens to impede access, the low-flying *A. pseudopunctipennis* readily entered the adobe dwellings between 7:00 P.M. and 5:00 A.M. to bite inhabitants. Light shy, this species was encouraged by the unilluminated interiors, and found convenient resting places in the humid, dark walls.

In most years malaria in Mizque was a seasonal phenomenon of the period of copious rains, warm temperatures, and high humidity. These conditions normally took hold by early December, though the lags in incubation placed the peak of malaria outbreaks from March to May. The period from June to October, with low relative humidity, cool nighttime temperatures, and scant standing water, was normally not a time of malaria transmission. With a life span of less than two months, a mosquito generation rarely extended through the dry, cool period in the imago stage. At temperatures of 0°C, the mosquito dies, but her eggs can lie dormant on dry earth in a state of diapause until the start of the next rainy season (Torres 1981). The parasites have much less temperature tolerance. *Plasmodium falciparum* ceases reproduction and thus transmission below 20°C; *Plasmodium vivax* does not divide below 16°C (Wernsdorfer 1980:44). This difference in temperature tolerance explains why vivax malaria was contracted at higher elevations than falciparum was (figure 4.6). As a case in point, one former hacienda, Viña Perdida, located between 2,371 m and 2,528 m, had a record of vivax malaria into the 1950s but no incidence of the more virulent form (ASNRA, 1957, 3102).

Above 1,800-m elevation, malaria can in most years only be seasonally contracted; below that altitude, temperatures rarely fall far enough to interrupt parasite reproduction if breeding sites were available. Paradoxically, malaria

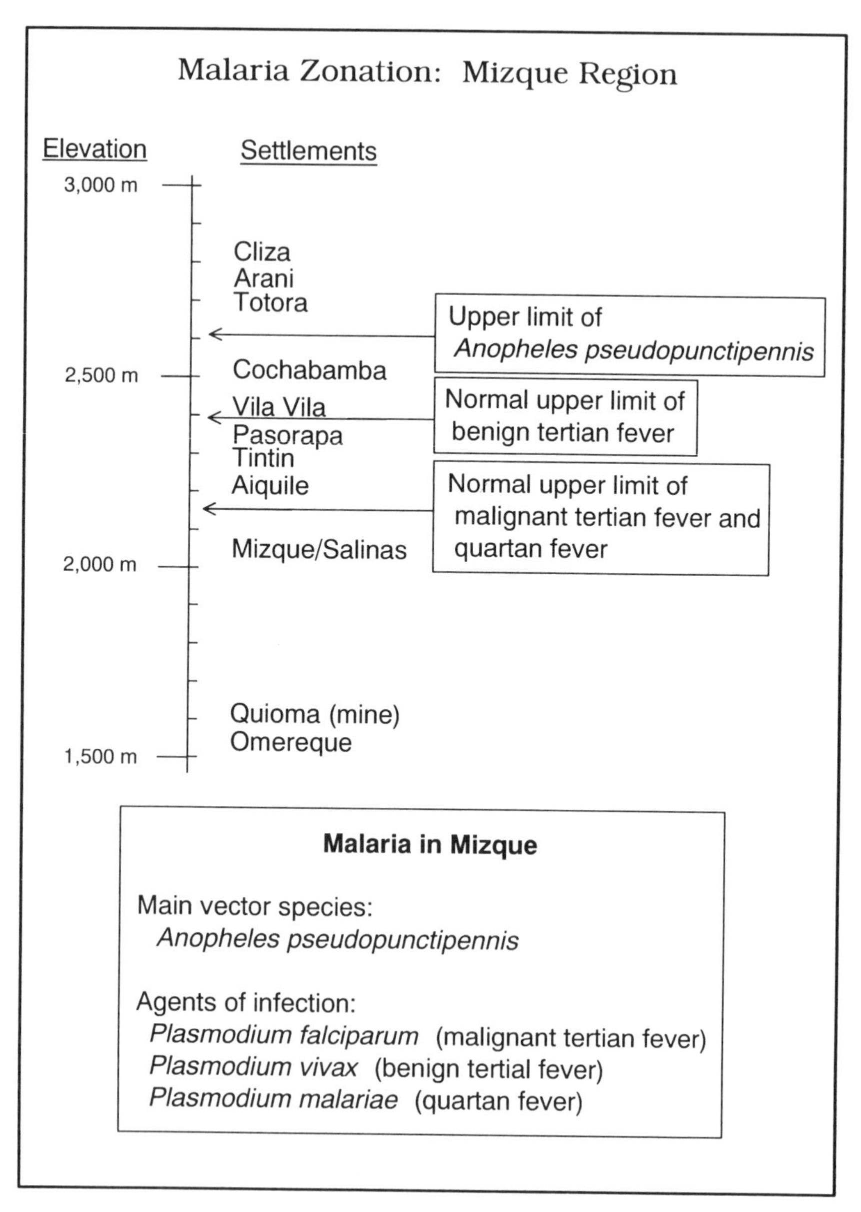

Figure 4.6. Elevation limits of the main mosquito vector and the three kinds of plasmodia carried by the mosquito as related to the main settlements in the Mizque region mentioned in the text

can spread when a drought extends into the rainy season. As stream channels contract and water flow stops, mosquitoes lay eggs in the stagnant pools. These conditions triggered a devastating malaria epidemic in the drought and famine years of 1877–1878. High mortality in Mizque and other valleys in Cochabamba Department, exacerbated by the famine created by crop failure, placed the Mizque region at its nadir of despair (*El Heraldo,* 4–26, 5–21, 6–11, 6–18, 1878).

Identification of the pathogenic complex permits speculation about the past dispersal of malaria. The lack of any real evidence for pre-Columbian malaria in the Americas points to its introduction from two different sources, southern Europe and West Africa (Dunn 1965).[6] Within western South America the coastal lowlands and many inter-Andean valleys etched below 2,100-m elevation acquired both tertian and quartan fevers within decades after the Spanish conquest (Jiménez de la Espada 1965, 1:210, 2:28, 35, 292; Guaman Poma de Ayala 1980, 3:1061). Malaria in these places could not have diffused by means of infected vectors moving from valley to valley across high mountain ranges. An infective *Anopheles pseudopunctipennis,* whose flight range is normally limited to within five kilometers of its breeding locale, cannot spread far under its own power. More plausibly, people brought the disease to Mizque. Individuals who came to Mizque either carried plasmodia long-distance onboard ships from the Old World, or they became infected with them upon disembarkation in the malarial coastal areas and then carried them inland in their bloodstream. Newly arrived in Mizque, these people infected previously clean mosquitoes with plasmodia, which, in turn, were rapidly passed on to other individuals. Spaniards, here by the late 1540s, were the logical bearers of *P. vivax* to Mizque. Subsequently, some of the African slaves brought to the region may have carried with them *P. falciparum;* and other blacks, *P. malariae.* The latter protozoan is well-known to be persistent, but even *falciparum* may live up to three years in the body. Its normally lethal outcomes to the unprotected make *falciparum* the likely culprit of the deadly epidemic of 1710.

Another modern understanding, one that is rooted in genetics, sheds light on the shift in population composition. The fact that Indians in endemic zones got sick with malaria and frequently died fits a racial pattern known elsewhere in the New World. Native peoples have lacked the antibodies to cancel or reduce the baneful effects of plasmodia in the bloodstream.[7] With time, however, some immunity would have developed among Indian residents of the valleys who survived the disease if the seasonal migration of unadapted outsiders from higher elevations had not inhibited the emergence of any such immunity. Periodically forced by need to work in Mizque, Indian laborers from the densely populated Cliza zone of Cochabamba Department characteristically returned to their high country towns with malaria (ANB, 1844; Viedma 1969:84; Melean and García Mesa 1905:28). Besides innate susceptibility, historical abuse of alcohol lowered discipline to respond positively to protective measures and contributed to their high rate of malaria contraction. The lack of spatial mobility for landless Indian workers, attracted to haciendas and therefore not free agents, also took its toll. Without the option of leaving at will for higher zones when malaria seasonally ravaged the valleys, native people inevitably contracted the disease. During the lethiferous malaria epidemic of 1877–1878, virtually all the remaining Indians of Mizque died (ANB, MI, 1878, 207).

People of Spanish ancestry who came to Mizque were less subject to malaria than Indians were. Several hundred years of Iberian exposure to the disease may have imparted some immunity (Urlegaza Uranga 1934). As in the New World dominions, "*calenturas intermitentes*" were the most common malady on the Iberian Peninsula (Hernández Morejón 1967:136). Endemic through most of southern and eastern Spain, malaria was also occasionally epidemic, as it was at Seville in 1785–1786 (Dominguez 1789). The relative immunity of Spaniards in Mizque can only be inferred; what is clear is that their position at the top of the socioeconomic pyramid gave them the option of moving elsewhere to avoid the disease. The vacuum was gradually filled by *mestizos,* who came to dominate the population, suggesting that the Spanish side of their mixed inheritance may have given them some resistance to malaria. Sanz's (1954:167) description of Mizque as a mestizo community was explained as the outcome of the devastation (*furioso estrago*) that malaria wrought on Indians.

As a group, people of full or partial African ancestry were most likely to be protected against malaria. Certain genetic polymorphisms gave blacks an advantage over other races in settling the New World tropics after the Spanish conquest, when malaria was introduced. Hemoglobin S, associated with the sickle-cell gene, which evolved in Africa where falciparum malaria is indigenous, offered a safeguard to a sizable minority of black people. More controversial is the view that a blood enzyme condition known as G6PD deficiency, which has a high frequency among West Africans, may also have imparted resistance to falciparum malaria. As for vivax malaria, certain blood-group antigens ("Duffy system") make people of Negro ancestry almost totally resistant to this form of tertian fever. As elsewhere in the tropical and subtropical New World (Friedlander 1977), this inherited protection enabled blacks to cope much better in hyperendemic valleys than Indians did until the introduction of chemoprophylaxis. Hacendados empirically realized this differential tolerance, but black slaves were expensive and yanacona Indians were cheap. Furthermore, black adaptiveness to the valles fuertes did not ensure their continued presence in the region. As haciendas were abandoned and also when slavery was abolished in the 1850s, people of African heritage disappeared from Mizque as they did from so many other Andean depressions where they had once formed a notable component of the population. In some cases, they migrated elsewhere; in others, their mixed-race descendants were gradually assimilated into the predominant mestizo component.

Twentieth-Century Malaria Control

A public health consensus in Bolivia on the link between malaria and mosquitoes came early in this century, but no real action was taken to control the disease until 1929. In that year, the new government antimalaria agency specifically targeted for clean-up the valleys of Mizque, the most persistent and disreputable focus of hyperendemic malaria in Bolivia. Indeed, blood-test examinations showed that all 800 inhabitants of the town of Mizque carried plasmodia in their blood (Flores 1929). Fear of malaria caused the famous European archaeologist Erland Nordenskiöld (1924:66–67) to avoid Mizque on his 1913–1914 trip, despite the possibility of finding prehistoric artifacts there.

Agency officials, headquartered in this isolated town rather than in a national or departmental capital, were given draconian power to implement a comprehensive plan which brooked no opposition from townspeople. The Andean tradition of the corvée was used to carry out clean-up tasks. Male residents of the valley aged 18–60 worked for no pay two days a year between 1929 and 1932 to drain land, fill swamps, straighten stream channels, cap wells, clean irrigation ditches, and remove the vegetation that had taken possession of the streets and lanes. Since fish eat mosquito larvae, fishing was prohibited. Dwellings were whitewashed and fumigated. Residents and visitors were required to submit themselves for clinical examination and, if malarial, to take quinine, which for the first time in Bolivia was routinely administered to those ill with malaria. With interdiction of the weekend sale of alcoholic beverages, moral renovation seems also to have been part of the campaign. In 1932 an additional allocation of government funds to carry on this work in Mizque was forthcoming (*La Razón,* 4-19-1932:7). It had finally registered that malaria was Bolivia's most pressing public health problem (Mendoza 1931). The authorities used the decadence and despair of Mizque as a symbol of what had to be done and the successes two decades earlier in Panama as an example of what could be achieved.

However, the outbreak of the Chaco War with Paraguay in 1932 distracted official attention from rehabilitating Mizque and focused it on the front line, where malaria was creating havoc with the military agenda. Bolivia's bitter loss of that conflict was due, in part, to a high incidence of malaria among troops of highland origin. After the war's end in 1935, returning soldiers sick with fever spread the plasmodia to uninfected mosquitoes in parts of Bolivia where it was previously unknown. In the years that followed, malaria pestilence struck down almost all the warmer intermontane valleys in this part of the Andes. In 1936 a malaria epidemic in Mizque made it a renewed focus of public health

Figure 4.7. Beginning in the 1950s, each dwelling, assigned a number by the antimalaria service (SNEM), has been periodically sprayed with DDT to kill mosquitoes.

attention. Then in 1941, an even more devastating eruption of *paludismo* hit the depressions and basins of Cochabamba Department including, for the first time, the city of Cochabamba (elevation, 2,550 m) (*El País,* 1-18-1942:3).

The virulence of the 1941 epidemic provoked an appeal by the Bolivian government for international assistance to deal with the malaria problem. In 1942 the Rockefeller Foundation, as part of the Servicio cooperativo interamericano de salud pública (SCISP), extended funds, personnel, and organizational skills for controlling malaria in the semitropical valleys of Cochabamba, La Paz, and Chuquisaca. The program, using Paris green to kill mosquito larvae and administering quinine to victims, was soon impeded by World War II. Beginning in 1947, DDT spraying and chemoprophylaxis using newly developed synthetic drugs produced striking results in Mizque. Compared with 1942, when 73 percent of the people examined tested positive for plasmodia, in 1951, no one carried the parasite. The partial elimination of the reduviid bug that spreads Chagas' disease was a side benefit of applying DDT inside dwellings. When in 1953 the Rockefeller Foundation phased out its program in Bolivia, the Pan American Sanitary Bureau and the World Health Organization filled the breach. In 1957 the Servicio nacional de erradicación de la malaria (SNEM) was formed as an agency in the Bolivian Public Health Ministry. By 1962 its systematic spraying program was so successful that officials predicted the imminent riddance of the disease (Moscoso Carrasco 1963) (figure 4.7). But that optimism proved to be premature, for beginning in 1970 resistance of plasmodia to chemicals led to sporadic malaria outbreaks in Mizque. Subsequently the malaria agency removed the word *erradicación* from its title. Expansion of the anopheline vector and the flow of human migrants enhanced malaria's reappearance as an endemic disease. Recrudescence of malaria has made more people in the valleys vulnerable than lived there a century ago.

Recent Land-Use and Settlement Changes

In the 1950s the valleys of Mizque underwent a major transformation when spraying programs successfully controlled the vector. Suddenly the 250-year-old hindrance to human settlement of these interlocking valleys was eliminated. At the same time, the Bolivian revolution proclaimed in 1952 began to implement its land reform program. The large haciendas that had monopolized the rural land use and structured local society were expropriated and parceled out as peasant-owned farms with an average size of 3.5 ha. Fewer than 10 percent of Mizque farmers are landless sharecroppers, banishing the yanacona tradition of landless peons. Free of the debilitating burden of malaria, agricultural production doubled between 1955 and 1967 in Mizque (Unidad sanitaria de Cochabamba 1968). Irrigation expanded to cover about 70 percent of the valley floors, permitting two, sometimes three, yearly harvests. Dominant crops are maize, peanuts, onions, garlic, potatoes, sweet potatoes, and fruit, especially citrus, cherimoya, and avocado. Lands for grazing and wood supply are held in common by reconstituted peasant communities. Several new periodic markets give evidence of the revival of surplus production. In the Omereque portion of the valley, land reform diversified land use beyond sugar cane and tobacco, which had been the main estate crops before agrarian reform (Guillet 1973). Establishment of a marketing cooperative encouraged peasants to concentrate on growing vegetables for sale.

The near elimination of endemic disease triggered a population increase of 32 percent in Mizque Province between 1950 and 1976. Newcomers filled the demographic vacuum left by malaria. More than half of the farmers in the valleys of Mizque are descended from families that were rooted in the densely populated Cochabamba Valley before the 1950s. That flux contrasts with nearby highland communities above 2,500 m, such as Raqaypampa, which have maintained indigenous traditions even though they too were once hacienda land (Calvo et al. 1994). Unlike the skewed pyramid of the 1930s, when malaria resulted in high death rates among the very young, over 50 percent of the population in the early 1980s was less than 20 years old. Structural changes in land tenure also created a much more dispersed settlement pattern in the valleys than was known before 1950. Beginning in the 1960s, towns acquired facilities they had never had before. Mizque obtained electricity, a piped drinking water supply, and a secondary school, and the people in Omereque built a modern church and rectory, boarding school, health center, and auditorium.

An improved transportation system could unleash more of the economic potential of the Mizque zone, especially if the irrigation systems were improved and extended. The valley floors are especially suitable for fruit and vegetable production. Cherimoya finds the optimal climate here for attaining its peak quality. Viniculture was modestly reestablished in the late 1970s after a two-century hiatus. Sustained development requires the containment of malaria, but mosquitoes and plasmodia now have partial resistance to synthetic chemical compounds. Malaria control efforts are also complicated by the latter-day realization that freedom from endemic disease is a trade-off for the toxic accumulation of chemical sprays and their nocuous effects on human and livestock health.

Contextual Perspectives

Inferences that relate to responsible agents, scale, and synthesis emerge from this study of an isolated Bolivian valley. Malaria's role in the demographic decline and retrogression of a once salubrious zone was clear already in the eighteenth century. Long before the ecological thresholds were tied to vectors and parasites, inhabitants figured out that malaria contraction was confined to the macrothermal valleys. The inverse correlation between malaria morbidity and altitude had become apparent in the seventeenth century. The association of malaria with standing water became suspect a century before scientists learned that the anopheline mosquito was the carrier and that its reproduction required a persistent film of water. Folk knowledge sometimes converges with laboratory findings.

Malaria is too frequently interpreted as a disease wrought by invasive biological elements without appreciation of it as a culture/nature gestalt in which humans have played a considerable part in its spread and intensification. It was likely that people coming to this region initially infected clean mosquitoes rather than the transmission originating as animal to man. Equally plausible was the human role in the amplification of mosquito breeding, which tightened the grip of endemic disease and promoted epidemic outbreaks. Soil erosion triggered sedimentation that led to river channel overflow and festering backswamps. Creation of a network of irrigation ditches, digging of wells, and the accumulation of water near habitations provided additional sites for vector reproduction. Domestic livestock constituted alternative sources of blood to sustain the requirements of large anopheline populations.

The effects of malaria went beyond the valley-upland dynamic to remodel the larger regional structure. By the mid-eighteenth century, Mizque had lost its urban character, Spanish inhabitants, and economic prosperity and was forced to cede its position at the apex of the central-place hierarchy to Cochabamba. Though changes since the 1950s have given the town a new lease on potential well-being, the departmental road system nevertheless leaves Mizque in a peripheral position. The demographic gap widens, and while Mizque has grown to about 2,000 inhabitants, Cochabamba has seen its population rapidly expand to more than 350,000. A historical problem defined by a mere 500 meters of elevational difference cast the die of regional formation.

Mizque suffered one of the largest and most virulent malaria histories in western South America. But one can extend this history to virtually all parts of the Andean region below 2,000 m, including the desert coast and the eastern jungles. The ecological and epidemiological perspective on settlement and land use of the hot country offers a logic for the transatlantic movement of African slaves to a land where an abundant supply of labor existed when the Europeans first came to the area. A similar process of demographic displacement occurred in widely separated areas in Bolivia, Peru, Ecuador, and beyond. A biological imperative operated for more than 400 years until malaria eradication programs began in the 1950s.

To a much greater degree than was previously fathomed, the verticality of vector-transmitted disease in the Andes elucidates human migration, population densities, and racial/ethnic change. Seen in an epidemiological light, slavery is thrown into clearer perspective as an institution than as simply labor replacement, which has been the stock interpretation of historians. The paradox of widespread poverty in zones with a well-endowed resource base becomes comprehensible. The ecological imperatives of the pathogenic complex in concrete locales merit special attention in reconstructing the cultural geography of Latin American settlement.

5

The Andes as a Dairyless Civilization

Llamas and Alpacas as Unmilked Animals

Why milking was absent in the pre-Columbian Andes is an enigma to some who ponder the anomalies of the world's culture history.[1] Two domesticated herd animals, the llama and alpaca, have been eligible candidates for a dairy focus that has never emerged. Understanding this lactary lacuna requires attention to the nature of these beasts as well as to the Andean past. But the greatest insight on this subject comes from outside the region itself. How milking emerged as a culture trait elsewhere in the world unveils the key to puzzling out its absence in the Andes.

Fiction has frequently crowded out the facts about South American camelids. Many people have blithely assumed that llamas and alpacas have been milked in the same way as cattle and goats. This presumptive trap is based on an unconscious Western conceit that milking is a "normal" practice and that milk is a healthful and delicious ambrosia for humans to ingest.[2] A worldwide array of 10 different domesticated ungulates that are milked fortifies a certain expectation that the llama and alpaca must therefore also be used in that way. Moreover, these two camelids have so many uses that it is natural to assume that milk has been one of them. Meat, fat, and blood are food. Fibers are woven into clothing, bags, and rope. Leather comes from the hides. Dung is a fertilizer and an indispensable fuel in the treeless Andes. Bones and the dried fetus have magical-religious value. These multiple uses have been elaborated over a long period of more than 6,000 years since the llama and alpaca were first domesticated (Wheeler 1984).

Domesticated New World Camelids

The llama (*Lama glama*) and the alpaca (*Lama pacos*) are closely related animals, although they are descended from two different wild species—the llama from the guanaco (*Lama guanicoe*) and the alpaca from the vicuña (*Lama vicugna*) (Wheeler 1995). When the two domesticated species are crossed, their offspring (*huarizo*) is usually fertile. All four South American camelids—two wild, two domesticated—possess the same karyotype (2n = 74), which indicates that they have not diverged nearly as much as different species in other mammalian families in South America, such as the canids and felids. Llamas and alpacas have many more anatomical and physiological similarities than differences. Although the adult llama weighs between 130 kg and 150 kg, compared with the alpaca at about 60 kg, their feeding habits and behavior resemble each other. Alpacas and llamas less than a year old are given the Spanish term *cría,* which is used in English as well. In most of this discussion, both species can be considered together. Their present distributions overlap, but that of the llama has always been the widest (figure 5.1). Both species have contracted in their range from Inca and early colonial times.

The Inca, who inherited from Andean predecessors their knowledge of camelid uses, made their own advances with these animals over the less than 500 years of their empire. The Inca and pre-Inca peoples selected and bred camelids according to color, size, and fiber quality (Wheeler, Russel, and Redden 1995). Given that level of human interest and control over breeding, use of milk as a source of nutrition might be expected to have occurred also. A lactation period of six months and a milk rich in vitamins, proteins, and fats would have enhanced their value to humans. Moro S. (1956) found that the fat content of alpaca milk fluctuated between 3 and 4 percent, figures somewhat higher than that of Davis (1955) and Jenness (1974) (table 5.1). More recently, Fernández and Oliver (1988:301) calculated the fat content of llama milk as 4.7 percent with a possible variation of 1.30 percent. Proteins, mainly casein, are more than twice as abundant in llama milk as they are in cow or goat milk. Other than that, llama and alpaca milks have organoleptic similarities to that of the cow, quite unlike their family cousin, the dromedary, which yields a strongly flavored and salty-tasting fluid.

Llamas and alpacas share certain behavioral traits that would have predisposed them to milking by their human keepers. Their docility resembles that of sheep. They allow humans to help in birthing, which occurs 350 days after conception. Keepers also sometimes intervene to facilitate copulation by tying a rope around the hind legs of skittish females. In spite of biological possibilities and human interventions, no milking pattern of camelids ever emerged anywhere in the Andes, either before the Spanish conquest or after.

Andean Milk Void

Although the llama and alpaca seem likely candidates, nothing points to the use of their milk in the Andes, past or present. Reconstruction of their domestication offers no evidence of Andean people's interest in that secretion

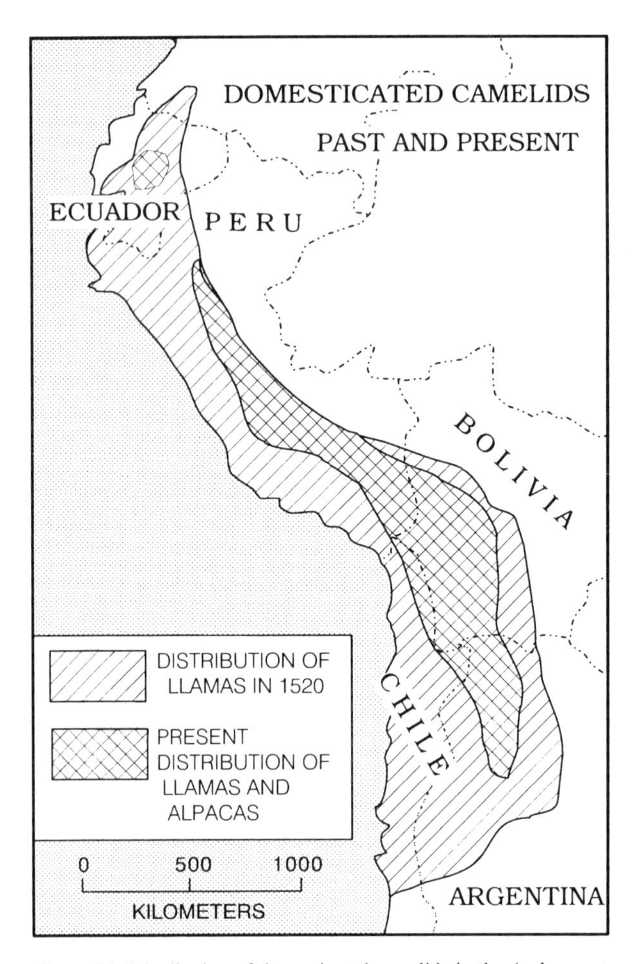

Figure 5.1. Distribution of domesticated camelids in the Andes, past and present. Coastal occurrence of llamas persisted into the seventeenth century.

(Browman 1989; Lavallée 1990; Lynch 1983; Novoa and Wheeler 1984; Wing 1986). Stylized drawings of camelids are abundant in Andean rock art, but none displays the salient udders that would offer clues about prehistoric uses (Berenguer and Martínez 1989:399). Large udders are not a genetic result of domestication; rather, they evolved through the practice of milking. No recorded Andean myths involve milk or milking (Urioste 1983:219). Nor was milking of camelids ever mentioned by the sixteenth-century chroniclers who wrote down detailed observations of Andean domestic economy. In the seventeenth century, Cobo (1956, 1:387) remarked that the "naturales destas Indias" neither used nor had knowledge of milk, whey, cream, butter, or cheese.

Quechua offers no lexicon of milk use, in contrast with its many terms to distinguish pelage types of llamas and alpacas (Flores Ochoa 1978). The words for human milk (*ñujñu*) and animal milk (*wilali*) suggest different conceptual spheres. Andean medicinal uses curiously exclude llama and alpaca milk at the same time that milk from the donkey and dog is included (Lira 1985:79–80). Three astute nineteenth-century travelers, Walton (1811), Tschudi (1885), and Middendorf (1893–1895), did not write a word tying camelids with milk for human consumption. Contemporary ethnographers who have studied llamas and alpacas in some detail have not commented on either of these two domesticates as a milk source to their owners (Calle Escobar 1984; Cardozo 1954; Flannery, Marcus, and Reynolds 1989; Flores Ochoa 1968, 1984; Orlove 1977; Sumar 1988; and Webster 1973).[3] Although I have found no evidence that it did occur, it remains possible that at some time or place episodic milking of a lactating llama or alpaca could have occurred. A hypothetical example might have involved a woman who failed to provide enough milk for her nursing infant and a pliant female camelid whose cría had died. However, there is no indication that the next critical steps of repeating that singular event and turning milk extraction into a regular practice has ever occurred anywhere. If milking had emerged over time as a cultural pattern, it would likely have survived the Spanish conquest and at some point would almost surely have been noted in the ethnographic record of the Central Andes.

Consequence for Genre de Vie

Had milking been put into practice, it conceivably could have triggered the development of nomadic pastoralism as an Andean way of life. Dairy products

Table 5.1. Percentage distribution of nutritional components in the milk of nine different domesticated ungulate mammals

Common and scientific names	% Fat	% Protein	% Lactose
Buffalo (*Bubalis bubalis*)	7.4	3.8	4.8
Camel (*Camelus dromedarius*)	4.5	3.6	5.0
Cow (*Bos taurus*)	3.7	3.4	4.8
Goat (*Capra hircus*)	4.5	2.9	2.9
Horse (*Equus caballus*)	1.4	2.0	7.4
Llama (*Lama glama*)[a]	2.4	7.3	6.0
Reindeer (*Rangifer tarandus*)	16.9	11.5	2.8
Sheep (*Ovis aries*)	7.4	5.5	4.8
Yak (*Bos grunniens*)	6.5	5.8	4.6

[a]The only included species not milked by humans

Figure 5.2. One way in which llamas are integrated into Central Andean agriculture is by pasturing them in harvested fields to establish the reciprocity of their eating the stubble and depositing dung to replenish soil fertility. Near Colcha (Cusco).

would have granted near self-sufficiency to high-altitude inhabitants above 3,800 m as they periodically moved in search of good pasture for their flocks. If domesticated camelids had acquired a dairy function, their herding would likely have also spread into the páramos of the northern Andes. However expansion in that direction was late, limited, and relatively unimportant. In the late fifteenth century, the Inca brought llamas as beasts of burden to their conquered territory in present-day Ecuador. Lofty zones were largely beyond the possibilities of agriculture, and the domesticated camelids had their greatest importance there. By and large these animals have been tied to an agricultural way of life which foreclosed in some ways their full potential in domestic economy (figure 5.2).

Milk animals arrived in South America from Europe beginning in the 1530s. Andean folk observed how Spaniards pressed the teats of a lactating cow, goat, or ewe to obtain a creamy liquid. However, that observation did not lead to a stimulus diffusion of the idea to include either the llama or the alpaca. Grasslands above 4,000 m, where camelids were abundant but which were environmentally difficult for cattle, would have been the most plausible zone in which that might have taken place. Andean failure to transfer contrasts with northern Scandinavia, where in historical times the Lapps (Saami) learned to milk reindeer, the domesticated mammal at their disposal, by following the example of dairy-loving Swedish farmers who came north with cows (Eidlitz 1969). Farther east in Siberia, other northern groups learned to milk their lactating does by imitation of the milkers of cattle, mares, and ewes to the south of them (Fohndahl 1989).

Pathways toward Understanding

The failure to milk alpacas and llamas and the anomalous situation that such a lack represents in world culture history have received no thoughtful assessment in the literature of the Andes. Even such an astute observer as Alexander von Humboldt (1970, 2:607) offered a flawed Eurocentric tautology to explain the lack. From Humboldt's perspective, New World camelids were simply not suitable candidates to provide milk.

Camelid Milk Quantities

Scarcity of yield is proposed as one major explanation for the failure to milk South American camelids. This hypothesis assumes that it was not worth milking the lactating female for the small quantities available. The line of reasoning also conveys the idea that after the cría has suckled, no fluid is left over in the udder for people to extract (figure 5.3). Considered historically, this argument is specious, for cattle, goats, and other milked mammals also had no surplus quantities of milk to surrender to humans in the early stages of their domestications. The earliest domesticated cows produced only trickles of fluid, not the astonishing milk production of genetically modified dairy cows. Given the minuscule yield in the early periods of milking, the motivation to extract milk on a regular basis was other than nutritional.

If milking did not arise spontaneously anywhere in the Andes or the New World, it is also quite certain that pre-Columbian peoples had no example from which they could borrow. The idea of fluid extraction did not spread to the Americas from any of the three continents—Europe, Africa, or Asia—where it had been known for thousands of years. Several possible pre-Columbian contacts between the Old and New Worlds have been postulated, but no one has ever seriously suggested that the dairy idea was among them. Andean people did not learn about this unusual source of human nutrition until the European introductions of the sixteenth century.

Characteristics of Camelid Behavior

Idiosyncratic llama and alpaca behavior is another reason proposed for the milking void. In this view, a lactating alpaca or llama does not permit a person to manipulate her teats, nor does she release the colostrum that triggers the consequent flux of milk for any but her own cría. But these hormonally con-

Figure 5.3. Llama mother and *cría* on the *puna* (Cusco Department, Peru). Although the female has four teats, the birth of twins is very rare. Photograph was taken in the dry season, when forage is of poor quality.

trolled defense mechanisms were known in other domesticated herd animals. In time either they were bred out of those species, or people learned how to coax the lactating female to let down her reserve, just as they would have in llamas and alpacas if they had started to be milked. Although milch animals have evolved progressively in the direction chosen by their human keepers, the process is far from complete. Behavioral reservations still occur with lactating mares, water buffaloes, and camels, which permit human removal of milk only in the presence of their young.

Lactose Intolerance

A third proposed explanation for the historical absence of milking alpacas and llamas is focused on a human physiological condition. Once weaned, most native New World people cannot digest fresh milk. Absence of the enzyme lactase impedes digestion of the lactose contained in the milk. Numerous other cultural groups that together make up more than half of the world's population manifest this same inability to digest milk satisfactorily as adults (Simoons 1970). If malabsorbers consume fresh milk, many of them become ill. In contrast, genetic selection through a long history of keeping dairy animals and drinking their milk in northern Europe, parts of Africa, and, it seems, parts of northern India has enabled the adult body to retain lactase. Other scientists have proposed that environmental factors were important in selection for this trait. But Holden and Mace (1997) have strongly reinforced Simoons' culture-historical hypothesis that adults with a high capacity for lactose digestion evolved in populations that kept livestock.

Nutritional investigations in Peru have found that native Andean people beyond a very young age may indeed suffer nausea and diarrhea if they drink fresh milk (Figueroa et al. 1971). Calderón Viacava, Cazorla Talleri, and León Barúa (1971) found lactose malabsorption in 63 percent of their Peruvian participants. Paige et al. (1972) registered malabsorption in 100 percent of persons over 12 years of age. In Bolivia, Aymara Indian schoolchildren showed lactose malabsorption in 55 percent of the 2–5-year-old age group and in 77.4 percent of the 11–15-year-old age group (Balarna and Taboada 1985). Psychological rejection of milk in the Andes does not reach levels found in China, where people consider it to be a repulsive, undrinkable secretion. However, for small Andean children, fresh or canned milk is an esteemed beverage. Lack of milk in the pre-Columbian adolescent and adult diets or its relative absence today has not been reflected in vitamin D deficiencies such as nutritional rickets. Other sources of calcium have been ingested in the Andes to compensate for that provided in milk. In some areas of the Andes, calcium-rich minerals—calcite (*katawi*), calcareous tufa (*hake*), and calcium carbonate (*khakya*)—are used to make sauces eaten with potatoes (Browman and Gundersen 1993). Ashes (*llipta*) of certain crop plants ingested as an accompaniment of coca chewing are another calcium source. Although no archaeological evidence is known for the use of the minerals, the use of llipta goes back in Andean culture history for 4,000 years.

Lactose malabsorption is not in itself a convincing explanation for the failure to have milked the two domesticated Andean camelids. All human groups are believed to have once been largely intolerant of milk sugars; as they adopted milking, those few who could digest it were at a selective advantage, until many generations later, they formed the majority. Furthermore, the gene that controls lactase production in adulthood has evolved in the Old World over six or more millennia, which is the length of time that llamas have been domesticated. Another indication that lactose malabsorption is not a persuasive barrier to milking can be found in the native Andeans' adoption of the colonials' practice. After the Spaniards' sixteenth-century introductions of the cow, goat, and sheep, the Indians saw these southern Europeans using fresh animal milk for babies when maternal supply was insufficient. Indians found that contribution to infant diet just as helpful (figure 5.4). In some places, adults have consumed fresh milk, but sparingly (Vokral 1989:176).

Most fresh milk extracted in the postconquest Andes has been converted into soft cheese. Spaniards in the early colonial period provided the initial

Figure 5.4. Indian woman near San Pablo (Cusco) milking a cow for her child, with the presence of the calf facilitating the operation

example for cheese making in the Andes (Vázquez de Espinosa 1969:261, 264, 324). In that process, fermentation largely changes lactose to glucose and galactose, so that only small amounts of the former remain in the cheese. That low lactose threshold enables many native people to consume it with little or no adverse physiological effect. In some places, the Indian peasantry adopted the making of unaged soft cheese. Most is made from cow milk, but in places, goat and ewe milks are used. Sheep cheese (*queso de Paria*) has long been associated with the Altiplano around Oruro. Bolivia today still produces 90 percent of the sheep's milk in Latin America. But many native Andean people do not themselves actually consume the cheese they make; instead they sell or trade it to mestizos. In the Andes, preservability and added value have made it feasible to transport cheese over quite long distances.

Religious Motives and Animal Transformations

The Old World Model

The Andean failure to milk ultimately is best understood by reconstructing how the practice of milking may have started in the world. Eduard Hahn (1896), the German geographer-ethnologist, deduced that Neolithic milking experiments had a religious, not an economic, motive.[4] In Hahn's view, humans did not domesticate cattle for the nutrition they provided to people.[5] Hahn and a later scholar, Erich Isaac (1970), surmised that the curved horns of the aurochs (*Bos primigenius*) evoked an epiphany of the moon goddess through their resemblance to the lunar crescent. Sacrifice of this animal species became a reenactment of the goddess's death, perhaps performed with each waning of the moon as the crescent appeared in the sky. This practice and its frequency would have required a constant supply of animals.

Capturing each aurochs from the wild could not be sustained. Live bulls and cows were kept in holding pens to breed. These matings of captives occasionally produced calves of a smaller size and a greater docility than their progenitors, and also with different variations in markings and hide color. Viewed as special, these variable descendants were allowed to breed. A gene pool that distinguished these bovines from the wild gradually formed, and phenotypic distinctiveness enhanced their sacred status. The hierophantic role of these new domesticates included pulling ceremonial wagons in religious processions, symbolically plowing the land in fertility rites, and undergoing milking as a ritual gift offering.

The earliest attempts at milking most probably took place only after domestication had created a pool of relatively manageable animals that allowed themselves to be milked in the presence of their calves. Opportunity to extract milk may have also come when the contents of the dam's overloaded udder became available to humans after the calf had been killed. The act of milking itself may have been a ritual. Priests were logically the ones to extract the fluid from the udder as part of a ceremony; they then could proclaim it to be a gift of life and fertility from the moon goddess.

Gradually the economic usefulness of cattle asserted itself for plowing fields, pulling wagons, and for the meat, milk, and hides which are so prized today. These, however, were later and secondary benefits of bovine manipulation. At its origin, it was religion that prompted the elaboration of the female principle. The moon goddess cult evolved out of the upper Paleolithic identification of female fertility as the giver of life (Walsh 1989). The emergence of agriculture and livestock husbandry gave concrete religious expression to it, for to ensure continued agropastoral prosperity, the cult of fertility had to be refined and codified.

The ancient prestige of milk as a cultic requirement provided an incentive to encourage the breeding of bovines for their docility and large udders. Over many generations as higher-yielding specimens proliferated, milk became commonly available and at some point was integrated into the food regime of the larger society (figure 5.5). As the religious purpose of cattle faded into the background, the nutritional benefits of meat and milk became a sufficient reason to keep these animals. However, cattle breeding focused on large milk

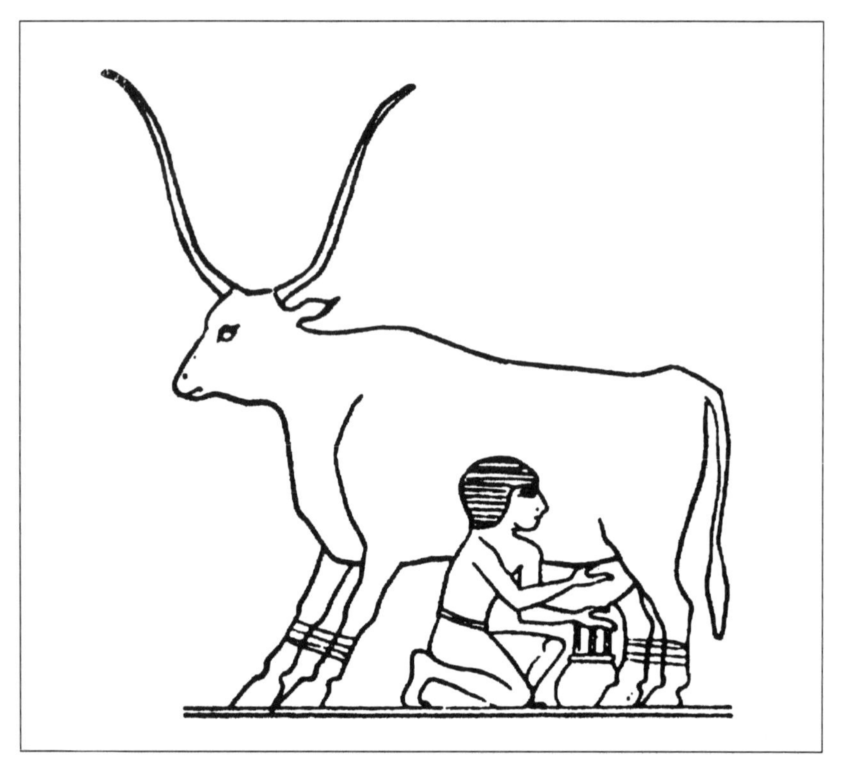

Figure 5.5. Milking of a cow with half-moon-shaped horns in Dynastic Egypt in the second millennium B.C. The tied feet indicate the imposition of force to accomplish the operation. For the Americas there is no iconography before the seventeenth century A.D. that shows extraction of milk from any animal, and animals that were represented thereafter were introduced domesticates.

output, and high butterfat content did not elaborate until the nineteenth century. Since then, so productive are modern dairy cows with their hypertrophic mammary glands that it is easy to see how the idea of milking in Western societies has been interpreted as a *raison d'être* for cattle domestication. If llamas and/or alpacas had been selected in that direction, they too would have eventually evolved large udders, long teats, and a milk supply well in excess of neonate requirements.

Thus milking arose as a practice in a particular convergence of cultural, historical, and geographical conditions. C. O. Sauer's (1952) idea that knowledge of milking was an unduplicated invention of a particular area at a particular time postulates diffusions from a Near Eastern center of origin. Archaeology points to western Asia as the broad zone of cattle domestication ca. 5000 B.C.

The earliest clear evidence of milking is found there ca. 3000 B.C., but another view, one based on genetic evidence, is that both types of cattle (*Bos indicus* and *Bos taurus*) derive from a single domestication event 8,000–10,000 years ago (Loftus et al. 1994). The earliest known representation of humans milking animals is dated back to the early Neolithic, in which a decoration on a libation vessel shows the back of a cow and a hand milking in the context of a statue of a mother goddess on a throne (Silistreli 1989). Iconographic evidence of milking from northern Africa suggests either a very early diffusion of the idea or a parallel zone of origin (Simoons 1971). The fullest evidence still points to the Near East, whence cow milking as a religiously inspired practice pushed westward to the Nile Valley and the Mediterranean and eastward to the Indus civilization (Simoons and Simoons 1968:234–58).

As the idea of milking diffused farther into more marginal environments, different animals were substituted where cattle did not thrive or were not important. In colder areas, ewes were milked; in semiarid zones, nanny goats. In deserts, camels were available for this purpose; and on the Asiatic steppes, mares where horses were the main domesticate. In the Indian subcontinent, milk was extracted mainly from cows, but in the high Himalaya, yaks were milked, and, conversely, in lowland areas, the lactations of water buffaloes were collected by human interlopers. By that process of substitution from a bovine model, milking spread widely as a cultural practice in parts of the Old World (figure 5.6).

Even though sheep and goats preceded the domestication of cattle, cows were the first animals to be systematically milked. Milk from cows has always been the most commonly available in most of the world and the kind of milk described in ancient texts as being a perfect and noble fluid. The sacred glow of milk is partly a legacy of a religiously motivated domestication of the animal and of the hierophanies associated with the ancient history of the practice.

New World Failure

If religion is the key to understanding the origins of milking, the pre-Columbian failure to milk llamas or alpacas must be understood in those same terms. Although it is not yet clear if their domestications had a religious motive, these ungulates have long had an indisputable religious importance in Andean culture and society (Gareis 1982). Andean adherents of Inca and pre-Inca religion made periodic sacrifices of these animals and used them in spiritual rhetoric (Brotherston 1989). The cults of the sun, thunder, Viracocha, and the moon each had their own camelid flocks (Santillan 1968:108). In this celestial pantheon, the moon was the feminine divinity and the wife of the sun (Silverblatt 1987:50). Called Mama Quilla (Mother Moon), she was given her own chapel in the Temple of the Sun in Cusco and at Quimquilla, near Huarochiri in the highlands east of Lima. There and at other sites of the Inca Empire, lunar cere-

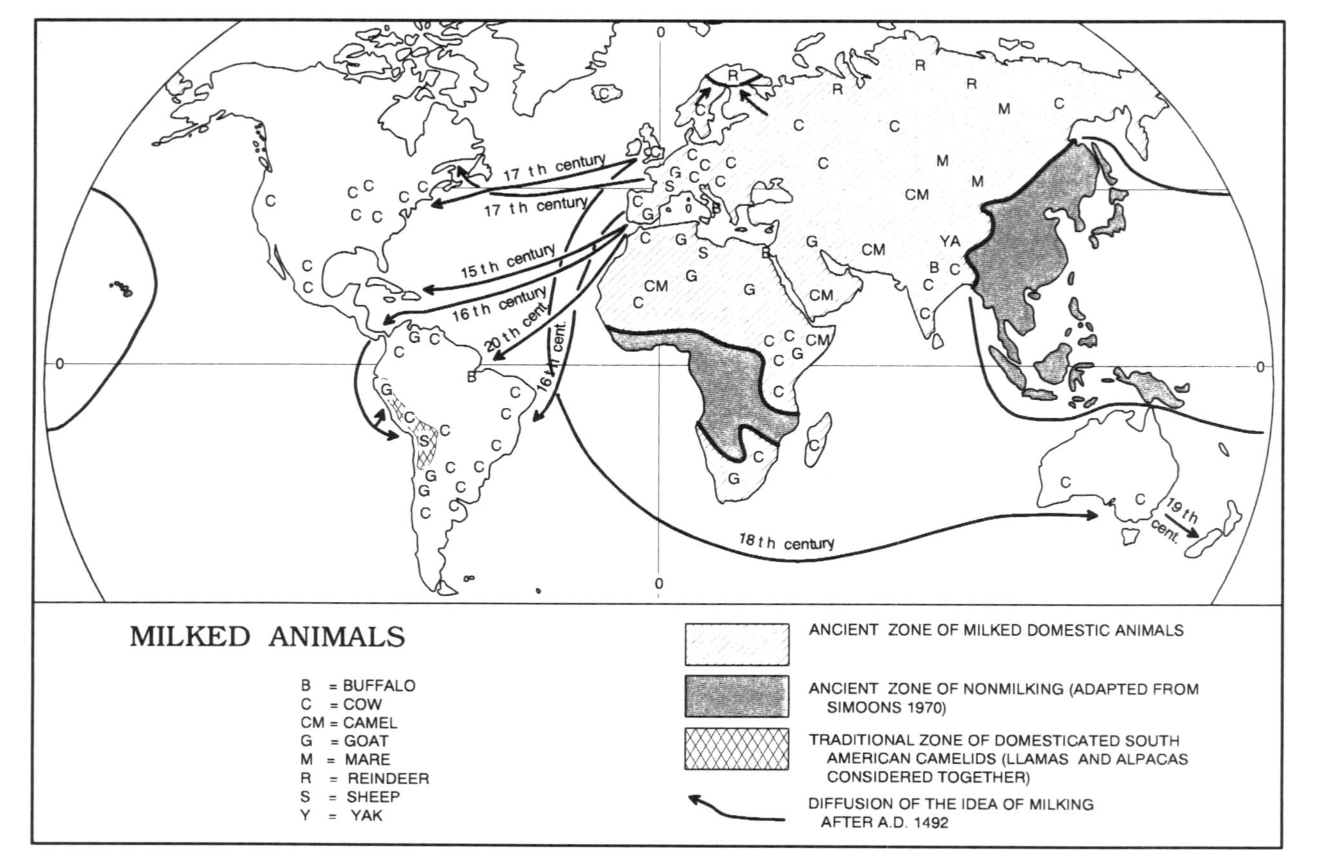

Figure 5.6. Traditional distribution of milking in the Old World and the post-Columbian diffusion of the milking practice to the Americas, Australia, and New Zealand (Sources: diverse ethnographic data; data on milking zones in Africa and Asia from Simoons 1970)

Figure 5.7. Recalling the Inca custom, this early colonial drawing shows a propitiatory sacrifice of a llama in the waning moon festival in March to guarantee the harvest. (Source: Guaman Poma de Ayala 1980, 1:214)

monies involved sacrifices of llamas and other animals (figure 5.7). However, none of the Andean rites of fertility, curing, or divination related to milk in any way. Instead, camelid fat, transportable and not readily perishable, acquired the key role in ritual. Use of fat is still seen in the "*pagos*" and "*despachos*" that individuals make to placate supernatural forces.

Other sacrificial rites of the past included smearing llama blood and eating cooked and raw llama meat (Guaman Poma de Ayala 1980, 1:228–29; Molina 1943:46–62). Although the moon cult was symbolically tied to the female principle and fecundity, milk was not an element in that symbolism. Unlike cattle in the Near East, the llama and alpaca had no apparent physical attribute that linked them specifically to the moon. Lacking horns on either sex, llamas and alpacas had no obvious lunar symbolism, so that Andean sacrifice of these animals had a more generalized context. No evidence occurs for the Inca sacrifice of crías, which would have been the window of opportunity to take milk from lactating dams without the competition of their offspring.

Totally removed from that learning curve, Andean people never, of course, sorted out the ways to stimulate a lactating dam to unleash the flow of milk from the mammary gland out of the teat. In the Old World, human manipulation of ungulate behavior and physiology was a key aspect in the milking process. Even today, cattle other than the highly evolved dairy breeds do not readily give up their milk to other than their calves. They must be familiar with the milker and either have their offspring nearby to trigger the milk-ejection reflex or undergo other human interventions. In Africa, cows are massaged, or the owner first blows air up their vagina to trigger the flow of milk out of the teat (Amoroso and Jewell 1963).

Idealist Perspective

Puzzling out the Andean milk void will always remain an approximation, but, once scrutinized, none of the classic reasons for that lack is persuasive. Pre-Columbian absence of the systematic extraction of Andean camelid milk did not have much to do with either a presumed inaptitude of these animals for milking or the physiological intolerance of their keepers. Milking does not depend on high levels of lactose tolerance in a population.

Instead, an Old World perspective on culture history shows the best pathway to comprehend this baffling absence. In that comparative light, the origins of milking as a practice had nothing to do with nutritional necessity, but were part of religion. Andean civilization did not establish any nexus between religious cult and lactating animal. If milk had acquired a sacred character, milking of either or both camelid species would probably have caught on as a human activity. The great stimulus for humanity to extract the contents of the udder was a metaphysical obsession that galvanized humans to pursue that extraction. It gave them the will to persist and the courage to intervene in the life of powerful beasts. Practical necessity as a mother of invention offers only the most shallow argument for the origins of milking or nonmilking. Not only the unfolding of lactic origins but also other practical uses and food prohibitions manifest the power of ideas in governing human material life. Cultural materialism offers only cheap, popular explanations devoid of fundamental appreciation of the motivational force of religion.

After the Spanish conquest, cattle, sheep, and goats were introduced by people also knowledgeable about milking and making dairy products. Competition from these Old World domesticates best accounts for the failure to transpose milking to the New World camelids. Camelids had no niche of their own that may have encouraged a stimulus diffusion. They thrive at high elevations, but so do sheep. The sedentary nature of Andean rural life also made milking much less central than it would have been if Andean highlanders had had a nomadic component and depended primarily on animal products. Species transfer of the idea of milking might have occurred if settlement mobility had imposed greater overall dependency on herds.

If divine propitiation explains the earliest origins of milking in the world, later proposals to extend the practice must have other motives. For alpacas or llamas ever to become dairy animals, a special rationale of health or novelty would be required. Such an eventuality would most likely occur outside the Andes. New uses of llamas and alpacas, including any sustained attempts to convert them into dairy animals, would probably occur in North America. The combined population of more than 50,000 animals and a spirit of innovation opens unorthodox possibilities. So far no one has launched or even proposed a plan to initiate a dairy option, which would require entrepreneurship, experimentation, and selection over a long period of time. To most it is still a bizarre, unworkable idea. Yet given the history of domesticated herd mammals, the milking of New World camelids is not an absurdity but simply still unexplored 6,000 years after their domestication.

6
Epilepsy, Magic, and the Tapir in Andean America

Understanding the enigmatic configurations of the nature/culture gestalt becomes a special challenge when an ancient pattern must be reconstructed from fragments of evidence and inferences. One example involves an improbable yet strong connection in Andean culture history between the tapir and human epilepsy. How such disparate phenomena as a wild animal and a neurological disorder could be related bespeaks the uncanny way humans have derived meaning from the biological phenomena surrounding them.

Epilepsy is a complex disease with several different manifestations, of which generalized seizures, also called tonic-clonic or grand mal, are the most characteristic. The onset of a grand mal is preceded by a sensory warning ("aura") which alerts the epileptic of the oncoming rush. Shortly thereafter, the epileptic falls to the ground as a burst of neuron discharges invade the brain. The body suddenly becomes rigid, and typically the victim lets out a sharp cry as air moves out of his lungs. The ensuing spectacle may include gnashing of teeth, dilated pupils, rolling eyeballs, and a face flushed purple. A minute or two of rhythmic jerking spasms and saliva foaming around the mouth may be accompanied by tongue-biting, genital erection, and incontinence. Having run its course, the seizure spontaneously ceases and the person again functions normally. In addition, a certain number of epileptics experience hallucinations and psychoses that include hyperreligiosity and mystical delusions (Glaser 1978; Trimble 1991).

Epilepsy in the Andes

Epilepsy has been a disorder in virtually all human groups, but its frequency varies. In the Andes, which are still inhabited largely by descendants of pre-Columbian cultures, epilepsy is reported to be four to six times as common as in North America (Senanayake and Roman 1993; Gómez, Arciniegas, and

Torres 1978). Its high incidence in western South America may be the continuation of a historical pattern, since heredity is believed to play a part in the incidence of this disease. Prehistoric evidence of epilepsy cannot convey just how common this disease might have been. An anthropomorphic ceramic from the Mochica period (400 B.C.–A.D. 600) offers tantalizing possibilities for neurological interpretation (figure 6.1). The chronicle of Guaman Poma de Ayala (1980, 1:107), written in the early seventeenth century, described and illustrated an epileptic attack of a royal Inca personage (figure 6.2). Chimbo Mama Cava, wife (*coya*) of the Inca Capac Yupanqui, suffered from periodic attacks of epilepsy, and for this reason the Inca divorced her.

Some cultures give epilepsy religious connotations; in Mediterranean antiquity it was called the sacred disease (Temkin 1971). Yet that association is surely much older and more universal than any written records. In the Paleolithic hunting culture, epileptics became shamans or spirit-diviners (Eliade 1964). These individuals were thought to have a divine calling, for their periodic but unpredictable seizures were seen as proof that the spirits spoke through them. Furthermore, ecstatic trances facilitated the soul-journey. In the throes of an epileptic fit an individual was thought to pass through the portals into another world. Apparent death was quickly followed by "return to life." That was surely evidence that the epileptic was an individual who was touched by the gods and could outwit death. People so afflicted claimed to have special powers. However, the ancients recognized two sides to this uncontrolled expression of spiritual energy: demonic (*morbus daemoniacus*) and divine (*morbus deificus*). In the latter category were the pythonesses of Apollo at Delphi, who gave out the answers of the oracle. They were recruited largely from among women who experienced convulsions, that is, those who had epileptic seizures.

As in ancient Greece, epilepsy in the preconquest Andes was a puzzling disorder that intriguingly was connected with religion. The chronicler Betanzos described the people who might enter the Inca priestly class. Some were known as *villackuna,* persons who underwent an "attack of madness" (Valcárcel 1978, 3:30). Blas Valera (1992:72) wrote of oracles in Quito, the north coast, Rimac, and Titicaca, whose priests (*watuckuna*) maintained a "pact with the devil" through periodic outbursts (*furor diabólico*). This indigenous link of religious practitioner and seizure survived the Spanish conquest. During the idolatry trials of the seventeenth century, Villagomez (1919:155) wrote that those chosen to be priest-curers had the "heart's malady" (*mal de corazón*) and were able to talk with the idols (*huacas*). He thus confirmed the pre-Columbian pattern of epileptics as a specific category of those who became diviner-curers (*huaca villackuna*). Spaniards perceived epileptic seizures to be the effects of the devil.

Also called *sonqoyoc* (heart men), these specialized priests were entrusted

Figure 6.1. The human face on this Mochica (ca. A.D. 600) stirrup pot shows whites of the eyes and protruding tongue. The puzzling conjunction of a dead head on a living body might be interpreted as representing a person in epileptic seizure. (Museo nacional de arqueología, antropología e historia del Perú)

Figure 6.2. Guaman Poma de Ayala's (1980, 1:106) drawing of the *coya* in an epileptic seizure

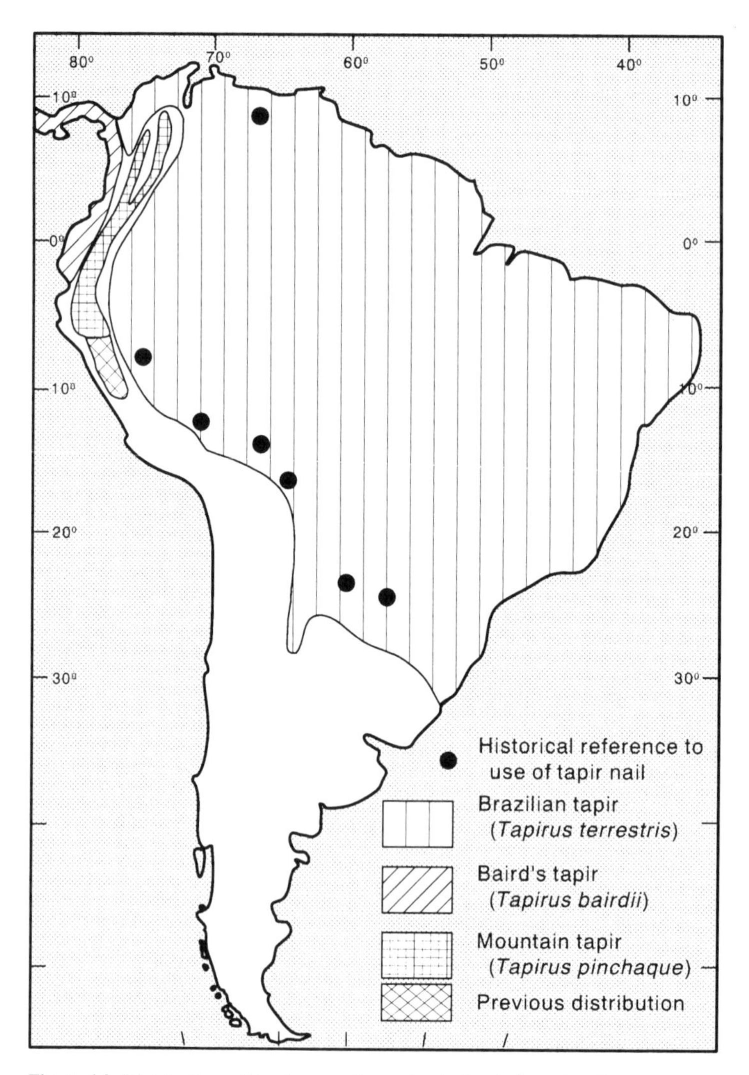

Figure 6.3. Distribution of the three tapir species in South America. Dots represent places where the use of tapir parts outside the main Andean zone is recorded in documents.

with aspects of ritual curing and animal sacrifices. These priests, who experienced seizures, particularly used in their rites the tapir, the South American animal that has had a strong tie with epilepsy for millennia. The zoogeography of the tapir provides a basis for understanding the lowland-highland dynamic of its curious role.

Tapirs of South America

The tapir is the heaviest mammal and the only perissodactyl in South America. Dispersed from the north into Central America in the late Pliocene, the animal subsequently deployed by way of the Isthmus of Panama over two-thirds of the South American continent. Three allopatric species occur in the Western Hemisphere (figure 6.3). Two members of *Tapirus,* almost hairless, are adapted to the heat and moisture of the tropical rain forest. Baird's tapir (*T. bairdii*) extends from its main range in lowland Central America and as far north as southern Mexico southward to the humid jungles of western Ecuador and Colombia. The Brazilian tapir (*T. terrestris*) is found over a vast territory from Venezuela to northern Argentina and west to the Andean foothills below 1,000 m (figure 6.4). A notable swimming ability facilitated the spread of this species throughout the Orinoco, Amazon, and Paraná drainage basins. The third and rarest species is the mountain tapir (*T. pinchaque*).[1] Found sparingly now between 2,000-m and 4,500-m elevation, this animal has a dense covering of hair that protects it against the cold (Schauenberg 1969).

Highland people had their earliest tapir encounters with the mountain species. It is presumably this animal of which Bruhns (1979) found a zoomor-

Figure 6.4. A nearly full-grown tapir (*Tapirus terrestris*) in eastern Peru shows the digits on his front foot.

phic representation in northern Peru. Now found only in Colombia, Ecuador, and Piura Department, Peru, this somewhat smaller version of *Tapirus* historically extended at least as far south as central Peru (Tschudi 1844). As human population density increased in the Andean highlands, intensified hunting and habitat loss reduced its former range and lowered its numbers ("Descripción geográfica . . ." 1793:178). At least within historic times, tapirs of the hot country have been much more abundant than their high-altitude congener. Forest-dwelling bow hunters relished their flesh, saved their feet as trade items, and turned their thick hides into war shields. Highlanders came in contact with the forest-dwelling tapirs when they descended to cultivate coca on the lower eastern slopes of the Andes. The unwary animals were ambushed in traps when they left the cover of the surrounding forest to browse on the precious coca shrubs.

Origins of Tapir Magic

The beginnings of tapir magic in the Andes plausibly go back to the time when hunting was an important subsistence activity in the highlands and *Tapirus pinchaque* was still an abundant quarry. Among the hunters were a few strange individuals self-selected to a shamanic calling. They were the ones who most closely observed this and other large wild creatures and learned all they could about their forms and habits. They learned that tapirs follow pathways in the forest, gather at salt licks, and seek out fallen fruit. This knowledge of tapir behavior made hunting them more successful at a time when human population had increased and availability of an easy quarry was less predictable.

Using mental imagery, shamans developed superior memory (Noll 1985: 450). That talent was combined with the urge to acquire repertoires of information on animals for both practical and spiritual purposes. Learning about prey species helped to ensure hunting bounty, but it also related to an intense desire to derive higher meaning for the human condition. In that quest, shamans sought personal transcendence by participating in the life of beasts and by identifying with the animals they killed. Becoming like an animal was a vital aspect of becoming the "master of the animals" (Ripinsky-Naxon 1993:8). In shamanic cultures the spirits of humans and animals are essentially the same. Metaphorical analogies were constructed between nonhuman and human characteristics. From these symbolic constructs shamans fashioned a highly personal system of rituals, beliefs, cures, and healing. The society sanctioned these unusual activities, for in them were seen the links between successful hunting and supernatural mediation.

Prerequisite to becoming a shaman was the soul-journey undertaken to discover elemental truths about nature and the self. For many shamans, that triplike trance was triggered by hallucinogenic substances; for others it could be induced by loud percussion and rhythmic motion. Still other shamans experienced epileptic seizures that signaled the journey to the other world. The epileptic condition has often been the defining motivation for a shamanic vocation, which provided these afflicted people with a special niche in an otherwise impossible world. In his study of primitive religion, Radin (1957) asserted that "neurotic-epileptoid" individuals predominated among shamans. Indeed, the world literature on shamanism points to these and other mental properties that are shared cross-culturally.

The tapir has been an important animal not only for the large amount of meat it has provided, but also for its unusually thick skin, which has been made into a variety of useful objects. As part of the search to organize what is knowable so as to supply interpretation to others, shamans noted the zoological peculiarity of the tapir foot. Unlike most mammals of the neotropical fauna, the tapir is odd-toed: its forefoot has four digits, whereas the hind foot has only three. The front digits vary in size; those on the back do not. Hunted and killed as game, this creature could be scrutinized for the greater significance of its unusual digitation. Lack of that organic symmetry may have placed it in the special realm of the supernatural as a manifestation of the charm of imperfection: "*der Reiz der Unvollkommenheit,*" as Nietzsche expressed it. The unusual can generate emotions, images, and delusions that are taken to be realities (Ackerknecht 1942:506).

Parallels, however fanciful, were pondered for larger significance. An anatomical marking of the mountain tapir has its own bizarre analogy with the epileptic human: the white hairs near the lips of the tapir recall the froth that collects around the mouth of the person during seizure. To shamans, the perceived epileptic condition of tapirs gave the animal numinous power and awe-inspiring qualities. Shamans in South America configured the tapir as a reflection of their moral concepts and as an archetype of their own behavior.

Shamans, possessors of secret wisdom and keepers of esoteric knowledge, extracted meanings from tapir anatomy and behavior. Tapir genitalia were anthropomorphized, creating the perception that the animal is a highly sexualized creature. Its outsized penis has made the tapir a mythological potent seducer of women. This "*macho de monte*" lurking in the forests has had a lascivious reputation among diverse native peoples (Langdon 1991:7; Mindlin 1995:23; Piso 1957:236; Seeger 1981:211; Zerries 1990:590–97). The tapir's mythic relationship with humans is a consistent part of South American folklore (Reichel-Dolmatoff 1985), but it also has its Mesoamerican equivalent where there is an abundant pre-Columbian iconography (Navarete 1987). In the context of shamanism, the tapir may be fused with the master of animals as a phallic forest spirit. The tapir represents a trickster figure with uncontrollable passion. Shamans also have often been viewed as tricksters, and early anthropomorphic figures of them frequently display a phallic feature. Between the world of pure essence and the realm of human desires, the tapir stands as a powerful symbol of the nature/culture gestalt.

The sexual behavior of the animal reinforces its epileptic associations. Tapirs in rut are vaguely reminiscent of humans in an epileptic seizure. Both signal the onset of their respective frenzies with a sharp penetrating vocalization. Among the diverse tonalities is a whistling sound almost like that of a bird. Early in the colonial period, Ordoñez de Ceballos (1614:36) observed that the noise made by tapirs in heat sounds like humans. In the fervor of sexual anticipation, tapirs randomly spray their urine, evoking the lack of human control over bodily functions during a generalized attack. The genital erection that sometimes accompanies seizures corresponds to the imagery associated with a tapir in rut. Part of this ithyphallic association may have been tied to the shamanic dream-vision which La Barre (1970:172) interpreted as fantasizing sex with an animal succubus. However farfetched these associations seem to be today, the tapir connected archaic cultures to other creatures in a way that went beyond predator and prey.

Tapir and Religion

The tapir-epilepsy-magic association possibly emerged at some point in the preagricultural period of Andean culture history before 2000 B.C. The triad survived the transition to the agricultural way of life that took hold by 500 B.C. Though farming and herding were then well established in the Andean highlands, hunting still contributed to subsistence. As local communities later consolidated into larger polities—Huari, Tiahuanaco, Inca—a specialized priestly caste replaced the lone shaman of a prior era. Caste members included epileptics, who conveyed their special knowledge of the cosmic order through a tapir cult. Contrary to Torrance's (1994:213) assertion, central Andean religion, with its huaca villackuna mentioned earlier, did not exclude ecstatic shamanism. The cult radiated from designated sites in the highlands, where its priestly caste cunningly supported itself by supplying amulets and carrying out ritual cures.

Archaeology provides an uncanny clue to this connection. Unlike many other creatures in Andean civilization, zoomorphic pottery of the tapir is very rare; the one published report of the possibility of a molded tapir suggests an uncertain identification (Bruhns 1979). Most intriguing was a ceramic drinking vessel molded in the form of a tapir foot, uncovered in 1967 near Ayacucho, Peru (figure 6.5) (Raymond 1979). Rather than reflecting the artistic whim of an individual potter, the vessel suggests a clear religious expression. Polychromatic decoration on that pot points to the tapir's association with epilepsy and religion. The recumbent avian figure, with a contorted appendage and unfocused eye, suggests the vacant stare and twisting movements that a person undergoes during seizure (figure 6.6). Birds are a prime metaphor for the mo-

Figure 6.5. Huari ceramic, dated before A.D. 800, molded in the form of a tapir foot, found in Ayacucho Department, Peru (Source: Raymond 1979, reprinted with permission)

bility needed for shamans to make soul-journeys to the sky spirits (Ripinsky-Naxon 1993). Possible use of this unusual object was as a container to imbibe potions ceremoniously as part of a tapir cult. No other pots of this kind have been recovered. However, Lathrap (1973:176) has argued that "a unique item in an archaeological sample represents [more] an established . . . pattern than it records a unique and idiosyncratic event."

The vessel is believed to belong to the Huari culture, which dominated the central Peruvian Andes during the Middle Horizon period between ca. A.D. 300 and 800. The model for this curious ceramic may have been the votive terra cotta pieces made of the human foot in the Tiahuanaco-Huari style (Stierlin 1984:143). Huari, an imperial state centered on Ayacucho, brought urban and religious concepts to the highlands, setting the stage for the Inca. Huari conquest of the Central Andes signaled the sway of a settled agricultural way of life. Hunting ceased to be an important highland activity. Huari agricultural productivity involved integrating diverse ecological zones. Religious practitioners became more powerful and sophisticated, and prestige pottery was made for the priestly caste. Some of these diviners-curers-magicians were epileptics who perceived the tapir as having its own epileptic condition. From

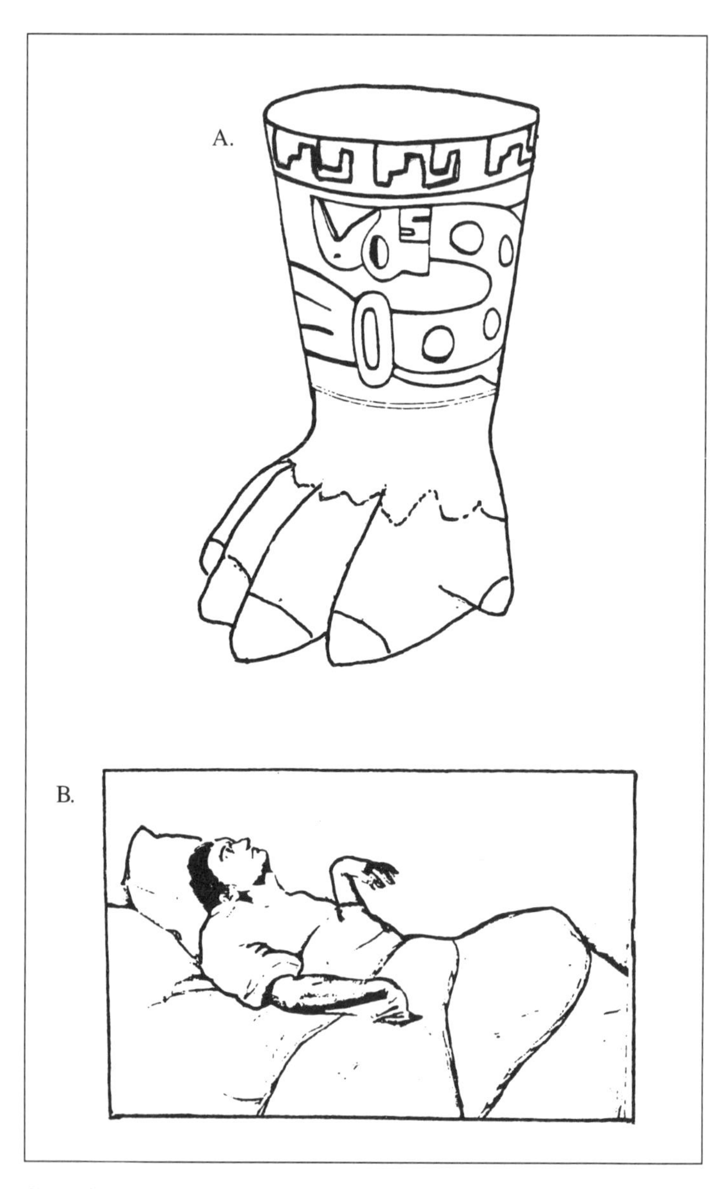

Figure 6.6. Decorative detail (A) on the same tapir pot as shown in figure 6.5 suggests a connection between a shaman (the bird symbolizing his soul-journey) in a seizure (from Raymond 1979) and (B) a person also in a face-up position in a state of seizure (Esquirol 1838).

that association, the tapir was given a spiritual status among members of the animal kingdom.

The Inca inherited the tapir cult from their Andean predecessors. Blas Valera (1992:51) recorded tapirs as having been among the sacrificial animals in the Inca period. A contemporary ritual in the western Amazon that involves tapir slaughter points to the symbolic power of this noble beast (Frank 1987). However, tapir sacrifice in the highlands, unlike that of llamas or guinea pigs, must have been uncommon. Even by Huari times, the southern limit of the mountain tapir had probably retreated northward. Keeping live specimens for sacrificial purposes could have been difficult because of the large bulk and narrow dietary range of either the mountain or the lowland tapir. Domestication, which could have solved the problem of supply, did not occur. Although tapirs of all three species readily tame when taken young and adjust well to a captive diet, they do not normally reproduce in captivity. The tapir cult was instead maintained by the sympathetic magic contained in certain parts of the dead wild animal.

Even before the Inca conquered the Andes, practitioners of the tapir cult looked to the boundless forests east of the mountains. There, another species of tapir, *Tapirus terrestris,* was still plentiful. It resembled the highland relative in most important aspects and seamlessly took over its uses and symbolism. Shaman-priests made forays down into the jungle to obtain materia medica, ritual paraphernalia, and sacred knowledge (Oberem 1974). They were the main intermediaries in obtaining tapir parts, but, in addition, several forest tribes traded jungle items to highlanders every year (Tschudi 1849:280–81). Much of this exchange of lowland and highland products occurred at annual fairs that were held into the nineteenth century (Miller 1836; Gade 1972).

In the Inca period, a tapir cult had its own idol ("Guagra Yanu") (Basto Girón 1957:70–71). One might extrapolate from the association of this cult with the idol that epileptic functionaries used seizures as a form of religious ecstasy in curing, divination, and sorcery. Tapir parts were key material objects in this magico-religious complex and functioned as storehouses of the numinous. Concretions found in the intestines of the animal were charms that were thought to keep epilepsy at bay and more generally to exorcise evil spirits. Magical power was attributed to the foot, with its unequal digitation front and back. Unlike the horse, which has a single hoof on each foot, the tapir's feet have several toes, each encased in its own small hoof. The hornlike dermal tissue that extends over the end of the digit is given names equivalent to *toenail* in both Quechua (*sillu*) and Spanish (*uña*); sometimes the words *claw* in English and *garra* in Spanish have been inappropriately applied to this part. Presumption of the powerful magic contained in the tapir foot could explain why an ancient potter molded not the entire body of the animal, but only its most pertinent part.

Post-Columbian Analogies and Parallels

Spaniards were the first Europeans to see the New World tapir. Never found in the Caribbean Islands, the animal was first described in 1509 on the Isthmus of Panama. Fernandez de Oviedo y Valdés (1959, 2:42–43), writing in the first quarter of the sixteenth century, offered the best early description, reflecting his remarkable talent for natural history.

In the Central Andes, the arrival of the Christian Spaniards destroyed most formal manifestations of Inca religion. Tapir magic, however, survived the conquest. Unlike the high priests of Inca state religion, the practitioners of this cult received a supernatural call to their office. Although the Spaniards did not tolerate public displays of its sacrifices, they were attracted to its curative aspects. Epilepsy was a disease of concern, and they called it by several names: *alfecería, perlesía, mal caduco, gota coral,* and mal de corazón. Its importance went far beyond the minority of people who actually manifested the disease. Spaniards thought epilepsy to be contagious, and thus its prevention was important. The apotropaic imperative explains why the feet, nails, and enteroliths were so widely valued through the colonial period. These parts came from both the mountain tapir and its lowland congener. Virtually all of Central and South America participated in tapir magic. Europeans in a wide swath from eastern Venezuela to the Río de la Plata associated the tapir with epilepsy (Caulin 1958:275; Lucena Giraldo 1988; Granada 1896; Roth 1915:368). Over the centuries, hunters have valued tapir feet as trade items that have had a ready market and thus often more value than the meat of the animal (Cutright 1940:18; Allen 1942:405; Ihering 1968:93).

Although the Spaniards did not know of the tapir before 1509, they used similar parts of other animals long before the sixteenth century. Stonelike concretions (*bezoars*) from the stomachs of herbivores, especially of the ibex or chamois (*Capra ibex*), had assigned functions in late medieval Europe. More than anything else, they were used as antidotes for poison, but also were considered valuable against gout (if taken with brandy), the pain of child delivery, bubonic plague, and for the prevention of epilepsy (Möckli-Von Seggern 1972:335). The magic of the ibex bezoar imparted medicinal value to other parts of the same animal: blood, fat, bile, horns, lung, liver, and rut gland. These elements were thought both to protect people from specific diseases and to alleviate or cure a specific health problem.

Spaniards who came to the Andes in the sixteenth century found indigenous people gathering enteroliths from native ungulates they hunted: vicuña, guanaco, white-tailed deer, guemal, pudu, brocket, and tapir. Bezoars from the tapir were specifically valued for "heart problems," that is, epilepsy (Calancha 1939; Guevara 1836:606–7; Jiménez de la Espada 1965, 2:11; Weddell

1853:188; Whitten 1976).[2] Spaniards, sharers of a belief in bezoar efficacy, markedly increased bezoar extraction from the above animals. The use of tapir parts was greatly reinforced, especially after sixteenth-century Spaniards set up a zoological parallel between two very different animals.

Analogizing Moose (Elk) and the Tapir

European Elk (Moose)

In their folk zoology and use of tapir parts, the Spaniards conflated two entirely different animals. They gave the same common names to the tapir as they had used for the moose (*Alces alces*), which in Europe is known as the elk. Not historically part of the Iberian fauna, the moose was nevertheless known to fifteenth-century Spaniards as an animal hunted in Scandinavia and Russia. Several names were applied to it: *anta* (or *ante*) or *danta,* terms of Arabic origin, were used much more than *alce* of Latin and, before that, of Greek derivation.[3] Spaniards also frequently referred to the moose/elk by the antonomastic name of *gran bestia.* An eighteenth-century Spanish dictionary defined *gran bestia* as "a mixture of camel and deer and as fast as a very heavy horse that suffers congenitally from epilepsy" (Real academia española, 1969, 1:600; my translation).

Until the nineteenth century, feet of the moose were traded throughout Europe as epilepsy curatives and talismans (Vélez 1613:177–78; Tissot 1770: 230). Specifically traded were the black and smooth hooves, two large and two small, which in the English of the time were called claws. In Spain the hoof was known as the *uña de la gran bestia,* "toenail of the great beast." To cure epilepsy, various procedures were followed. The afflicted grasped a piece of the nail in his left hand and placed it in his left ear (Vélez 1613:177). It was also rubbed on the chest. To prevent epilepsy, a piece was hung around the neck so that it touched the flesh; a piece of the hoof was carried in the collet of rings worn on the finger; or the pulverized moose/elk hoof was taken in infusion (Pennant 1974:21). Prevention was important; in Spain as elsewhere in Europe, epilepsy was considered to be a communicable disease. Ancient Assyrio-Babylonian concepts had the notion that epilepsy was a contagion, and the idea persisted through the Middle Ages into the Renaissance. Fourteenth-century curers classified it as one of five contagious diseases (Kanner 1930b:119).

Attribution of the moose/elk foot or part thereof as effective in the treatment of epilepsy derived from fanciful associations.[4] It was long believed that the moose could cure epilepsy because the animal itself suffered from the same disease (Johnstonus 1755, 1:93).[5] Count Buffon (1780, 4:348) explained that this bizarre notion came from the animal occasionally falling suddenly when pursued. In fact, a precipitous collapse occurs when the fleeing moose trips on its fore feet; it has nothing to do with epilepsy. The roots of this belief are still unknown, but it is very possibly tied to an ancient shamanic interpretation of this animal.

A corollary of the legend once so ardently believed had the moose able to cure itself of an epileptic attack by inserting its left hind hoof into its left ear (La Martinière 1671). That belief may have been a false extrapolation based on a moose observed trying to rid itself of a parasite. The greatest magical value lay in the rear left hoof, based on the principle of *similia similibus curantur.* Pomet (1725:246), a purist on such matters, warned that fraud in the commerce of moose/elk parts could be avoided if the left hind foot were sold still attached to the lower leg with its hair intact. The moose horn also was valued as an antiepilepticum, perhaps extending the magic to a more generally acquirable part. Into the late eighteenth century, pharmacopoeia described the use of moose feet against epilepsy (Schneider 1968:30). But by then skepticism had surfaced. Count Buffon (1780, 4:349), who himself reportedly suffered from seizures, discounted moose feet as having any therapeutic value.

Tapir Transfer

In tropical America, Spaniards came into contact with an entirely different fauna from that known in Europe. The two continents share no native genera or species. The sharp zoogeographic separation did not, however, prevent them from analogizing the Old World moose and the New World tapir. Similarities between the two now seem far-fetched, but to many sixteenth-century Spaniards, whose knowledge of the moose came from hearsay, the resemblance was real. Both the moose and tapir have a short tail and a neck shorter than the head. Heads of the two creatures bear a certain similarity in their length and nose parts: the tapir's prominent proboscis recalls the almost trunklike appearance of the moose (figure 6.7). Small eyes and poor vision also characterize the two animals, which feed mainly at night. Both have thick hides. Forest-dwelling browsers, lowland tapirs and moose are fond of standing in water and foraging on aquatic plants, and both are strong swimmers. The male tapir, like the bull moose, shows erratic and sometimes aggressive behavior during the mating season.

Objectively, however, the dissimilarities between *Tapirus* and *Alces* make those parallels superficial. They belong not only to separate families (Tapiridae vs. Cervidae) but also to different orders (Perissodactyla vs. Ungulata) with all the evolutionary divergence that implies. The physiognomic differences are substantial. Although each is the largest beast on its respective continent, a full-grown moose of Europe weighs at least twice as much (ca. 600–800 kg) and stands twice as tall (2 m at the shoulder) as the full-grown tapir. Large palmate antlers set the male moose apart from the hornless tapir. The front hooves of a moose are larger than its back hooves.

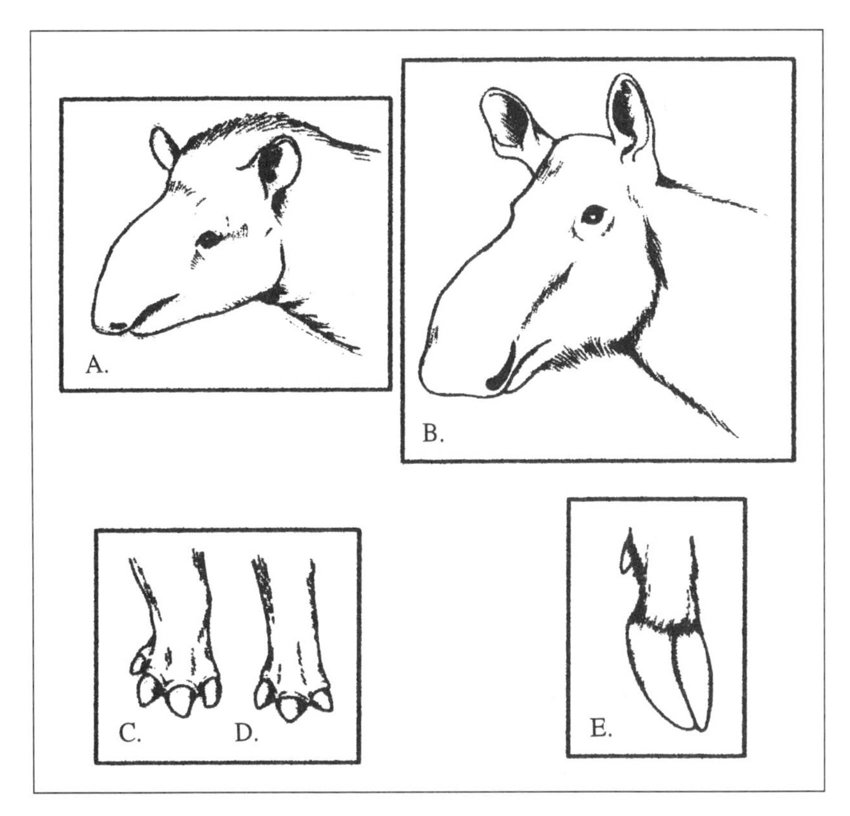

Figure 6.7. Comparison of the head of the tapir (A) with that of the moose (B)—both have an overhanging lip. However, the once highly valued feet of each are dissimilar. The tapir's front foot has four digits (C); the hind foot has only three digits (D). The moose's large cleft hoof (E) is very different. (Drawings by Betsy B. Howland)

To understand how people historically viewed such dissimilar animals as being similar, it is important to recognize a different ordering of knowledge. Before the mid-seventeenth century, the known was based not only on accurate description of observed characteristics. Spoken or written sources of information that were part of naive traditions—fables, legends, commentaries—were also considered to be valid sources of information. Also, signs and names were then integral parts of the things themselves (Foucault 1971).

When the Spaniards arrived, the tapir already had an established cultural association. Validation of the tapir's ability to prevent or cure human epilepsy may have stemmed from a belief that the animal suffered from the same disease and was able to cure itself (Valdizán and Maldonado 1922). The Europeans reinforced it by confounding the tapir of the Neotropics with the moose or elk. Some moose folklore may have been transferred to the tapir. As with the moose, the efficacy of the tapir to combat epilepsy was said to be greatest in its rear left hoof (Cobo 1956, 1:369; Vargas Machuca 1892:153–54). Spaniards transferred several names for the moose to the New World Tapiridae. *Anta, danta,* and *gran bestia* became the common terms for this animal in South America.

This wholesale transfer must be understood in terms of sixteenth-century concepts, whereby similitude of use and certain superficial appearances grouped animals together. In that context, transmutation of names seemed natural, although not without eliciting some disagreement. Fernandez de Oviedo y Valdez (1959:42–43), who was one of the most acute natural history observers of his day, protested against the application of the term *danta* to what he saw as an entirely different animal. Notwithstanding his objection, the name transfer caught on, perhaps out of a Spanish reluctance to adopt native names. For 200 years they persisted in calling llamas by the Spanish names for sheep: *ovejas* or *carneros de la tierra.* By the early seventeenth century, the three Iberian names of anta, danta, and gran bestia had displaced the indigenous Andean names of *huagra* (transcribed or phoneticized also as *guacra* or *guagra*) and *ahuara* even among native peoples (González Holguín 1952). However, a hybrid term for the tapir, *sacha vaca* (*sacha* = wild in Quechua; *vaca* = cow in Spanish), is widely accepted in Peru and Ecuador. The various name transfers to the tapir point to how part of the early enterprise of natural history was constituted through language.

The allure of tapir magic was also reinforced by a European belief in amulet therapy, which is tied to the sixteenth-century rise of the *virtuosi* who insisted on the reality of occult qualities (Eamon 1994:138). Its roots lie in the Pythagorean theory of occult properties immanent in plants, animals, and stones. Natural magic reached its height of acceptance in the seventeenth century, after which it declined in Europe and, more gradually, elsewhere in areas of European settlement.

Spaniards who went to the New World relied on a variety of talismans to protect against certain diseases, among them epilepsy. A piece of the tapir nail, the whole digit, or even the entire foot was carried around the neck to prevent seizures (Marcoy 1874, 1:543). Some persons slept with the nail or foot on their torso over the heart (*Documentos* . . . 1929). As a curative for those who had the affliction, in western South America the tapir nail was pulverized and taken internally with a liquid. Spaniards and Indians were involved in trading *uñas de la gran bestia,* an expression derived from a Spanish folk remedy using the toes of an entirely different animal (Real academia española 1969, 1:600; Velasco 1977, 1:97). Entrepreneurial Spaniards, imagining a lucrative trade in tapir foot if the popular mind confounded it with that of the moose, must have been the most enthusiastic promoters of that name transfer. Prevention was much more lucrative than cure, for the former covered everyone. The special magic of the tapir hind foot may actually represent a Spanish idea borrowed from the moose in Europe. Shaman-diviners (*hechiceros*), acting as guardians of indigenous and colonial traditions, have kept this piece of folklore alive over the centuries.

Tapir and Epilepsy Lore Today

Nails, feet, and intestinal concretions of the tapir are still used in Andean folk medicine. Itinerant medicine men, of whom the Callawaya of Bolivia are the best known, include those items in their repertoire of materia medica (Girault 1984), yet tapir parts have also been sold in the folk-remedy stalls (*hampi katu*) set up at highland markets and at harvest fairs.

The array of animal products offered in these markets attests to the long-standing interconnectedness of the three very different natural realms of western South America. Pre-Columbian people had similar access to remedies and ritual substances from ecological divergent sources. Seashells, dried starfish, and powdered sea lion tusks are among the remedies brought from the arid littoral to the west. Dried llama fetuses, alpaca fat, condor blood, and deer horn come from highland sources. The eastern lowlands contribute, among other things, grease of the spectacled bear, jaguar fat, dried snake, and tapir stones, feet, and nails. In the contemporary Andes, sympathetic association has broadened the scope of tapir magic. Amulets, originally restricted to the foot, hoof, or part of the nail, now also might be bone, tooth, a piece of hide, or even blood (Estrella 1977:116). The tapir's application has also widened; many people today especially in Ecuador view its parts as generic good-luck charms for success in love or business.

Scientific explanations for epilepsy that dawned in nineteenth-century Europe have reached larger Andean towns. But in rural communities, epilepsy is still understood to be a cardiac, not a neurological, problem.[6] Its Quechua names of *sunqo nanay* ("heart illness") and *sunqo chiriyay* ("heart gotten cold") convey the folk etiology: seizures are believed to result from heart stoppage which occurs when the *aya waira* (wind of death) blows over the victim and chills that organ. Other traditions are also maintained: in some Andean communities an epileptic is self-selected to be a folk healer (Sánchez-Parga and Pineda 1985:530).

Larger Meanings

Starting as the private imagery of a nameless shaman, Andean tapir magic has endured in the Andes as a mythic dramaturgy for certainly two millennia and probably twice that. Its longevity derives from its connection with epilepsy, a condition that imparted religious authority. Power of possession was translated into a vocation of divination and curing. Epileptic priests were naturally considered to be experts in preventing and curing.

Tapir magic possibly became entrenched in Andean culture history when some epileptic shamans became priestly specialists. They exploited their neurological condition and the unpredictable ecstasy associated with it for larger cultic purposes. They then became the archpromoters of the tapir cult. In the colonial period, the religious associations vanished, but the tapir as a chthonic animal persisted as folk knowledge. Indeed, it became stronger than ever when grafted onto the preternatural beliefs of Spaniards. The *arcana naturae* that was so much a part of European medicine in the Middle Ages and Renaissance fit seamlessly into tapir magic.

Some perceived anthropomorphized behavior of the tapir established a connection with epilepsy. And tapir amulets were efficacious: at least 90 percent of those who wore them would never get the disease. The idea that epilepsy is contagious was refuted in the eighteenth century when it became clear that this neurological disorder is restricted in any given population to a minority of genetically marked or brain-injured individuals. The curative use of tapir parts did not disappear after that realization gained ascendancy. Tapir magic was seen as having an apparent efficacy in halting a seizure in those afflicted. The termination of a seizure minutes after it starts was there for all to witness. That it requires no intervention was an idea that came later.

Today in the Andes the triad of the tapir, epilepsy, and magic is no longer generally known. Most people have either totally forgotten or retain only a faint memory of this relict of Andean folk medicine and shamanic cosmology. More than a simple vestige from an irretrievable past, the triad has a larger significance in Andean culture history. The link of idiopathy with an element of the wild biota mediated the passage from a hunting existence to an agriculturally based civilization. Second, the moose/elk transfer from European folk medicine to the tapir-epilepsy story relates to the efficiency of diffusion but also to the psychic unity of mankind. Tapir magic suggests both the interchangeability of human and animal that is part of the evolution of spirituality. In Jungian terms, this tapir association might also be seen as a vestige of our animality that has lost itself in the nebulous abyss of time (Jung 1964:12). Old World and New World pre-Columbian parallels are not to suggest simplistically that similar ideas are generally imbedded in the human psyche, but that the shamans' calling follows a pattern through many generations of human experience that goes beyond any one culture.

7
Valleys of Mystery on the Peruvian Jungle Margin and the Inca Coca Connection

Human settlement of the upper and lower peripheries of the vertically staggered Andes was once greater than it is today. The high puna and the hot forested country were both inhabited in ways which reflect the fact that the rural Andes had more people before the conquest than at any time since. Yet, the Inca occupation of the jungle valleys continues to pose a special conundrum for those who study that event, for Quechua-speaking highlanders were out of their element in the low valley environment. A broad distinction can be made between the sedentary farmers and herders of the cool lands and at much lower elevations the *chunchos* (barbarians), who lived in much less complex societies that practiced slash-and-burn agriculture.

That broad adaptational separation into different geographies and cultures was a useful generalization at a continental scale (Troll 1931; Gade 1996). It explains much of the trade between highland and lowland peoples in Andean culture history (Gade 1972). However, from a regional or local perspective, the cultural and environmental frontier loses its validity, for when studied at a large scale, the upper limit of the tropical forest turns out not to be a cultural boundary at all.

Preconquest cultures of highland origin had settled certain jungle valleys within a day's reach of cool plateaus. Despite basic reservations that mountain inhabitants had about this danger-filled territory so different from their own, they were drawn there for several different reasons. Several resources lured them to lower elevations. Gold was panned from streams in the Andean foothills, and woods for tools and weapons were cut from the forest. Certain crops also grew in that zone; chief among them were cotton, capsicum pepper, and fruits, but especially coca. Coca cultivation was of the greatest long-term significance for Inca presence, although not all forest settlement featured economic extraction. Huayna Capac built Machu Picchu as his own private estate to capture the warmth and lush green ambiance so removed from cold, brown

Cusco. In the first tumultuous decade of the conquest, the tropical forest also offered a place of refuge for Manco Inca and his entourage. In 1539, Manco established a settlement in the warm Pampaconas Valley of the Vilcabamba region in a last-ditch attempt to stay free of Spanish control radiating from Cusco.

Interest in finding evidence of high cultural achievements in the jungle stems from knowledge that indigenous highlanders once controlled some of it. Rumors of a putative "city" called Paititi motivated forays east of the cordilleras beginning already in the sixteenth century. Did such a place actually exist, or did native people fabricate it to keep the Spaniards with their greed for gold off balance? One Peruvian writer viewed Paititi as a metaphor for hidden wealth and not a real place (Angles Vargas 1992). For true believers, the overgrown ruins of Paititi still await discovery somewhere beyond the last major range of mountains. Paititi seekers have set their sights as far afield as the Llanos de Mojos in the Río Beni drainage of eastern Bolivia. There, productive agriculture based on an extensive complex of ridged fields led to a high level of prehistoric achievement (Levillier 1976). However, most questers and also many local legends believe Paititi to be an Inca site close to the Andean foothills (Gilt Contreras 1960). The discovery of Machu Picchu in 1911 and the rediscovery of Vilcabamba in 1964 greatly stimulated the possibilities that more "cities" lay hidden under the jungle growth. Over the past four decades explorers have focused their efforts in the hilly but forested upper drainage area of the Río Madre de Dios system (Neuenschwander 1963; Iwaki Ordóñez 1975; Aparicio Bueno 1985; Deyermenjian 1988, 1990, 1995).

Location of ruins densely overlain with vegetation may involve two distinct kinds of plant cover. The cloud forest (ceja de la montaña) is bone-chilling, dripping wet, and supports few mammals, birds, or people. Farther down, below 1,500 m, the rain forest (*selva alta*) takes over. It is hot and unhealthy, but more useful for agriculture than the ceja. In the Inca period, the slopes and valleys of both forest types were referred to as Antisuyo.[1] After the conquest, that same territory was called the Provincia de los Andes. Later in the eighteenth century, the meaning of the word *Andes* shifted to cover the cordilleras, valleys, and plateaus of the highlands. *Montaña* then gained ascendancy as a term for the thickly forested expanse of the piedmont in the Amazon drainage. *Yungas,* a parallel term of preconquest origin, referred to the hot country. Along the eastern front, that zone extended from ca. 2,300-m to ca. 800-m elevation. *Chaupiyungas* (chaupi = half) was the cooler zone above 1,200 m. Two other words, *cocayungas* and *pallayungas* (*pallay* is "to pick coca"), referred to the coca-growing realm which began ca. 1,800 m. Today *Yungas,* when capitalized, is primarily a regional term for the eastern slopes of La Paz Department, Bolivia.

By whatever name, the forested country at lower elevations has had many negative connotations. Highland people who venture there have always had to deal with the hot temperatures, high humidity, clouds of biting insects, threatening beasts, and unfamiliar food. Most of all, the jungle has invoked terror for what cannot be seen. Two diseases, one native, the other introduced, have been contracted in the hot, wet climate. Since long before the conquest, mucocutaneous leishmaniasis (*juk'uya*) took its toll in this forest region (Gade 1979). The grotesque facial disfigurations that characterize this disease in its advanced state have long been a prime symbol of the insalubrity of the tropical forest. In an attempt to control its contraction, Inca coca workers traveled to the fields for short periods. After the conquest, expansion of coca cultivation led to a disregard for previous safeguards. Early colonial chroniclers and Viceroy Toledo himself were appalled by the high incidence of this disease. Toledo established a commission to investigate its causes, cure, and prevention. He then issued a series of labor edicts for the coca valleys. One set the maximum allowed work stint in the yungas at 24 days, which approximated Inca practice. Coca field owners failed to respect the short work rotations, the violation of which was said to have exacerbated Indian mortality. Also, the Europeans who came with the conquest brought malaria to these valleys. Paroxysms of fever (chucchu), apathy, weakness, and jaundice affected a large number of those who went to the eastern lowlands. More lethal and easily contracted than leishmaniasis, malaria was a constant endemic threat that occasionally occurred in virulent epidemic form. Malaria may have been the cause of two epidemics, those of 1589 and 1720, which killed off a large number of people in this zone. Indigenous people had no genetic immunity to it. Shunning valleys below 2,000 m was the only certain way to avoid contraction until the 1950s, when synthetic drugs and DDT greatly reduced malaria incidence.

Coca Cultivation East-Northeast of Cusco

Even before the Incas controlled the Central Andes, highlanders cultivated coca themselves when they could not reliably acquire the quantities they desired by means of trade with lowland tribes. By ca. A.D. 800, the Huari people were getting coca from the Apurímac Valley, east of Ayacucho, either by trade or by cultivation (Raymond 1992). Farther south around Cusco, highland people first got coca by exchange or as a tribute item from lowland tribes. Besides cotton, coca was the only major plant on which the Incas depended that could not be produced in the cool highlands.[2] It was thus logical that they would eventually grow it themselves to assure its supply. Within a 200-km orbit north, east, and southeast of Cusco, a dozen warm valleys provided a suitable environment for the cultivation of the coca shrub. Details of site, situation, and trail linkages must be laid out to understand this distribution at the jungle-highland interface. The work of Renard-Casevitz, Saignes, and

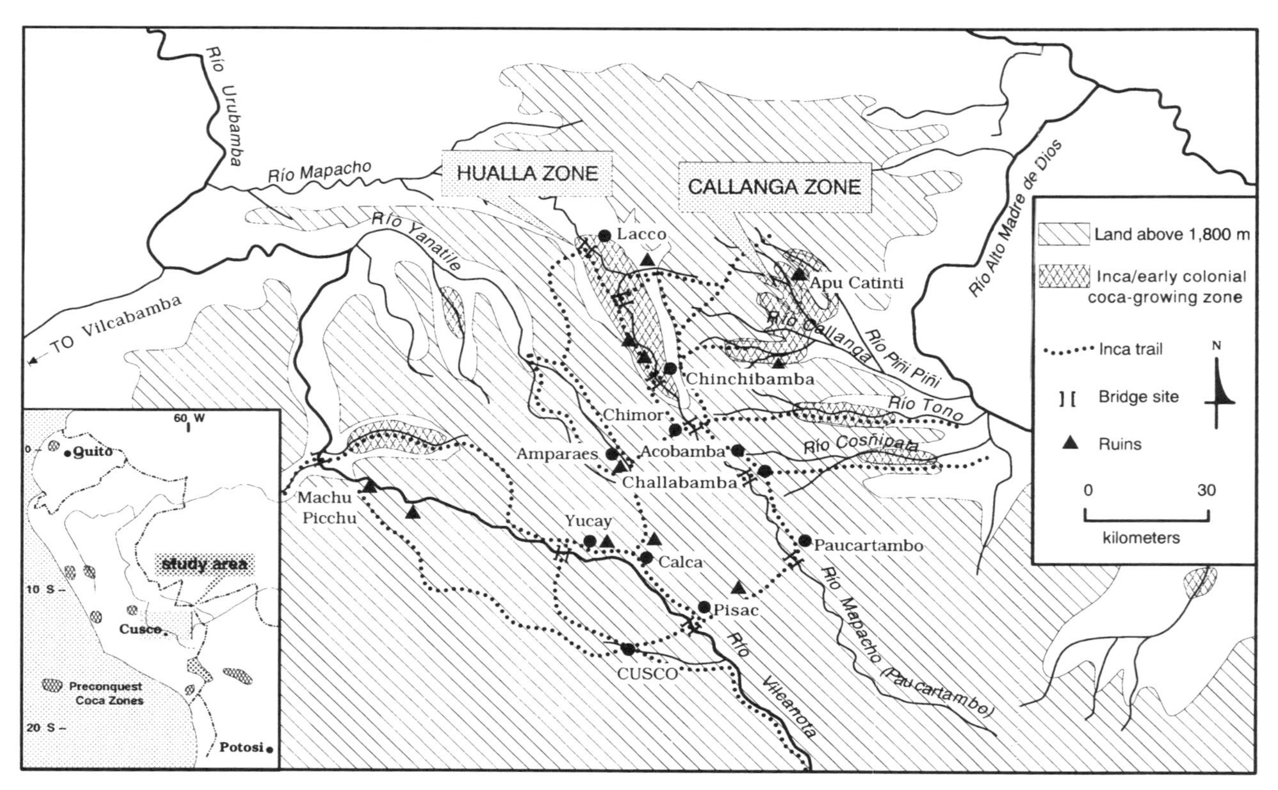

Figure 7.1. Inca and early colonial coca zones north and northeast of Cusco, highlighting Callanga and Hualla

Taylor (1984) forms a case in point for the failure to come to grips with the geographical reality of highland intrusions into the forested zones.

Two main drainage basins holding cocayungas can be distinguished: the main coca-growing area lay directly east of Cusco in the upper tributaries of the Alto Madre de Dios and the other in the Mapacho or Paucartambo Valley. Headwaters of the Madre de Dios form three main parallel valleys: Piñi Piñi (also called the Toaima until the nineteenth century), Tono, and Cosñipata. In each of them, subtributaries form smaller lateral valleys, where coca was also grown (figure 7.1). A second area was grouped in the lower reaches of the Urubamba drainage; this was broken into a series of valleys, among which was the Vilcabamba, Lares, Mapacho, and the main depression of the Urubamba itself. Whether coca cultivation in each of these valleys began before or with the Incas is still unknown. In some of them, forest tribes cultivated coca as a tribute item to the Inca until the Inca themselves took over its cultivation. Sarmiento de Gamboa (1960:254) related that in 1471 Topa Inca deployed troops to defeat forest people and that their victory set the stage for highland people to be sent as mitimaes to grow coca for highland needs.

Beginning in the early sixteenth century, the supreme Inca, Huayna Capac, took personal ownership of numerous coca fields. Labor for coca's harvest came from his servants (yanaconas) in the highland valley of Yucay (2,800 m) near Cusco, which Huayna Capac had taken over as his own royal property. There, with labor from native people of the zone and transplanted migrants from as far away as Quito, he built and maintained palaces and acquired terraced gardens and fertile maize fields.

Some of his yanaconas from the Yucay Valley became short-term migrants (*cocapallas*). They traveled to the warm valleys below the eastern escarpment once a year for three weeks in staggered contingents to pick, dry, and pack the coca leaves. In the Inca period, this group included Huayna Capac's yanaconas from Yucay; their service substituted for the tribute that others in the Inca realm had to pay. The brief work period was a health precaution to lower the risk of contracting leishmaniasis, which has been a disease with strict ecological parameters. Permanent workers (*cocacamayos*), older than those who came periodically to pick coca (Santillan 1968:106), were responsible for laying out new coca fields from the cut-and-burned forest, making new seedbeds, weeding the plots, fighting infestations of leaf-cutting ants, and weaving the coca baskets. They also had their own fields of subsistence crops and coca. Disease contraction may have been an expected outcome of their role; however, some of these people may have been resistant or immune. To take further advantage of a warm growing environment, they also cultivated other crops, especially capsicum pepper (ají).

After the fall of the Inca, the Spaniards discovered how important coca was to native people and also what a profitable trade it involved. Spaniards began to acquire coca-producing properties in 1548, but its production did not become a Spanish monopoly, as has frequently been reported, for kin-based communities (ayllus) continued to control the coca lands they had before the conquest. Under either Spanish or indigenous control, the same work arrangements and dual labor system persisted. Permanent workers often came to these valleys to escape the onerous mita requirements imposed on their home villages. Members of conquered forest tribes also labored on the plantations to fulfill tribute obligations that had originated in the Inca period. Short-term laborers in the cocayungas were members of highland communities. Some of them also owned patches of coca shrubs of their own (Sarabia Viejo 1989:183). Peasants from ayllus in the high country descended into the valleys to pick and pack (Cook 1975:202; Villanueva Urteaga 1982:266).

Yucay was more distant from the coca fields but had a special link inherited from the preconquest period. In return for his key role in the conquest, Francisco Pizarro was awarded as his encomienda the Yucay Valley and two nearby communities (Villanueva Urteaga 1970; Rostworowski de Diez Canseco 1970). Called the Marquesado de Oropesa, it included the best climate and soils for growing floury maize in the Andean highlands. A strong symbolism in this transfer occurred when the conqueror took over Huayna Capac's royal reserve and yanaconas and assumed a title of nobility. After Pizarro was killed in 1541, his young mestizo son Gonzalo inherited the repartimiento. Upon Gonzalo's death at age 11, the encomienda reverted to the crown. To lure Sayri Tupac, the son of Manco Inca, out of his Vilcabamba refuge, Viceroy Mendoza in 1558 awarded him the repartimiento and the title that went with it. On Sayri Tupac's death in 1560, his daughter, Beatríz Coya, who had married the Spaniard Martín García de Loyola, acquired the repartimiento.

Yucay peasants remained involved in coca tasks. At the height of the coca boom in the late sixteenth century, around half of the 3,500 Indians in the Yucay encomienda migrated to the eastern valleys for short periods, as their custom had been since the days of Huayna Capac (Villanueva Urteaga 1970:94). About 400 different migrants from Yucay deployed into different valleys for work detail about every 120 days. But as coca demand dropped in the seventeenth century, so did the need for workers. After Sayri Tupac's direct line came to an end in 1741, the royal lands in the cocayungas were sold to individuals. The two-century tradition of Yucay people seasonally moving to the hot valleys to pick coca leaf ceased.

Pathways of Movement

Calca-Amparaes-Chimor was the major route to the east until the Pisac-Paucartambo raod superseded it in the eighteenth century. Calca was a key town on the way to the coca fields (figure 7.2). Cocapallas stopped there at the shrine (huaca) of K'umurumiyoq to seek divine protection from tropical diseases in

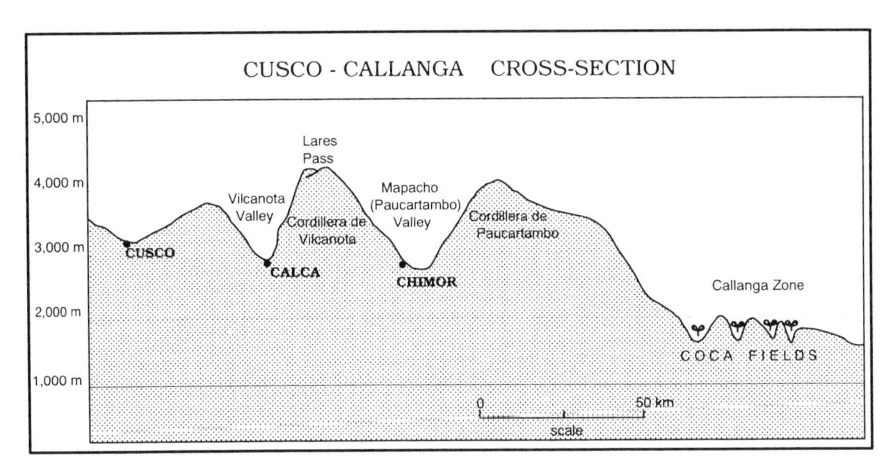

Figure 7.2. Cross-section from Cusco and from the Vilcanota (Yucay) Valley to the coca zone of Callanga, showing relief traversed

the yungas (Gutiérrez Pareja 1984:80). From Calca the road went up and over the Lares Pass at 4,400 m to Amparaes. A full day's trip from Calca, Amparaes was the site of a *tambo,* where travelers to the jungle spent the night. Extensive ruins attest to Amparaes' being an important settlement in the Inca period (Flores Ochoa 1985:270). Located near the convergence of several roads and having abundant pasture, Amparaes became a place to raise pack animals. Into the colonial period, a large number of llamas were bred and traded there. Though superseded quite early by mules, llamas have not yet disappeared as pack animals.

In the Inca period and for about four decades after the conquest, Indians carried the leaf on their backs up the steep trail to the point where it was transferred to llamas (Ramírez 1936:37). A 1575 viceregal decree prohibited Indians from carrying out coca on their backs unless it was from their own *chacras* (Sarabia Viejo 1989). Numerous by then, mules replaced humans in carrying coca to the waiting llamas. These camelids normally carried two baskets of coca, each weighing 11.5 kg when filled. Coca brought in baskets from the fields was unloaded at warehouses (*depósitos*) and repackaged for long-distance shipment. Storage centers were holding places in a dry, cold atmosphere before the leaf was redistributed elsewhere. They also become control points for taxing purposes.

Long llama trains required three months to make the trip from Paucartambo to Potosí (Glave 1983:49). On the return trip, pack trains transported preserved foodstuffs of chuño and *charque* (sun-dried meat) as provisions. Much of this food came from highland communities in Paucartambo (Zimmerer 1996b). The permanent work force in the cocayungas could not grow enough food to

Figure 7.3. Part of the stone-inlaid portion of the Inca road crossing the Cordillera de Paucartambo between Chimor and the coca-growing country on the eastern side (Photograph by Gregory Deyermenjian, reprinted with permission)

feed temporary workers, who preferred in any case their traditional highland diet. Mules gradually replaced llamas in the long-distance movement of coca. Mules moved much faster on the trails than llamas and also were able to carry loads six times heavier than what llamas could carry. They could also twist their way down steep paths into the hot valleys with no ill-effect. These advantages outweighed the fresh fodder and water they required and the damage incurred to trails and bridges from their greater weight and sharp hooves.

From Amparaes, the main trail led 25 km and more than 1,000 m down to the Río Mapacho at Chimor. Chimor was a fragile but crucial link between Calca and the cocayungas. A 14-m-long hanging suspension bridge of plaited plant fibers loomed 60 m above the water. Pack animals loaded with coca gingerly picked their way across the swaying structure on a surface 1.5 m wide. Wear-and-tear of a year's use required replacement with fresh branches plaited and laid. Before and after the conquest, several nearby ayllus repaired this bridge as a regular tribute obligation. In 1956, steel cables replaced the vines.

From the Chimor bridge, trails led east and north. Eastward, three different Inca paths were built to the cocayungas of the Alto Madre de Dios drainage: Chinchibamba to the Callanga and Piñi Piñi; Acobamba to the Tono; and Challabamba to the Cosñipata and Pilcopata. All three of them started at the Río Mapacho, arched up and over the last cordillera, and then descended sharply through cloud forest into the coca-growing valleys (figure 7.3). These same trails were used after the conquest. Shelters (*bohíos*) set up along the way provided refuge from the rain; when wet, coca leaves begin to ferment, which makes them worthless.

The Tono and the two headwater streams of the Toaima/Piñi Piñi, including its tributaries, the Callanga and Pitanga, formed the main core of Inca coca cultivation. Three Inca towns were established in the Tono Valley (once also called Cocaytono or Tonoyanavire). These settlements survived the conquest, and in the late sixteenth century each held between 300 and 500 people (Pärssinen 1992:394). A church, hospital, priest, and judge were located in the Tono Valley (Matienzo 1967:175, 188). A trail connected the Tono and Pilcopata valleys.

For two centuries after the conquest, the Tono and Piñi Piñi and their tributaries maintained a substantial population of highland people. Poor health conditions instigated labor conflict (Maúrtua 1907b, 2:221–30). By the nineteenth century, the old towns, most of the coca fields, and the hospital and church had disappeared. Bovo de Revello (1848:26) noted vestiges of the settlement of Tono located on the borders between Haciendas Chaupimayo and Huainapata. On Hacienda San Miguel he found a bell inscribed with the date 1561, which had belonged to the church in Tono. San Miguel, the once-flourishing coca property in the early nineteenth century, moved its product out on the Acobamba-Tono road. The first Spanish intrusion led by Pedro de Candia in

1539 took this trail. The travelers Herndon and Gibbon (1854, 2:42–46) and Markham (1991) used that route into the jungle. Faustino Maldonado in 1860 and von Hassel in 1904 also used this trail. The trail between Challabamba and Cosñipata-Pilcopata also served people and products in the Tono Valley, to which there was a 15-km spur. Movement along this major trail involved a perpendicular descent via switchbacks 25 km long that began near the base of an extinct volcano, Apu Cañachuay, also a revered huaca. Mules on this trail could carry only half the normal weight: 68 kg (6 *arrobas*) rather than 136 kg. It was so narrow in places that only one mule could pass (Miller 1836:178).

Other Inca roads led northward into the Lares and Mapacho valleys. Göhring (1877:103) was among the first to refer to the "existence of a wide stone-inlaid road on both sides of which were buried many utensils belonging to the time of the Incas." Hassel's (1907:366) 1902 expedition also found trails.

Two Lost Valleys of the Cocayungas

Two other zones east of Cusco, Callanga and Hualla/Lacco, had coca cultivation under the Incas (see figure 7.1). Scarcely known today, their importance spanned the conquest only to fall into near oblivion in the late colonial period. An aura of the lost, hidden, and remote has surrounded them for many years. No motorized transport on land or water is yet possible from the highlands to these two places; one must enter by walking or riding horseback on rudimentary trails. In the lowland forest, numerous navigable streams create far less isolation. Difficulty of access to verify the configurations on the ground explains why many maps of Peru or parts thereof have not shown either place. Maps that do have these names often put them in wrong places.[3] Antonio Raimondi (1965), Peru's most indefatigable explorer, did not reach either of these two valley zones in his nineteenth-century travels.

Although Callanga and Hualla lie within 60 km of each other as the crow flies, they are drained by rivers which flow in opposite directions. The Callanga's waters flow into the eastward-moving Piñi Piñi and then into the north-northeasterward-moving Alto Madre de Dios, which is one of the tributaries of the Madeira. The Mapacho drains into the lower Urubamba, which in turn feeds into the Ucayali. The Ucayali disembogues into the Marañon before those same waters course into the mighty Amazon itself. The basic hydrography east of Cusco has taken a long time to reveal itself. Although Miller (1836) showed it correctly on his map sketch, the first atlas of Peru had the Río Paucartambo flowing into the Madre de Dios (Paz Soldan 1865). Not until 1903 was it confirmed that the town of Paucartambo and the lower Mapacho Valley were located on the same stream and that it flowed into the Urubamba. Variable names for the river have caused confusion. Mapacho is the pre-Hispanic term still extant. Two other names came into use before they were seen as the same river: Paucartambo is the toponym used farther up the river, whereas Yavero is applied more to the zone near its juncture with the Urubamba.

Both Callanga and Hualla/Lacco were Inca economic outposts where coca and other tropical crops were grown, processed, and brought out on spurs of the Inca road system. Those who have written the most about Inca roads—Hagen (1976), Strube Erdmann (1963), and Hyslop (1984)—said nothing about these extensions. Stretches of them are still intact. Without vehicular access, these two zones languish from the high cost and logistical difficulty of getting products to market. Although mules are the best pack animals for these parts, mule breeding and driving (*arrieraje*) have largely disappeared. Horses bring in supplies and carry out agricultural products, but compared with trucks, this mode of transport is expensive and slow. Without regular pack trains, Callanga and Hualla/Lacco are more isolated now than they were in the late Inca and colonial periods.

Callanga

Callanga or some variant of it has been a toponym in the Piñi Piñi headwater zone since at least the Inca period.[4] At different times it has been attached to five kinds of places—river, valley, zone, town, and estate—in that area. Nystron (1868), a Swedish-American engineer, was among the earliest to publish a map showing the Callanga to be a headwater of the Río Piñi Piñi. The map of Jorge von Hassel (1898:299), a German engineer contracted to investigate the selva margins of Cusco, also showed the Río Callanga as well as other tributaries of the Piñi Piñi.

Nevertheless, Callanga's exact location continued to be elusive. Llona (1904) wrongly placed the "valle de Callanga" as a southern, not northern, tributary of the Piñi Piñi. He showed a populated place called Callanga near the junction of the Callanga and Piñi Piñi rivers. Pío Aza's 1926 map demarcating Dominican mission territory had "Cuyanga" (*sic*) as a synonym of the Río Yucaria, which is, in fact, a different stream. The American Geographical Society map at the scale 1:1,000,000 perpetuated these hydrographic confusions (AGS 1938). Two recent Peruvian atlases show Río Callanga but not the names or arrangements of its tributaries (INP 1963–1970:527; IGM 1989). Until large-scale topographic maps of the region become available, uncertainty about the configuration of the Callanga will continue.

Bafflement about Callanga's location and even doubts of its existence as a real place were for many years particularly frustrating to biologists. Around 1890 Ivan Garlepp, a collector of South American mammals, brought back to Europe specimens whose source was labeled simply as "Callanga." Among them were the first primate obtained from this part of the Amazon Basin (Thomas 1921; Ceballos Bendezú 1981–1982). Those collections, if accurately labeled, suggest that a primary forest covered Callanga in the late nineteenth century. Abandonment of the *cocales* around the end of the colonial period would have permitted near total forest regeneration in the intervening six decades. More like trees than shrubs, the wild-growing coca plants that Deyermenjian (1988:19) found in 1985 as part of the understory in the tributary valley of the Mameria are probably descendants of the coca shrubs once cultivated.

Ecological change may also explain faunal shifts. A particularly beautiful water strider, *Velia helenae,* was found in the Stockholm Museum with "Peru Callanga" indicated as the place of collection. This insect species and other biological material labeled as from Callanga had no parallel elsewhere. To be considered the place where earliest scientific description was made ("type locality"), Callanga's real existence as a place first had to be verified. From 1935 onward, an accomplished biological collector, Felix Woytkowski (1978:160), was obsessed with finding Callanga. After several false attempts, this Polish expatriate finally reached the valley bottom in 1953, where he found that enigmatic name of Callanga given by the people who lived there. He described it as a "bottomless well surrounded by sheer walls of rock." That confirmation satisfied the world of science that Callanga was indeed a real place. Woytkowski was nevertheless greatly disappointed not to have found *Velia helenae.* An anthropogenic transformation could have changed the local biogeography once again.

Callanga was an important coca-growing zone of the late Inca and early colonial periods. At any one time, no more than 500 people of highland origin could have worked there, of whom fewer than 100 were permanent residents. Most workers came for short periods of time as part of their mita obligation and then returned to their home communities. The largest group of cocapallas came on a six-day trip from the Yucay Valley (Cook 1975:202). Stone-faced terraces in Callanga, now covered with forest, were constructed for growing coca. Only such a long-living perennial plant yielding a product of high value would have justified terrace construction in this zone. Highland people, not forest tribes, had the technical experience and social organization necessary to construct them. Presence of stone terraces and remains of stone buildings as well as smaller artifacts of highland provenance offer persuasive evidence that the Incas used this area before the Spanish conquest (Deyermenjian 1988:25; 1990:75).

It now appears that Callanga was the center of a zone of several valleys that the Inca had occupied for the express purpose of growing coca. Deyermenjian (1990, 1995) and his associates discovered ruins of Inca terraces in nearby depressions from ca. 1,900 m to 1,300 m above sea level in the valleys of the Siwas and Pitama rivers, which join to form the Callanga. To the north, the Río Mameria, an upper tributary of the Río Nistron (= Nystron), which in turn flows into the Piñi Piñi, contains undisturbed evidence of Inca occu-

Figure 7.4. Ruins of an Inca building of rough-cut stonework in the remote valley of Mameria in the Piñi Piñi (Toaima) drainage, where the Inca grew coca (Photograph by Gregory Deyermenjian, reprinted with permission)

pation (figure 7.4). An Inca fortress was built on a hill site, Apu Catinti, in the Mameria drainage (Maúrtua 1906, 6:11). An Inca garrison there would have protected coca workers, coca fields, and trails from incursions of forest-dwelling tribes. Bovo de Revello's (1848:25) description of the hill site as containing the ruins of two "large cities," was unsubstantiated by Deyermenjian (1990:80), who found no evidence of settlement in 1986 or 1989 high on Apu Catinti. He discovered at its base only the remains of a 10-m-long structure with walls 2 m high, but nothing there suggested to him a city or town of any sort.

Callanga's lifeline to the outside world was the Inca-built spur from Chimor. The trail wound through cloud forest on both the east and west sides of the Cordillera de Paucartambo. A causeway was built across a wide stretch of marshy puna. The bottom of the Callanga gorge was reached by way of a steep, six-hour-long descent from 3,300 m to 1,500 m. This particular entrée to the jungle did not stop with the Spanish conquest. In 1542 Pedro de Candia reached the Alto Madre de Dios by way of the road through Callanga (Matienzo 1967:167). Large amounts of coca continued to flow out on it from Callanga and Piñi Piñi for about a century after the conquest. Highland Indians from the Yucay Valley labored in the coca fields owned by Francisco Pizarro. His brother Hernando Pizarro owned other coca chacras nearby in the "valle de Toayma," downstream from Callanga. That Callanga was a recognized place in the early colonial period is suggested by its listing (sometimes called Atacallanga in the records) as an *anexo* in the reducción of Challabamba (Maúrtua 1906, 2:141).

With the decline of silver production at Potosí in the late seventeenth century, that city's population, wealth, and coca consumption fell correspondingly. The reduced demand for leaf could be met from the much closer Yungas of La Paz. Shipments of coca from the Cusco region to Potosí stopped. The coca fields in Callanga were abandoned, which, together with the ravages of endemic malaria, reduced the motivation to stay there (Villanueva Urteaga 1982:258). Other discouraging factors were a coca disease that killed many plants, and hostile forest Indians, who were seen as potential pillagers of coca haciendas at any moment of the day or night. Interethnic conflict was often blamed for abandonment (Oricaín 1906:374; Raimondi 1898:258; Markham 1991:112). However, Nystron (1868) disputed the accuracy of reports that chuncho pillaging was the cause of the nine abandoned haciendas in the Tono and Piñi Piñi drainage basins. As abandoned coca fields in Callanga regenerated into forest, families of forest tribes returned to their ancestral territory to fish and hunt again. At times they collided with highlanders at the few remaining haciendas, but many reports of confrontation seem to have been invented.

With scant traffic and no maintenance, the Chinchibamba-Callanga trail became partly obliterated by fallen trees, mudslides, rockfalls, and the advance of wild vegetation. At some point in the eighteenth century, the path fell into obscurity. Its rediscovery dates from the Amazonian rubber boom, when a government-sponsored exploratory party sought new transportation connections between the highlands and the head of river navigation in the selva. An expedition led by Ramón Chaparro in 1894 accidentally came upon a *camino* into the Callanga Valley (Robledo 1899:440). Hassel (1898) mentioned this trail as overgrown but well preserved, yet did not include it on his map.[5] Subsequent maps also failed to show this trail.

In the late nineteenth century, Hacienda Callanga was established in this remote valley. This huge spread of 16,000 ha covered the valley depression and slopes above. Its center of operation was near the junction of the Callanga–Piñi Piñi and Nistron rivers (12.59°S., 71.13°W.), at 1,500 m above sea level. Distillation equipment was brought in to produce rum made from sugar cane, and to the scattering of people on the adjoining plateau, *Callanga* became a generic word for *aguardiente* (Banco de la nación 1969). Other agricultural products—coffee, peanuts, citrus, bananas, papayas, and manioc—are largely uncompetitive. Remoteness made the hacienda unprofitable, but another problem was dissension within the Bueno family, who for decades had owned this property but were preoccupied with their landholdings elsewhere.

Callanga's connection with the highlands became tenuous. The Chinchibamba trail was no longer passable by then and the Acobamba-Tono trail was in poor condition. The sole access route to the Paucartambo yungas east of Cusco shifted to the Challabamba-Cosñipata trail. Then, beginning in the 1920s, the Challabamba trail was widened and graded as a one-lane vehicular road (*carretera*). That link offered the most direct route to the jungle from both the departmental capital in Cusco and the provincial capital of Paucartambo. The highway was also near the large sugar-cane hacienda of Sven Ericcson, a Swedish engineer who was the moving force behind the road construction (Mörner 1994). Movement of people, vehicles, and taxable goods in and out of the jungle could be conveniently controlled at the Paucartambo bridge. That heavily used highway was later extended deeper into the forest. By the 1960s, vehicles could reach Salvación, a military base beyond the confluence of the Piñi Piñi and the Río Alto Madre de Dios. With total attention focused on the Challabamba-Cosñipata road, Callanga was ever more marginalized. Two days of travel from Callanga were required just to reach the foot of the escarpment to begin the steep climb to Paucartambo.

Gualla/Hualla/Lacco

The lower Mapacho is another valley of mystery that fell into obscurity and whose recent history is still one of isolation.[6] As the crow flies, it lies 165 km from Cusco; overland it is more than twice that distance. Hualla/Lacco refers here to that part of the Mapacho Valley between 1,700 m and 900 m above sea level. The valley progressively widens into an amphitheater-shaped basin about five kilometers across, where abundant land for agriculture coincides with mild temperatures. Upstream to the south, the river courses through a canyon with few possibilities for settlement or agriculture. This gorge long formed a barrier ("*cinco leguas de despoblado*") to movement and settlement downstream (Villanueva Urteaga 1982:264).

Vertical differences also characterize the Lacco Valley (known in the colonial period as Gualla or Hualla) laterally from its lowest point near the Río Mapacho to the heights on both sides. A macrothermal climate on the floodplain changes rapidly to warm temperate as the land rises in elevation on both sides. On the surrounding upper river terraces, cool temperatures have favored potato growing and, beyond 3,500 m above sea level, pastoralism on the high grassy slopes. The whole zone has had crustal instability within historic times as manifested in earth tremors, cinder cones, opened smoking vents, and hot springs.

Settlement history of the lower Mapacho is recorded under the name of Hualla.[7] Highlanders grew coca there as one of the distant points in the Cusco orbit for its production. Hassel (1898:303; 1907:367) was the first to record the well-preserved bank terraces, remains of stone dwellings, *chullpas* (mortuaries), bridges, and inlaid trails. Polentini Wester (1979:64) was convinced that this once densely inhabited stretch of the Mapacho was the fabled Paititi.

A fortress and temple were among the constructions high on the western ridge of the valley, not in the depression, below which had been cleared for growing coca. Most agricultural terraces, presumably for growing coca, were located above the elevation of 1,300 m. In the depression below, besides coca, capsicum peppers and fruits were grown as secondary products by those who had coca as their primary concern. Farther down the valley, wild materials, including copal, Peru balsam, wild cacao, and folk remedies of plant and animal origin, were collected from the forest.

To move tropical products out, Hualla was connected to the imperial trunk roads which integrated the zone into the economic and social realm of the Incas. Guaman Poma de Ayala (1980, 3:994) wrote in the early colonial period of a "*camino real*" that went into the jungles of "Anti Guaylla." One trail partly followed the heights above the west side of the Mapacho Valley and connected with one that continued westward over the divide to Amparaes. A second trail crossed the cordillera into the Lares Valley and from there went up to Amparaes. To the east, an Inca trail that followed the crest of the Cordillera de Paucartambo turned down into the valley to the bridge crossing of Chimor. An Inca trail also connected the lower Mapacho with the canyon of Callanga.

Remote yet accessible to Cusco, these trails gave the region strategic advantage during the struggle against the European invaders. In 1537 Manco Inca first sought refuge with his army in Gualla after he had abandoned his hopes for a safe highland redoubt (Murua 1987:245). Manco spent several weeks assessing Gualla as the potential center for his Inca state-in-exile before deciding to set up his resistance headquarters in the Pampaconas Valley of Vilcabamba. In that jungle redoubt, he was farther from the center of Spanish control in Cusco and had a strong defensive position in case of invasion. Gualla's entire territory was the personal possession of Huayna Capac. After the conquest Gualla (also spelled Hualla), like parts of Callanga, became a repartimiento of Pizarro, who derived economic benefit from the coca produced there. This connection explains why Gualla was one of the 24 repartimientos of the corregimiento of the Yucay Valley. In Callanga, part of the corregimiento of Paucartambo de los Andes, individual coca fields, not the whole valley, belonged to Pizarro, just as Huayna Capac had owned them in the late Inca period.

Gualla had two major tribute responsibilities: producing 1,140 baskets of coca and delivering them in three yearly installments to Potosí. Proceeds in silver from the sale of 700 baskets of coca went to the *encomendero.* The income from 256 baskets supported the priest in charge of their catechization. Profits from 144 baskets went for judge salaries, and money that came from 40 baskets was given to the two headmen (*curacas*). Moving this coca to Potosí required a pack train of 190 llamas on each of the three annual trips. In Gualla the cool grassy heights far above the valley provided favorable conditions for a large number of pack llamas. As elsewhere in the Central Andes, burning the forest to create grassland favored the expansion of grazing. Llamas were a key to transport in this zone. Loaded with coca for the long journey ahead, camelids took the Inca road near the crest of the Cordillera de Paucartambo to their destination in Potosí. Potato fields at these higher elevations supplied chuño for the voyage and food for coca workers down in the valley.

Gualla did not escape Viceroy Toledo's program to nucleate native people into villages.[8] Although it lay isolated beyond the cordillera, administratively Gualla became an annex of Lares.[9] Often the priest had no hosts, wine, or candles for mass. In 1649 an earthquake destroyed the village of Gualla (Villanueva Urteaga 1982). With their water supply obliterated, 40 surviving families dispersed their dwellings to individual coca fields. Even though the church was destroyed, a Spanish priest continued to serve these people of highland origin in addition to 16 other families at Suyo, five kilometers down the valley. The land beyond Suyo was unpopulated, although Polentini Wester (1979) located remains of a trail suggesting that the preconquest geography was somewhat different. In the lowest reaches of the Mapacho, unlike in the Madre de Dios drainage, highlanders and the forest Indians (*infieles*) remained in separate realms.

As elsewhere in the Andes, Gualla's population substantially decreased during the early colonial period. The 1575 census commissioned by Viceroy Toledo counted 622 residents in Gualla, of whom 190 were tribute-paying males 18 years and older. Not included were the temporary workers who picked the coca. But by 1602, the number of *tributarios* in Gualla had fallen to 70, which translated into a total population of not many in excess of 300 (Cook 1981:226). They labored under a heavy tribute burden at the same time as the demand for coca faltered. Producing zones much closer to the mining centers could supply coca more cheaply, leaving the most isolated and distant zones without a market. Gualla turned mainly to subsistence farming or other crops. In the early colonial period Suyo, part of Gualla Parish, became a large property of a Spaniard who planted orchards there. In the eighteenth century Alcedo (1967) described it simply as a town in the province and corregimiento of Calca and Lares, without any reference to its prior role in producing coca. Neither the 1786 nor the 1792 population enumerations listed Gualla (Vollmer 1967). It is unclear whether its population had been wiped out or no census taker had reached it.

In the late nineteenth century Lacco became the focus of settlement attention in the lower Mapacho Valley. Lacco originally referred to the first upriver bridging site, a crossing that also marked the upstream limit of canoe navigability. Lacco was adopted as the name for the gargantuan property which had its center of operations at 1,400 m above sea level near the eastern side of the bridge (Mendizábel 1902). The entire 100-kilometer-long valley zone was also called Lacco after Hacienda Lacco, which covered much land on both sides of the Mapacho. Coca was long the major product. To guarantee a labor supply for the hacienda, the owner leased large parcels to *arrendires* on which they grew their own subsistence and commercial crops. Eventually these arrendires gained property titles to those parcels, which in the usage of the time were also called haciendas. Their emergence can be best understood as a response to labor shortages.

Unclaimed state-owned lands in Lacco were also available for cultivation and settlement in exchange for a twice-a-year payment to the state for each hectare used. Highland peasants who petitioned for *denuncios* in Lacco typically migrated there after the harvest was completed in the highlands. After three to four months on those jungle plots, they returned to their home communities in time to plow and sow their family chacras. The tight balancing act of farming in two widely different localities contributed to instability of landownership. Rights to the land were sometimes forfeited for nonpayment of taxes. In 1961 Lacco had 45 denuncios with an average size of 100 ha.

Isolation has affected the kind of land use, especially until the 1970s, for only agricultural crops of relatively high commercial return could withstand the high cost of transport. Cacao, coffee, coca, tea, *achiote, palillo* (*Escobedia,* a dye plant), and *cube* (*Lonchocarpus,* a species yielding rotenone) were planted earlier in this century. But the harvest and shipment of any of these products depended on commodity prices from year to year. Subsistence field crops from Lacco normally found no market, although during the cinchona and rubber booms, Lacco provided agricultural staples to workers who collected them. When the Lares road was completed to Quellouno in the 1970s, Lacco became less remote. It now lies within two days of Calca—one day on the trail, the other on a truck from the roadhead.

The area of Hualla and Lacco has held no long family histories. In the Inca period, the long-term mitimaes and short-term cocapallas were transplants from distant places. New workers came to take the place of those who had died or fled. In the colonial period, the labor force had the same duality of permanent residents who had been outsiders (*forasteros*) and seasonal migrants who picked, dried, and shipped the coca. Epidemic diseases, especially yellow fever and malaria, took a heavy toll, but only in the twentieth century did the insalubrity of the area become generally recognized outside the zone. Malaria epidemics, especially the notorious outbreak of 1933, devastated Lacco and all other valleys of the Vilcanota-Urubamba drainage below 1,800-m elevation. Heavy mortality and the terror that malaria inspired left Lacco largely devoid of people for a decade. Malaria continued as an endemic disease there through the 1950s. A third of the population was infected in 1957, an incidence which was lowered after a malaria eradication program was put into place around 1960.

Health measures and available land for settlement under the denuncio system again attracted migrants from the highlands. Of the 768 people counted in 1968, 62 were landowners; 130, lessees (arrendires); and 184, laborers. Now, aside from a poorly attended school, the small dispersed population lacks government services. Without a central village or a critical mass of people, social and regional isolation has been intense.

Reconstructing the Past

Although their exact origins in time cannot yet be dated, both ruins and historical records confirm highland occupation in the forested valleys east-northeast of Cusco. Regeneration of natural vegetation, obliteration of trails, and the disappearance of villages impart elements of mystery about the location and size of preconquest settlements in the jungle. The unknown and the tantalizing possibilities about Inca settlement have spurred explorers to follow leads in search of a "lost city" in the jungle. Although two precedents exist in this century for such a discovery, the chances of encountering ruins of other arresting sites would seem to be slim. Existence of Paititi as a large settlement rests on a very tenuous foundation. Nonetheless, a fuller story of the Inca cultivation of tropical products will tell us much more about their presence in the jungle. Inca preference for using stone in building terraces, dwellings, forts, and trails provides a real basis for reconstruction of this aspect of Andean culture history and historical geography.

Different pieces of evidence about Callanga and the Mapacho point to the following configuration. Preconquest settlement in these low zones had coca as its prime motive. Each zone combined a good growing climate and a fairly low health risk. High leaf quality was assured by daytime temperatures between 18° and 25°C, cool but frostless nights, and rainfall in excess of 1,500 mm/year. These conditions favor the development of thick leaves high in alkaloids and flavenoids (Acock et al. 1996). If temperatures are too hot, the plant tissue of *Erythroxylon* is low in desired substances, tastes bitter, and becomes brittle when dried (Plowman 1984). Coca grown at 1,300 m is twice as potent as that harvested at 600 m above sea level. Stone terraces between 1,900 m and about 1,200 m in Callanga and the lower Mapacho suggest that the Inca had empirically discovered the most suitable altitudinal range in which to grow the best coca. Presence of certain plants as surrogate indicators of climate, such as the pacay tree (*Inga reticulata*), could have pointed to the appropriate locales for growing the best coca leaf. Moreover, migrant leaf pickers who came to the valleys could reach the coca fields on access trails that did not entail risk of leishmaniasis contraction. Once in the valleys, workers were within walking distance of higher elevations that put them beyond the possibility of that disease risk.

These precautionary measures were abandoned two decades after the conquest. The Inca pattern of coca growing began to change in the 1550s, when a surge in leaf demand was brought on by the huge new market created at the silver mining camp of Potosí. Entrepreneurial Spaniards commandeered land and labor in the cocayungas to profit from high leaf demand. Greed took precedence over concern for the health of native people, who constituted the labor supply. Coca quantity became more important than quality. Early colonial cultivation of leaf in the eastern valleys expanded beyond its Inca distribution into lower elevations, where four harvests a year, rather than just three, could be delivered. Warmer climate meant more leaf production and greater profits, but also greater mortality from disease, about which the chroniclers are so clear. But the expanded production was not to last, for in the late seventeenth century, the coca boom faded in the warm Cusco valleys. Permanent and temporary populations shrank as demand decreased and production went into a steady decline. Other areas, especially the Yungas of La Paz, provided cheaper leaf grown much closer to Potosí. Hualla and Callanga lost their prior importance and even their continuity as known places. Isolation, property fluidity, harsh tribute demands, population instability as a result of migration, and disease buried their pasts as much as the jungle growth did that subsequently reinvaded. The sense of mystery about both zones is closely related to their ambiguity as identifiable places. It is nevertheless clear that a pre-Hispanic infrastructure was carefully put in place that has not been equaled since.

8
Guayaquil as Rat City

The vermin and weeds that have accompanied human settlement almost everywhere in the world carry tales of easy dispersal, rapid establishment, and ecological mayhem. Many of these unwanted organisms have become established in environments to which they were specifically preadapted, but others have such broad tolerances that they are distributed nearly universally. Among the rank cosmopolitans is the rat, a commensal genus almost as biologically successful as humans. So common are rats in the world's urban areas that it is difficult to grasp how a purported social-cultural geography of animals in the city could ignore their presence (Philo 1995).

Ecuador's largest city and main port, Guayaquil, has a rat geography with commensality at an extreme. Rats are everywhere in that important tropical vortex of human activity. They swarm in shuttered markets and slaughterhouses. In the city's poor neighborhoods rats and people share living spaces, indoors and out. In tall downtown buildings rodents can be heard gnawing through the night. They scuttle through hospitals and darkened cinemas. Their proliferation has multiple impacts on human health and property.

Guayaquil exemplifies a common observation that rodents infest port cities because that is where they arrived as passengers from overseas. Docking ships have delivered immigrant rats of varying genetic constitution, and warehouses provide them shelter, food, and a base from which they can invade the rest of the town. In Bombay, Lagos, New York, and other coastal cities, commensal rats typically have been viewed as a public health issue, not as part of the urban ecosystem. The ecology of rats and their cultural-historical dimensions in much of the world remain unexamined. A gentle prod toward correcting this omission has come from Wagner (1970:28–29), who has directed his intellectual curiosity into some dark corners of Latin America and elsewhere. He stated that ". . . it would be revealing to look at the population of vermin as distributed alongside and intermingled with the human population in various countries. What is the implication of the distribution of the rat or the cockroach population or of the population of the zopilote buzzard or of pariah-like dogs for the state of communities in Latin America?"

In Guayaquil, as in many other cities, rats and humans are the two most numerous vertebrate creatures. The former are dependent on the food left by the latter, an ecological relationship that is close to parasitism. Given that connection, it is remarkable that man and rat coexist with such minimal direct confrontation. Rodent stealth and nocturnality ensure that rats generally are successful cohabitors with people. Their strategies in that regard have very possibly been honed by behavioral evolution during the period of their anthropogenic associations. Commensal rats are not easy to categorize in the continuum from wild to tame. Often they are referred to as domesticated species, presumably to distinguish them from the native fauna. Yet *domestic* usually implies an animal genetically changed by conscious human selection, which is not the case in these rats. Others have described these species as wild. In some oceanic islands, notably New Zealand and Madagascar, these rats are found in a truly wild setting apart from human habitation, because they have been able to move into an unoccupied ecological niche or to compete successfully with native organisms. But that does not describe urban rats any place in the world. Aside from those exceptions, the term *wild rat* is meaningful only when these animals are compared with albino laboratory rats, next to which the free-living pigmented rats seem to be positively savage. Urban rats are peridomestic creatures which have acquired an evil reputation (Canby 1977). Frequent hyperbole of rat size relates to the negative way in which humans perceive them.[1] Yet their omnivorous habits, lack of an estrous cycle, and fondness for copulation make them more like us than we want to admit.[2]

Spread of the Rat to the Americas

Both the rat and its European introducers have been in the New World for around five centuries.[3] The Americas have many species of native rodents, but none that belong to the genus *Rattus*. Humans have assisted rats in moving abroad, but only as unintentional passengers; few stowaways in the history of the world have had such a wide impact. Until measures were taken to prevent unwanted organisms from boarding and to fumigate those found, virtually all overseas voyages from Europe undoubtedly carried rats.[4]

Three commensal rodents—two kinds of rats (*Rattus rattus* and *R. norvegicus*) and the house mouse (*Mus musculus*)—showed the same uncanny ability to sneak aboard, consume a range of shipboard edibles, and then, when land was near, decamp at the propitious moment. If nutrients were abundant, control lax, and predation low, rat numbers on board ship substantially increased on transatlantic voyages. A four-month-long trip and a gestation period of only

22 days enabled rats to engender at least two generations before the destination was reached. The few rats that lurked aboard when the ship set sail multiplied into hundreds by the time it reached its New World destination. Ship cats did not effectively control rats. Although good mousers, most cats are unsatisfactory ratters. Moreover, the cats themselves sometimes ended up in the cooking pot.[5] Ship captains occasionally tried to keep rats in check by offering a prize to the crew member who captured the largest or the most (Flores 1985:366).

Ship rats and crew members competed for the same food. Hardtack (*bizcocho*), made from wheat flour twice baked to reduce the moisture content, was the staple of early colonial sea voyages. Grain-lovers above all, rats were happier nibblers of hardtack than were the crew who ate the tough, tasteless blocks out of necessity and habit. Other shipboard victuals included fresh meat (available only early in the voyage), salt pork, salted fish, rice, broad beans, and chickpeas. Cheese was carried for those times when starting a cooking fire was impossible. Garbage was tossed into the sea, so the rats, denied their normal food source, raided provisions stored in wooden barrels and canvas sacks. They were relentless in getting to food and water. Pottery jars and iron-hooped casks holding food were gnawed near the base. Live animals put on board as trip rations for humans or as trade items were sometimes bitten until they bled to death. Water jars were gnawed near the top; sometimes rats fell in, drowned, and fouled the drinking water meant for human use.

Ship rats could consume enormous quantities of food. Vázquez de Espinosa (1623:88–89) described the voyage of the *Nuestra Señora de Candelaria* in 1622 between Vera Cruz, Mexico, and Havana, Cuba. He recorded that a large number of rats had eaten much of the hardtack, flour, legumes, and meat meant for the long voyage back to Spain. At the port of Havana the ship was cleaned of over 1,000 rats and then reprovisioned. But near Bermuda, after two months of sailing, the remaining rats had again multiplied and devoured so much of the food supply that the ship again had to be reprovisioned, this time from an accompanying ship in the convoy. To make sure the ship would make it back to Spain, the crew and passengers killed thousands of rats.

Lack of sanitation and commensality went hand in hand on board these sailing vessels. Rats, people, and domestic animals shared tight spaces and notoriously filthy conditions. Except for the captain, crew members scarcely washed, never changed their clothes, had no beds to sleep on or table from which to eat (Martínez 1983:108). Vomit spewed the deck during bouts of seasickness. Unsanitary conditions enabled vermin to multiply. Rats, lice, cockroaches, and the vile stench had to be tolerated. That intimacy increased possibilities for the spread of rat-borne diseases, most notably bubonic plague. However, there is no record of this disease on board ship or in the New World colonies in spite of its presence in Spain into the seventeenth century.

If provisions ran out for the crew, rats lost their normal repulsiveness and came to be seen as edible. Magellan's crew might not have circumnavigated the globe in 1519–1520 had they not eaten the ship rats to avoid starvation while crossing the Pacific Ocean. Only sailors with money got this prized item; the mangy corpse of a rat changed hands for half a year's pay (Pigafetta 1972:65).[6] The pursuer and the pursued also have occasionally been reversed in maritime history. Garcilaso de la Vega (1960:438) recounts a case of a sick old man nearly devoured by ship rats.

The Black Rat: Pioneering Vermin

The black rat (*Rattus rattus*) was the earlier of the two main commensal species in the genus to have reached the Americas.[7] It is one of about 50 *Rattus* species in the world, of which 7 live in close association with people (Nowak 1991:786–87). *Rattus rattus* appears to have first come to Spain by ship during the Roman period in the second century B.C. (Reumer 1986). An enterprising voyager, the black rat hitched a ride on Columbus' first voyage in 1492. Excavation of the admiral's ill-fated first settlement of La Navidad, built in part with the shipwrecked remains of the *Santa María,* has yielded the jaw of a black rat (Deagan 1987:675). Fernández de Oviedo y Valdés' (1959, 2:31) statement that rats came in the earliest of transatlantic ships is vindicated by this radiocarbon-dated find. The overworked adage about the adeptness of rats in leaving a sinking ship would seem in this case to have applied.

In western South America, the same species of rat arrived five decades later than in the Caribbean. Thick isthmian jungle hindered the rapid dispersion of rats to the new port of Panama, although over time they did reach the Pacific coast, probably carried on muleback, hidden in supplies. Well before the end of the sixteenth century, they had overrun Panama City (Garcilaso de la Vega 1960:437). Although Pizarro, the conqueror of Peru, sailed from Panama, this site was not the original source of these cunning rodents in the viceroyalty of Peru. Zárate (1947:486) recalled that a ship sponsored by a provincial bishop, Gutiérrez de Carbajal, brought the first rats directly from Spain to Peru through the Straits of Magellan. Four harrowing months after setting sail from Seville in early 1541, that vessel was the only one of three that made it to the Pacific side (Morison 1974). Gómara (1946:277) corroborated the timing of this introduction when he said there were no rats in Peru until the time of the first viceroy of Peru, Blasco Nuñez Vela, who administered between 1544 and 1546.

Most Spanish chroniclers were not interested in where rats came from or when they arrived. The zoologic wisdom of the time considered rats to be among the organisms that arose, *mirabile dictu,* from the stinking mud of swamps. This idea, which dated from Aristotle and was promoted by St. Augustine, explained that putrefaction triggered spontaneous generation (Porta 1957:28). José de Acosta (1962:55) further explained that rodents fell into the category of *animalejos imperfectos* because their innately flawed qualities were the result of spontaneous generation. Two seventeenth-century observers elaborated much the same idea. Vázquez de Espinosa (1969:16) carefully distinguished between a "perfect" animal such as the deer, whose presence was attributed to having been on Noah's ark and whose distribution involved migration, and the "imperfect" rats and frogs, which emanated from rotting matter. Bernabé Cobo (1956, 1:321), normally a sound observer of biota in the New World, wrongly concluded on the basis of its Indian names that the rat was native, thus corroborating Aristotelian abiogenesis. To most observers of the time, that putative mode of reproduction provided a plausible explanation for the omnipresence and abundance of rats and mice. Into the eighteenth century, this line of thinking hindered sound perspectives on the biogeography of this peripatetic animal (Real academia española 1969, 3:497). Oviedo (1959, 2:31) reflected the contradiction between the received idea and actual observation. He expressed no doubt that spontaneous generation accounted for rats in the New World, yet at the same time noted that individual rodents were either male or female, that they reproduced through coitus, and that rats were brought from Spain. Vázquez de Espinosa (1969:16, 237–38), traveling widely in western South America in the early seventeenth century, also seemed to have been puzzled by the discrepancy between accepted Aristotelian dogma and observed reality. Repeating the notion that rats are products from the earth's putrefaction, he nevertheless later mentioned that the rats and mice he saw in the Río Cauca region of Colombia were descendants of those brought in ships from Spain.

The Peruvian mestizo El Inca Garcilaso de la Vega (1960:438), born in the sixteenth century, was on the right track. He noted both the abundance of rats on ships and their establishment in coastal areas, implying that places reached by ship were the earliest to acquire rats. Although the reliability of Garcilaso's accounts on some other topics has often been called into question, his observations accurately account for rat spread. Rodents disembark in several ways. They scamper down the mooring ropes. They sequester themselves in boxes and barrels that are unloaded from ship to dock. They can also swim to shore. Hendricksen's (1983:87) claim that rats can tread water for three days and swim half a mile in the open ocean seems far-fetched, but jumping ship before it sank was a useful instinct. Many rats surely drowned in the fray, but just one pregnant female reaching shore was all that was needed to establish a new rat population. Their invasion of port towns can be attributed to desperation to get off the vessel. Propelled by hunger and thirst, they decamped more quickly than the human travelers.

Rats increased exponentially at their coastal points of introduction and later expanded into the hinterland to find new food sources. On occasion human conveyances inadvertently moved them to new places. Already by the late sixteenth century their proliferation in the New World caused problems. Fernández de Oviedo y Valdés (1959, 2:31) mentioned how this "aggravating kind" (*enojosa casta*) of rodent was found in farms and towns through the islands and mainland of the Spanish Indies. Gómara (1946:277) claimed there were so many rats in Peru that they destroyed entire crop harvests, did not let Spaniards sleep, and frightened the Indians. Garcilaso de la Vega (1960:437) commented on their fecundity, the damage they caused in Peru, and the need for rodent control. The idea that rats should be controlled emerged in sixteenth-century Europe, and it became well established in Latin America by the seventeenth century (Konkola 1992). As in Spain, arsenic sulfate (*rejalgar*) was the customary ratsbane until the eighteenth century, when the much more effective white arsenic replaced it. One rat-catching method was the baited basket, still used today in the Andean countries. Dogs were more effective as ratters than the cats introduced from Spain. An autochthonous solution in hot climates was the rat-eating snake placed in granaries and houses (Cobo 1956, 2:354).

The Brown Rat as Latterday Immigrant

More than 200 years after the black rat arrived, its wily congener, the brown rat (*Rattus norvegicus*), reached the Western Hemisphere. Like the black rat, the biological origin of the brown rat was as a wild species in Asia, from whence it spread to Europe and on to the New World. However, the brown rat did not gain a foothold in western Europe until about 1730, more than a millennium after the black rat had filtered into southern Europe around the second century A.D. In eastern North America, the brown rat disembarked around 1780. Not until 1851 did this species appear in California, where it arrived by way of the port of San Francisco. Along the west coast of South America the brown rat does not seem to predate 1870.[8] Which vessels carried the first brown rats overseas will never be known. Sea voyages have accounted for perhaps 80 percent of the worldwide dispersal of this species, so in that regard, it is as much a shipboard rat as its black cousin is. Guayaquil could have received its first brown rats from ships that sailed from Spain, where this species had been known since about 1760. But open commerce in the late nineteenth century diversified the sources from which Guayaquil could have received it, including North America.

The presence of two different species for more than a century complicates the rat geography of Latin America (figure 8.1). Historically they often were not distinguished, for to the layperson, a rat is simply a rat. They have not always even been differentiated from mice, which reflects, in part at least, a semantic conflation of European languages. Neither the Spanish, nor the Portuguese, nor the Italian language clearly distinguished between rats and mice. Historically, the rat in Spanish was "a certain kind of mouse" (*ratón*) (Vélez 1613:125; Real academia española 1969, 3:496).[9] In Latin, *mus* designated both mouse and rat; in medieval Spanish, *mur* was used for both (Corominas

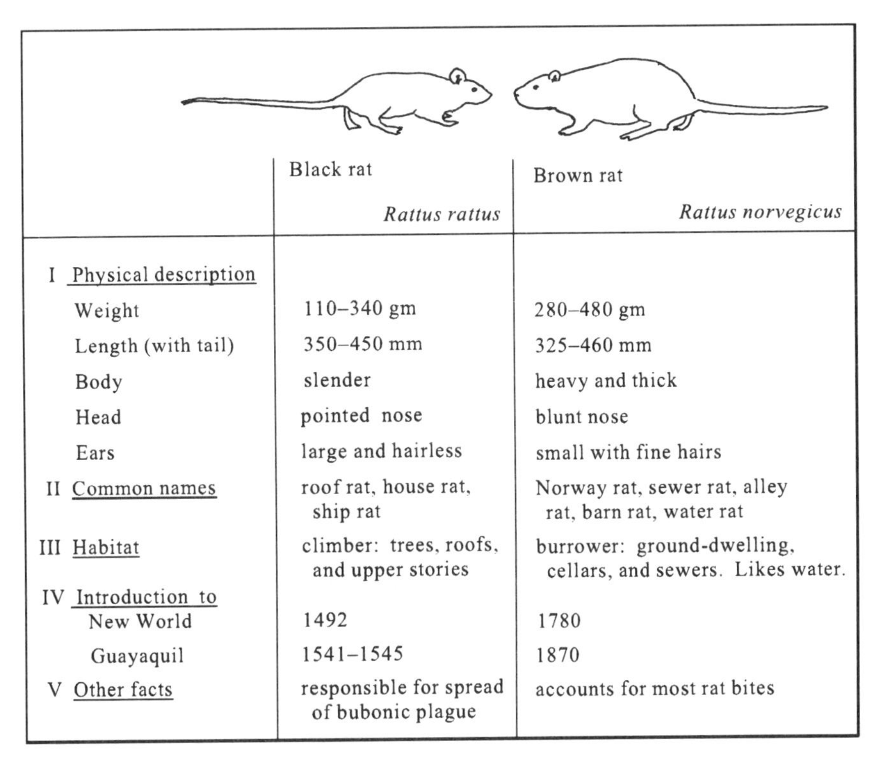

	Black rat *Rattus rattus*	Brown rat *Rattus norvegicus*
I Physical description		
Weight	110–340 gm	280–480 gm
Length (with tail)	350–450 mm	325–460 mm
Body	slender	heavy and thick
Head	pointed nose	blunt nose
Ears	large and hairless	small with fine hairs
II Common names	roof rat, house rat, ship rat	Norway rat, sewer rat, alley rat, barn rat, water rat
III Habitat	climber: trees, roofs, and upper stories	burrower: ground-dwelling, cellars, and sewers. Likes water.
IV Introduction to New World	1492	1780
Guayaquil	1541–1545	1870
V Other facts	responsible for spread of bubonic plague	accounts for most rat bites

Figure 8.1. Comparison of the two commensal rat species that have invaded Guayaquil and most other places in the world (Source: Pratt, Bjornson, and Littig 1976:6)

1954:478–80). Linnaeus placed *Mus* as the genus for the house mouse and both species of commensal rats: for the black rat, *Mus rattus,* 1758; and, for the brown rat, *Mus decumanus,* 1778, meaning "huge mouse." These binomials persisted until 1881.

Vernacular terms for the rat in western South America have included native words. *Pericote* has been used in South America for rat, but in some contexts it also means a mouse or any kind of wild rodent (Hildebrandt 1969:298–300).[10] The Quechua word *ucucha,* which originally applied to wild native rodents of the Andes, was later extended to those introduced.[11] In the late seventeenth century, Velasco (1977, 3:193) classified five kinds of ucucha, three of which were native rodents. One of these three was an *Oryzomys* species, a rice rat indigenous to the coastal region. It ate Indian crops long before the Europeans arrived and was enough of a pest to have been the subject of Mochica iconography. The two introduced species on Velasco's list were the *huasi-ucucha,* which corresponded to the house mouse (*ratón casero*), and the *yana-ucucha,* which

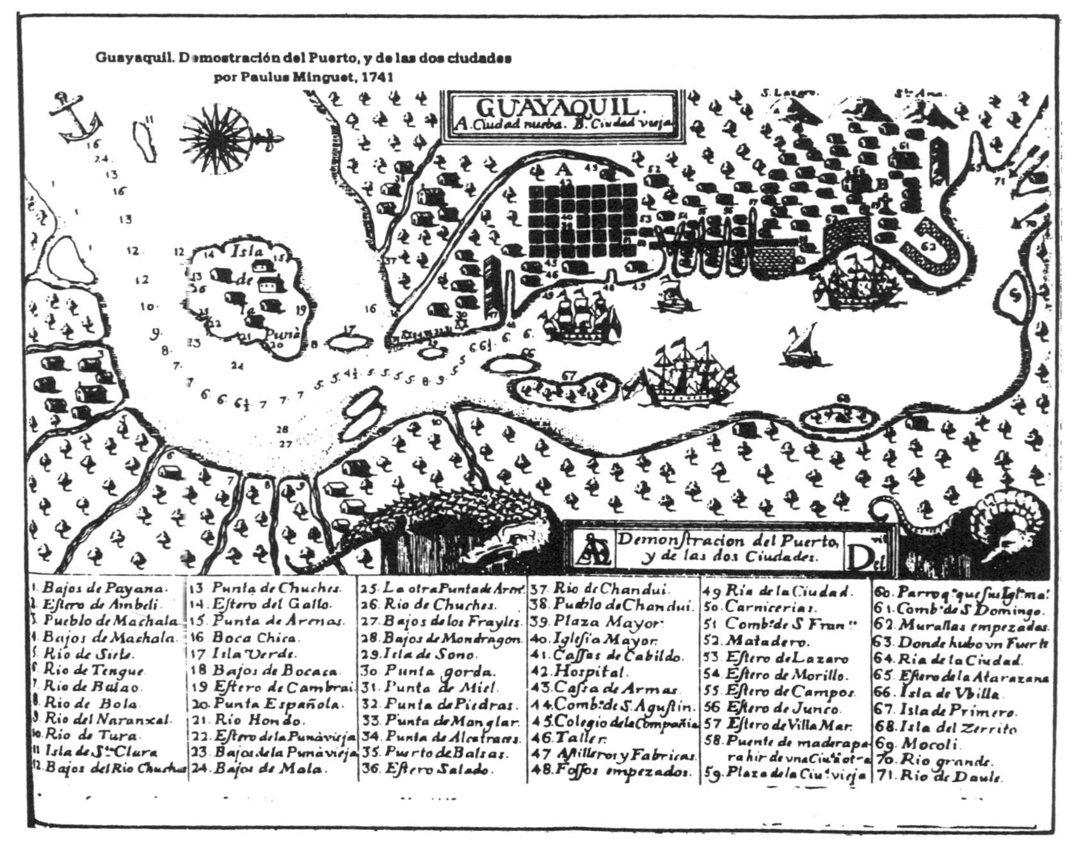

Figure 8.2. Guayaquil as shown on a cartographically stylized map of the eighteenth century (Source: Alsedo y Herrera 1987)

referred to the black rat; the brown rat had not yet been introduced. Although several native rodents, including *Oryzomys* spp., have commensal tendencies, the two introduced *Rattus* species of Asian origin and European provenance are the ones that have caused much destruction in rural and urban Latin America. The seventeenth-century writer León Pinelo (1943, 2:78) wrote of rat hordes laying waste to standing crops on the coast of Peru and supernatural efforts to control them through excommunication and pronouncements of anathema.

Rats and Colonial Guayaquil

The town of Santiago de Guayaquil was founded in 1534, but its site was subsequently moved to a hillside above flood levels. After 1693, population increase forced the settlement to expand from its cramped hillside location back to the poorly drained flats along the Río Guayas, 55 km from the open ocean. Shipbuilding, using timbers from rich forests in the basin, was the main activity during much of the colonial period (figure 8.2). By 1735, the Spanish crown had sponsored the construction there of 176 substantial vessels. Guayaquil also had a soap works (*almona*) that depended on animal fats brought from surrounding livestock haciendas. Port activities, driven mostly by smuggling, became important in the eighteenth century. Goods-laden ships, carrying more passengers than just the human sort, circumvented the strict mercantilist policy of Spain.

Black rats first arrived in Guayaquil shortly before 1550, seemingly on supply ships coming from Peru (Lizarraga 1987:61). They contributed to the squalor and insalubrity of a benighted place wedged between the flood-prone river to the east and five tidal creeks (*esteros*) to the west. The town turned into an amphibious hellhole in the season of drenching rains, hot temperatures, and suffocating humidity from January to May. Standing water provided countless breeding places for clouds of mosquitoes and gnats (*manta blanca*), which made Guayaquil nearly intolerable to many visitors. Also living in these waters were caimans with an appetite for human flesh (Jiménez de la Espada 1965, 1:128–29).

Indoor fauna caused unremitting torment. Mosquitoes, cockroaches, ants, termites, scorpions, spiders, beetles, moths, centipedes, and rats pullulated in these man-made spaces. Gnawing and scurrying rodents attracted reptiles, venomous and otherwise, which slithered around the premises looking for prey. The sixteenth-century chronicler Lizarraga (1987:61) opined correctly that the many snakes in houses were attracted by the large rat population, which built nests in the thatch and rafters. These and other uninvited "*bichos y sabandijas*" shared the bed and dinner table as if they were part of the décor. As elsewhere in the early colonial period, Guayaquil's heavy infestation of creepy-crawlies was explained by the heat and humidity of the location. Following the classical line of thinking accepted as late as the eighteenth century, these conditions caused rapid putrefaction, which begot lower forms of life from the swamp mud (Alsedo y Herrera 1987:55). Rat proliferation in Guayaquil had, of course, nothing to do with putrefaction and everything to do with superabundance of rations and convenient shelter. Streets and plazas were full of fetid garbage. Dead and decaying animals left on the public right-of-way offered an endless repast for buzzards (*gallinazos*) during the day and rodents at night.

Guayaquil's oppressive climate, noxious vermin, frequent fires, and pestilential conditions gave to the colonial descriptions of the city a persistent theme of calamity. Its filth was notorious even by eighteenth-century standards. Alcedo (1967, 2:176) castigated Guayaquil for its squalid streets (*calles puercas*). Epidemic and endemic disease followed this port city through the colonial period and into the independent Ecuador of the nineteenth century. The American scientist Orton (1869:29) disparaged it as "one of the most pestiferous spots on the globe." From the 1540s to the 1950s malaria wrought special havoc as Guayaquil's main endemic disease ("Descripción de la Gobernación de Guayaquil 1973:72). Yearly outbreaks (*anuales pestes*) of malaria killed and debilitated many residents (Baleato 1887; Brandin 1955). Just as in the Roman Empire, paludismo was intuitively connected to the swamps, but not until the end of the nineteenth century were mosquitoes proven to be the vectors in its transmission ("Relación de Guayaquil" 1994:499). Yellow fever first appeared in 1842, when the town had about 23,000 people, and for a century thereafter caused many deaths. Smallpox, tuberculosis, measles, diarrhea, and a number of unidentified diseases contributed to Guayaquil's high mortality, unhealthy population, and death-trap reputation, which continued over more than three centuries.

Before the twentieth century, rats in Guayaquil were sources of aggravation through propinquity, though they were not yet understood to be carriers of disease. Ulloa (1990:238) found hordes of these rodents in the houses of the city. When at night they emerged from their hiding places, their strepitous audacity kept him from sleeping. Worse than that, Ulloa noted rats as a probable cause of fires, for they knocked over lit candles to eat the wax. Another eighteenth-century observer, the French scientist La Condamine, likewise found Guayaquil to be insufferable with its rats, mosquitoes, and reptiles (Gómez 1987:92). Around the same time, Francisco Requena mentioned the rat infestation of Guayaquil (Laviana Cuetos 1982). Velasco (1977, 3:198) near the end of the seventeenth century observed Guayaquil's flourishing rat population closely enough to use it as an example to disprove a larger proposition of Count Buffon, who claimed that interclimatic transfer of animals from the midlatitudes to the tropics resulted in depauperate individuals; that certainly was not the case with rats.

Guayaquil maintained its reputation as a rat-infested city through the nineteenth century. The American scientist Orton (1870:29) did not mention rats, but he did point out some of the other disgusting fauna that penetrated everywhere: ". . . intolerable mosquitoes, huge cockroaches, disgusting centipedes, venomous scorpions, and still more deadly serpents, keep the human species circumspect. . . ." Like so many travelers before him, the Englishman Edward Whymper (1892:391) had to deal with vermin-ridden lodgings when he passed through Guayaquil in 1879 on his way to climb Chimborazo. Putting the best face on his predicament, Whymper philosophized that the "harsh gnawings of voracious rats were subdued by the softer music of the tender mosquito."

The idea of doing something about all this filth and these vermin goes back to the late eighteenth century. The grossly unhygienic conditions on which rats thrived in Guayaquil raised concern when an interest in municipal health and sanitation swept the Spanish Indies around this time (Clément 1983). Discussions were held about sanitation at least as early as 1771, but it was not until 1794 that a concrete plan to clean the city was actually proposed. Not until 1820 was municipal garbage collection actually begun (Hammerly 1987:139).

The development of garbage collection inconvenienced rats only a bit, for they had facile access to stored foodstuffs. After 1720 cacao became the economic lifeblood of the city. For two centuries Guayaquil led the world in processing and exporting this commodity. Between 1774 and 1809, the annual number of shiploads of cacao increased from 50,000 to 100,000. The cacao fruit—opened, seeds removed, and set out to dry in preparation for shipment—attracted rats, who especially like to eat the mucilaginous pulp that surrounds the aromatic seeds. In addition, much rice was grown in the region, and imported wheat was ground and stored in dockside mills for distribution to bakeries. These omnivorous feeders also ate the offal of the slaughterhouses and the fats destined for the soap works.

Black rats do not ordinarily dig burrows because their original nesting habitat was in tree branches. Guayaquil's traditional house construction provided ample opportunities for above-the-ground nests. As late as 1860, Guayaquil's dwellings were primarily made of plant materials. Bamboo or wood were cheap and can withstand the periodic tremors which were thought to make stone or brick dangerous (Hassaurek 1967:4). One-story dwellings had siding of the split culms of *caña guadua* (*Guadua angustifolia*) with roofs of wild banana thatch. In low-lying areas, houses of this local bamboo species were atop posts because of fluctuating water levels, but that did not impede rodential access. More affluent people lived in two-story wattle-and-daub (*quincha*) houses with thatch or tile roofs. Typically the upper floor was used as the residence and the ground floor as a warehouse or commercial shop; rats happily nested inside or in the roof. Quincha and tiles, which the cabildo tried unsuccessfully to impose in the late eighteenth century to reduce the fire hazard, did not discourage brother rat. Curved tiles and holes in the quincha made convenient nesting and hiding spaces.

Rats had easy ingress to institutional buildings in Guayaquil, which, with the exception of one church, were also made of wood. Sailing vessels under construction in Guayaquil harbor during the colonial period also had a natural attraction for rats. Timberwork of the masts provided possibilities of secure shelter for the naturally climbing *Rattus rattus.*

Beginning in the late nineteenth century, Guayaquil's rat population received a new element that filled an ecologically open niche. *Rattus norvegicus,* which evolved in the marshy meadows of Manchuria, showed its hereditary affinity for water in Guayaquil. Streams, ditches, ponds, and pools favored their proliferation. These rats need to drink often, and Guayaquil met this requirement. Casual methods of waste disposal made their food quest effortless. As settlement expanded, collected garbage was dumped on mudflats of esteros for the next tide to carry away. The brown rat found unlimited sustenance in these garbage-strewn places.

Clay soils of Guayaquil suited the brown rat's burrowing instinct. Had the city been built on sand, this species of rat might have been much less numerous. Human contact came when the rat's low-lying burrows were inundated and it was forced toward other sites. Sewers, drains, and cesspools provided food, shelter, and dampness. If food was available and predation low, brown rats also moved into buildings. Basements were preferred, but in Guayaquil only large commercial buildings had them. Ground floors then became rat space.

Rats in an Urbanizing Environment

With precarious buildings, wandering livestock, and refuse-filled streets that alternated between being quagmires or dustbins, depending on the season, nineteenth-century Guayaquil held its reputation as "one of the most disagreeable cities in the world" (Miller 1968:91). Pineo (1990) documents more than half a century of disease and death in Guayaquil up to 1925. The rapid gain in population after 1870 and a 40 percent increase in area made Guayaquil the largest city in Ecuador by 1880. Urban infrastructure was only gradually installed. Piped water dates from 1887, and the first sewage lines were built in 1892, though the system covered only 17 streets. Not until 1926, when a foreign firm was contracted to extend it, did the sewer system serve most of the city as then defined. Beginning in 1903 and continuing for many years after that, another kind of public work project was to fill in the tidal creeks with rock, soil, and garbage. This effort created dry land, building space, and obviated the need for bridges.

Rat space expanded in tandem with settlement growth. When the cacao boom collapsed in the 1920s, ruined ex-farmers from the surrounding Guayas Basin settled in the city. Those who were able to, moved to slums in the city center; the overflow constructed split-bamboo dwellings on the swampy fringe. Rats also gravitated to manufacturing plants, especially those processing organic materials. The sugar refineries and the brewing, tanning, and flour-milling establishments, which were attracted to the city for its work force and port, all had parasitic rodent fauna. Rats also infested retail food outlets (*pulperías*) and thoroughly took advantage of the municipality's system of renting out spaces on public rights-of-way to processors of agricultural products. This practice continued well into the 1930s.

The discipline and sense of order needed to discourage commensality has always been in short supply in Guayaquil. Andean cultural tradition may have played its part. Among migrants of highland origin, the custom of keeping unpenned guinea pigs as meat animals raised on kitchen scraps increased tolerance for other kinds of rodents in dwellings. Like medieval Europe, nineteenth-century Guayaquil did not dispose of wastes properly. Raw sewage, offal, and household refuse festered in the mephitic heat of this tropical city. The citizenry had not yet accepted the germ theory of disease. Municipal garbage collection was desultory. Scavenging buzzards, dogs, and pigs were allowed to consume waste matter, or it was removed and dumped in vacant lots, roadsides, quarries, and tidal creeks or used as fill (*relleno*) in the outlying parts of the city.

Health Impact of Rats in Guayaquil

More than simply a nuisance, rats in Guayaquil were responsible for the most serious health crisis the city faced in its 450-year-old history. Yellow fever killed more people, but bubonic plague, one of the most virulent infections of people known, had the potential of total devastation. The plague in late medieval Europe provided the world with perhaps the most horrifying example of a rapidly spreading contagion. As early as 1903 officials excluded certain ships from docking on reports received that the bubonic plague had broken out in neighboring coastal Peru (Izquieta Pérez 1903). But in 1908, despite precautions, plague entered Guayaquil. Outbreaks elsewhere in South America occurred between 1899 and 1909. It was part of the third plague pandemic that spread to all six continents from its apparent center of original infection in China (Warf 1992).

That the disease vector was carried on rats had been known to science only since 1898. Most plausibly, the first infected rats reached Guayaquil on a boat loaded with flour from the Peruvian port of Paita (García Rizzo 1970a:29). Long's (1930) assertion that Oriental rat fleas (*Xenopsylla cheopis*) infected with the deadly bacillus *Yersinia pestis* were brought to Guayaquil, not on rats, but in jute bags from Calcutta, India, remains doubtful, for such a means of introduction has not been confirmed for anywhere else. Guayaquil never had the pneumonic form of bubonic plague, which can spread from person to person without rats or fleas.

In Guayaquil, the plague outbreak was closely connected to rats from the beginning. In February 1908 a large number of rats died of plague in a riverfront warehouse. From there, plague was reported in a neighborhood of chicha shops, in which many bubo-infected dead rats and guinea pigs were found. As fleas abandoned the cold corpses of dead rodents for nearby warm human hosts, the disease spread through the city. The Oriental rat flea was the only insect vector known to have spread bubonic plague in Guayaquil. Its reproduction there found favorable environmental conditions. Guayaquil has no month in which the daily mean temperature rises above 26.4°C; temperatures above 30°C shorten the life of an adult flea.

Between 1908 and 1930 more than 8,000 cases of bubonic plague were recorded in Guayaquil, and many more are said to have gone unrecorded. Though the mortality rate was 40 percent, far fewer people died of *bubónica* than died of yellow fever, which claimed 8,000 Guayaquil residents in 1842 alone. Traumatized, the city enlisted international sources of money, medicine, and technical expertise. Plague disappeared from the city in 1930, only to reappear five years later, when infected fleas and rats arrived from the Sierra on the railroad. A third, much more enzootic, phase of the disease occurred between 1939 and 1965. In that period only 212 cases and three deaths were reported in Guayas Province (Jervis 1967:422).

Guayaquil's bubonic episodes reveal several patterns. Overwhelmingly the disease was contracted within the enclosed spaces of dwellings, where humans, rats, fleas, and bacteria intermingled. Second, children disproportionately contracted the disease. Of the 1,718 plague patients analyzed in the city between 1912 and 1915, 735 were between the ages of 1 and 15. Blacks and mestizos, who in this city have always constituted the poor, contracted plague and died from it at more than twice the rate of white people (Pareja 1916).

Out of the bubonic years came a greater understanding of Guayaquil's rat composition and microdistribution. Health officials organized live-capture programs to identify species, the degree of infestation, and the kind and number of their fleas (Eskey 1930:2,088). *Rattus norvegicus* constituted 75 percent of the rats trapped, suggesting that the brown rat had mostly displaced the black rat within half a century of its arrival. That *Rattus norvegicus* was the main plague carrier in Guayaquil conflicts with Moll and O'Leary's (1945:1) assertion that only *Rattus rattus* was enough of an indoor commensal to be able to spread bubonic plague and that when *R. norvegicus* killed off most of the black rat population in the eighteenth century, the disease disappeared from Europe.

Interspecific competition explains the microbiogeography of rats in Guayaquil. The more prolific, omnivorous, and aggressive brown rat had largely taken over in Guayaquil. In encounters between the two species, a fundamental biological antagonism asserts itself, and the stronger brown rat invariably dispatches the black. The two do not interbreed.[12] However, congeners can coexist in the same building if pressure on food resource is low. Where that occurs, brown rats occupy basements and lower floors, whereas black rats find shelter in the false ceilings and attics and on roofs. Through the twentieth century the brown rat has triumphed. In a 1965–1967 study, almost 95 percent of the rats trapped were *R. norvegicus* (García Rizzo 1968). In New York, Boston, and Paris, that species displacement is complete (Manson 1996).

Sanitation and Rat Control in Guayaquil

Rat control efforts beyond the immediate household do not predate the twentieth century. In the long history of bubonic plague in the world, measures against rats to deal with that disease did not predate 1896. Guayaquil benefited from what had been learned in that past decade. Bubonic plague spawned stringent measures to try to reduce rat numbers. In 1904 health officials in Guayaquil realized their city was vulnerable to the advance of bubonic plague from Peru and tried to take some precautions to stop it. Preventive vaccination began in 1909, when between 12,000 and 15,000 people received Linfa-Haffkine vaccine in a nine-week period. This vaccine, developed in 1897 a few years after Yersin discovered the plague bacillus, proved to be largely ineffective. Vaccination resulted in side effects and offered protection for only a few months. Not until after a living antiplague vaccine was developed in 1935 was a substitute to Linfa-Haffkine available. The prophylactic effort included soothing a nervous public. The health board asked banks to disinfect bills with autoclaves, even though bubonic plague cannot be transmitted by handling paper currency.

Primary effort was directed toward rodent eradication through killing rats and denying them food and shelter (Dirección de sanidad pública, 1923). Deratization campaigns using traps and poisons were labor intensive and always allowed some wary animals to survive and breed anew. More effective was cyanide gas fumigation of ships, trains, train stations, warehouses, markets, and slaughterhouses. To deny rats food and harborage, a cleanliness crusade was also organized. The city council provided a list of sanitary regulations, many of which were directly tied to rat control. Each neighborhood was to form a brigade to tidy up public and private spaces. These teams removed garbage and other refuse, rags, papers, and useless old objects from tenements, patios, and stores. They also cleaned drains, sewers, ditches, and latrines. A municipal decree prohibited keeping pigs, goats, and chickens inside houses. The health department recommended daily cleaning of the walls and floors of churches, schools, barracks, and theaters with a solution containing mercury bichloride. Sweeping with a dry broom was prohibited.

Public persuasion was another dimension of the antiplague effort. In 1909 the municipality recommended that floors be mopped several times a day with disinfectant to keep fleas at bay. Poor households received this solution without cost (*Gazeta municipal* 10-1-1909). Posted were instructions to remove standing water in and around dwellings, for rats must drink frequently. Removal of open containers of water led to an unexpected decline in yellow fever, for they had been prime breeding places for *Aedes* mosquitoes.

Festering mounds of garbage, for so long part of the city, were removed. To eliminate the rat heaven at the municipal garbage dump, organic and inorganic refuse was burned. To this end, an incinerator was installed in 1904, although the high cost of firewood forced the city to shut it down six years later. Sheds and warehouses connected with shipping and fishing activities were torn down as a way of reducing the wharf rat population along the banks of the Guayas. Markets were cleaned and reorganized. Wooden structures around the main market were declared to be "*criaderos de ratas*" and deemed to be so infested that they were all destroyed and a new market was built on a different site. A few pieces of private property were expropriated and destroyed for the same reasons. In the most rodent-dense *barrio* of the city, Quinta Pareja, the existing buildings remained, but no new construction was allowed between 1909 and 1935.

Rat hordes in the urban fabric led to concern for long-term solutions. Guayaquil's building code was revised in 1914 to discourage murine invaders. The only wood allowed on new buildings was for doors, windows, casings, and moldings. Wattle-and-daub construction (quincha) was forbidden. It was thought that quincha not only encouraged rat nests, but also provided hiding places for tetanus spores, bedbugs, and the triatomine insect that spreads Chagas' disease. Rules concerning interior construction were designed to prevent spaces for rats. Double walls, fake ceilings, attics (*tumbados*), and spaces under the ground floor were banned. The new building ordinance also outlawed two roofing materials, thatch and curved tiles, which conceal rat nests. Flat tiles, sheet metal, and slate were acceptable.

Special ratproofing requirements were promulgated for businesses and institutions. Pasta factories, bakeries, rice hullers, and flour mills were especially targeted, for grain is normally the most desirable nutrient to both rat species. All retail entities in which food was present—restaurants, sweetshops, dairies, grocery stores, inns, hotels—were required to take measures to keep rats off the premises. Schools, convents, churches, asylums, theaters, barracks, hospitals, and clinics were required to be ratproof, using any means at their disposal.

By 1928 bubonic plague was clearly in remission, and the city lost interest in the war on rodents. But the sharp decline in the problem could not be attributed to wide compliance with antiplague decrees, ordinances, and recommendations. Construction restrictions and ratproofing guidelines were too draconian for many people. Individuals complained that cement floors caused rheumatism and that the exposed beams were unaesthetic when double walls and false ceilings were dismantled. Long (1930) believed that plague persisted in Guayaquil more than two decades longer than it should have because house construction materials and design encouraged so much rat infestation long after decrees had been issued. Why bubonic plague finally disappeared from Guayaquil is unclear. Extermination campaigns over the years killed hundreds of thousands of rodents and for a short time made dents in their population. But the flea was not eliminated, and rodent numbers surged again after rat control stopped. Human immunity to *Yersinia pestis* would not have substantially increased in only three decades of exposure. The dominance of the brown rat, which does not share habitational space with humans nearly as much as the black rat does, may explain the remission and then evanescence of bubonic plague in Guayaquil.

Historical Role of Rat-Instigated Fires

Guayaquil has had numerous engulfing fires, most of them of indeterminate origin. In the city's first 250 years, 10 conflagrations occurred (Alcedo 1967, 2:176). Those of 1692, 1707, 1764, 1896, and 1902 largely destroyed the city. Major fires in 1801, 1804, 1812, and 1917 left gaping holes in the urban fabric. Guayaquil's historical impecuniosity has been blamed on this devastating succession of fires ("Relación de Guayaquil" 1994:499). The flammability of traditional building materials explains the city's special vulnerability. Exact causes are notoriously difficult to reconstruct in most cases, but among the culprits are rats. They can start fires by knocking over lighted candles, and nowadays do so by biting through insulation around electrical wires.

Modern Guayaquil as Ratopolis

The end of the twentieth century finds Guayaquil still swarming with rats. Neither the trauma of bubonic plague nor an impressive list of ratophobic legislation has galvanized the collective will to keep the number of rats in check. Sophisticated rodent-control procedures used in industrialized countries have scarcely been implemented. This lag does not correspond with Guayaquil's ready acceptance of so many other innovations.

Rat Numbers and Density

The actual size of Guayaquil's rat population is a matter of conjecture, for no citywide rodent census has ever been attempted. Some local health officials use a rough guide of 5 rats for every person, which would amount to a population of 10 million for the whole urban area. By contrast, New York City uses a working figure of one rat per person. High levels of infestation become apparent where rats are seen during the day, a pattern that results when food competition forces some of them to forage in the open. For every rat glimpsed, 10 lurk unseen below ground or otherwise hidden in the immediate area.

Other evidence of rats are holes in the ground from 8 cm to 17 cm in diameter; no other urban animal makes that kind of opening. Inhabited burrows are located within a 40-m radius of a dependable food source. Rats do not travel long distances to forage. Inside buildings, rat infestation is betrayed by holes of similar size in the walls or floor. Characteristic scats and grease marks along walls provide additional clues to infestation. High-pitched chittering and a scraping sound emanating from inside walls announce their presence. A particular musty odor in a hallway or room can be a subtle sign of brother rat inhabiting the premises.

Overall, the city could have about 800 rats per hectare. Food availability is the key to their relative densities. Shantytowns, which have fewer than 200 persons per hectare, may contain between 1,000 and 1,500 rats per hectare. Shantytowns greatly extended after 1950. The metropolitan population multiplied almost five times, from 258,996 in 1950 to 1,119,344 in 1982. The modern bridge across the Río Guayas opened much land for new settlement, as at Duran, which became part of metropolitan Guayaquil in 1979. The settled area increased more than 12 times, much of it on land that the municipality had taken over from surrounding haciendas. In 1944, the city of Guayaquil acquired Hacienda Guasmo through eminent domain. Its mangrove-covered terrain was ill-suited for human constructions, and no city water, electricity, or sewage was provided. Nevertheless, the municipality and its leading politicians made a profitable business of renting and selling unimproved parcels beginning in earnest in the 1960s. Unlike in Quito, efforts to direct urban growth in this energetic but chaotic city through master plans have never succeeded. Free-wheeling enterprise, corrupt politics, and an individualistic survival ethos have long driven many urban decisions in Guayaquil.

To make this land habitable, organic and inorganic fill material, together with quarried rubble, was dumped on it to raise house lots and lanes above the water level of the rainy season. Individual landholders added trash and garbage beneath their dwellings, thus disposing of waste and further filling depressions. Elimination of standing pools of water in that fashion has benefited control of malaria, but rats took advantage of this practice to construct burrows and find food.

Garbage has been a major source of rat food in the shantytowns, which long have had no regular collection. Through the 1980s, only 490 metric tons of the 1,226 metric tons of garbage generated in the city were collected (Municipalidad de Guayaquil 1995:1). The unremoved excess continued to be used as fill or dumped on public spaces such as in tidal creeks or along roadsides.

So deficient was municipally run garbage removal that a crisis ensued, resulting in a total change in the service. In 1993 a private contract was negotiated with a Canadian firm that now organizes, collects, and disposes of 92 percent of the city's garbage in its own landfill. Yet garbage problems remain in the shantytowns, where collections are not daily or door-to-door (figure 8.3). Since the large trucks cannot negotiate narrow muddy lanes, bags of gar-

Figure 8.3. Accumulated garbage is prime rat food in Bastion Popular, a shantytown in the north-eastern part of metropolitan Guayaquil.

bage accumulate at central collection spots until removed. In that period dogs and rats make evening visitations to these bags. Rat dinner is also available in the garbage strewn in vacant lots, waterways, and parks and along road-sides. Weedy vegetation, which grows rankly in this climate, provides good rat cover. Most dependable from a rodent's point of view is the feed meant for chickens and pigs raised in the house patios of outlying neighborhoods. Rodents harbored there invade nearby properties that do not keep small livestock.

City-controlled markets have among the highest rat densities in Guayaquil (figure 8.4). The 52 wholesale and/or retail distribution points throughout the city sell primarily bulk foods. The two largest, the Mercado del sur near the river and the Mercado central in the core, each have more than 400 sellers inside and many others outside. Thousands of buyers go there in the morning and early afternoon for their daily provisions. Smaller markets serve neighborhoods throughout the city. Rat infestation compromises the cleanliness of all these food dispensaries. Municipal health officials rate 85 percent of these 52 markets as being in a "bad state of hygiene" (Municipalidad de Guayaquil 1994). A dozen fumigations of certain markets over the past 15 years have killed many rats. Yet, deratization campaigns of this sort do not solve the problem. Rat populations rebound within three months of the end of any extermination campaign.

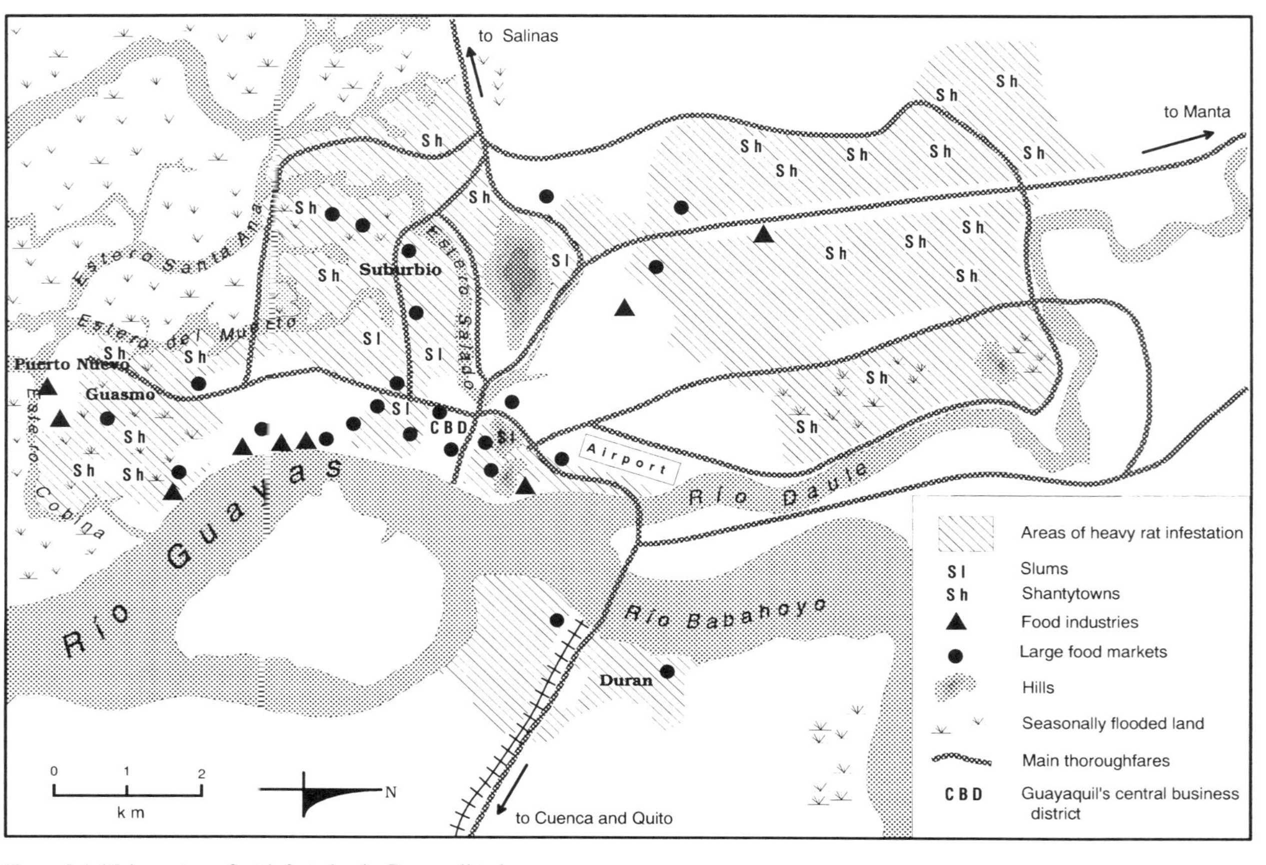

Figure 8.4. Main centers of rat infestation in Guayaquil today

In most markets, permanent stalls holding bulk foods are virtually impossible to keep free of rodents. Resident ratpacks contaminate substantially more than they eat. Neither the seller nor the ordinary buyer is scandalized when rat feces or hairs are found in the rice or nibble marks are found on the vegetables. Food carts licensed by the municipality are a mobile extension of market selling. On any one day, between 6,000 and 7,000 *ambulantes* set themselves up on the public right-of-way with produce and/or prepared foods. Rodents consume edibles that remain on the cart overnight as well as the garbage discarded from them.

Nighttime rat visits to stored provisions are a problem well beyond markets. In many households, rats regularly gain access to kitchens and pantries. Commercial food establishments offer an ideal setup for rodential larceny in the tranquil hours from late night until dawn. In food-processing firms, rat colonies can be expected if a serious control program is not in place or if the building is not ratproofed. These four-footed intruders are said to have posed problems at one time or another in almost all of the 500 declared and 300 clandestine food-processing firms in Guayaquil. About 15 large enterprises in the first category practice regular fumigations to destroy rats. However, infestation is often publicly denied, and information on the extent of this kind of contamination in Guayaquil is not released.

The historical port of Guayaquil on the banks of the Guayas near the heart of the city was well known to have some of the major concentrations of rats in the urban area. In 1973 the commercial port function was transferred to a new facility on an estero, not the main river, eight kilometers south of the center. To this facility many goods are brought in ratproofed containers, but bulk materials in warehouses are still accessible to rats, and wheat, Guayaquil's most important food import, attracts them. At the new port rats do not swarm in numbers as large as those in the historical port. At the remaining wharfs on the banks of the Guayas, rats still feed on garbage left by the boats that tie up there. Nothing in Guayaquil indicates that rat numbers have decreased since the bubonic episodes. Rather, rodent populations have grown in tandem with human increase, for each new person requires more food and generates added food wastes that sustain rats.

Rat Harborage

Rats are everywhere in Guayaquil and are most numerous where food is most plentiful. Outdoor rats are overwhelmingly *Rattus norvegicus,* which dig burrows to create a complex of tunnels with multiple exists (Lore and Flannelly 1977). *Rattus rattus* lives mainly in building interiors, where it is more likely to be heard than seen.[13] It can survive out of doors as well and nests mostly in trees, especially palms, found in parks, roadsides, and on residential property.

Indoor rats are found in all kinds of building uses. In Guayaquil, both poor and well-to-do people live mainly in single-family dwellings; only 13 percent of *guayaquileños* live in apartments. Rats occasionally invade middle- and upper-class homes, but their presence there is not normally tolerated. Persistent rat infestation in residential Guayaquil is in lower socioeconomic neighborhoods, which take in most of the city. Seventy percent of city residents are poor and must live either in inner-city slums or in shantytowns (Rojas and Villavicencio 1988). With substandard housing occupying most of the metropolitan area, it is easy for rat colonies to get established in them or nearby. The dilapidated inner barrios, such as Las Peñas, have a long history of infestation, but much more extensive and populous are the outlying *parroquias* of Febres Cordero, Ximena, and Tarqui, which collectively compose the massive shantytown of El Suburbio. Guasmo, located at the south edge of the city, is the most recently settled shanty zone. Large areas of El Suburbio and Guasmo have no piped water, flush toilets, sewer lines, or telephone service. However, electricity, garbage collection, bus service, schools, and health clinics are now generally provided. Fill dumped to create dry ground includes a mix of materials that leave passageways and living space for rats.

In the worst slums the old wooden and quincha-walled structures are in an advanced state of disrepair and are almost without exception full of rodents inside and out. In the outer neighborhoods, 80 percent of the dwellings are made of caña guadua alone or in addition to cement block, wood, or adobe. Guadua is an inexpensive and easy-to-use construction material for popular housing in Guayaquil (Parsons 1991). At the same time, its precariousness facilitates rodent infestation. In less than a hour, a rat can gnaw its way through this bamboo. When used unsplit for support poles and rafters, the hollow culms, 14–20 cm in diameter, provide passageways and hiding places for rats. When household economies permit, brick or cement block replaces caña guadua. Poured concrete floors, which inhibit, though do not prevent, the creation of ratholes inside the room, are considered an important improvement in these neighborhoods. Rats find more secure shelter in dirt patios, where they can hide amid the clutter of stored building materials and find food in animal pens.

Guayaquil's sewage disposal system, which covers only about 40 percent of the city, harbors an indeterminate rat population. Part of that system has storm sewers in addition to waste-water pipes. Prime living sites for sewer rats are the disused pipes that lead to abandoned or torn-down buildings. Even though Guayaquil has few home garbage disposal units connected to sewers, rats find edibles in the waste materials that move through the collectors and flow untreated into the Río Guayas. Sewers also offer them a means to move undetected into buildings by way of toilets and drains. The low water pressure of Guayaquil is blamed for the ability of rodents to advance from the collector pipes into domestic plumbing. However, toilet-bowl ingress, the subject of harrowing tales, is not a common occurrence. The grate in the patio connected to the city sewer provides an easier entryway.

Guayaquil has no systematic rat control program. From time to time, the city does exterminate rats in public buildings and markets when their numbers get out of hand and the political will is engaged. No municipal programs exist for the deratization of private property. Individual households in prosperous neighborhoods endeavor to eliminate rats by the use of traps and poisons or by hiring commercial exterminators. Lax municipal enforcement of building codes greatly encourages rats. Unhygienic conditions of dwellings built on land rented or sold by the municipality and without basic services promote rodent proliferation. Sanitary inspection of commercial establishments is desultory, and violators are rarely held accountable. Ironically, rodential mischief has created gaps in the documentary history of the city by shredding municipal archival documents through the centuries.

Government does not deserve all or even most of the blame. Factory and retail business owners consider eliminating rats to be not worth the expense, and only large operations have rodent control programs. In poor residential areas efforts are minimal, and rats readily move between households. People are habituated to their presence, and coexistence creates a certain level of tolerance. Many people do not react with alarm, disgust, fear, or hate in an encounter with a rat. But if the opportunity presents itself, inhabitants will normally kill a rat with the same aplomb as they show when dispatching a cockroach.

Rats, Human Health, and Property

Rats as Disease Vectors

Rats in Guayaquil have continuing if unquantifiable effects on human health and property. As host or reservoir, rats are partly responsible for spreading several diseases besides bubonic plague. Leptospirosis, an acute febrile illness caused by a pathogenic spirochete, is a serious rat-related affliction that can lead to kidney or liver failure. The *Leptospira* parasites live in several domesticated and commensal mammals, which eject them in their urine. Humans pick up the pathogen through the skin mainly from contaminated water or by ingesting contaminated food and drink. Leptospirae can survive for long periods in the myriad stagnant puddles and moist soil of Guayaquil. The connection of these parasites with rats was first appreciated when it was shown in Japan that a high proportion of the human patients suffering from this disease had been in contact with rodents (Ido et al. 1917).

In Guayaquil this malady, once called *fiebre aduanera* (customs fever), has probably occurred for a long time. Its symptoms of nausea, vomiting, and jaundice were often confused with yellow fever. After the spirochete was identified in humans, the researcher Noguchi found that a high percentage of the commensal rats he examined in Guayaquil were infected with it (Valenzuela 1932:4).[14] Between 1979 and 1994, of 3,990 presumed cases of leptospirosis reported in the province of Guayas, 86 percent were in the city of Guayaquil

(IHMT, 1979–1994). Of this total, 371 cases involved *icterohaemorrhagiae,* the serotype most characteristically carried by rats. On that basis, health officials consider rats to be the prime carriers. Research elsewhere has confirmed the role of rats in the spread of leptospirosis (Thiermann 1977). Livestock, which can be vaccinated against the disease, became less culpable than rats in its spread.

Rats are hosts to a number of other zoonotic diseases in Guayaquil. Murine typhus may have occurred in the city, but its symptoms of high fever, chills, and nausea can be confused with those of other diseases. Zinsser (1934:243–46) asserts that it was carried to Latin America from Spain, where he identified as murine the typhus epidemics that occurred in the sixteenth century. This rat-borne form of typhus has been epidemiologically confirmed in the world only since 1928, when it was found that the feces of the Oriental rat flea (*Xenopsylla cheopis*) can contain *Rickettsia typhi* bacteria. As with bubonic plague, murine typhus is transmitted when rats invade buildings and come in close human contact.

Salmonella, which causes typhoid and paratyphoid fever and gastroenteritis, is spread when people eat food contaminated with rat droppings. Enormous amounts of food are contaminated by rats in the city, but to what extent their droppings are infected with salmonella bacteria is not known. Typhoid fever was once a major killer in the city and is still fairly common.

Another zoonosis, Chagas' disease, is contracted when a triatomid bug transmits trypanosome protozoa into the human bloodstream. These blood-sucking insects also bite domiciled animals, which then become trypanosome reservoirs. Health authorities in Guayaquil believe that rats play an important role in the cycle of transmission. Various species of this bug are implicated. In Ecuador, the rat is the primary host for *Triatoma dimidiata* inside houses (Arzube-Rodríguez 1968). In Panama, where more than 50 percent of the rats examined were infected with *Trypanosoma cruzi,* that connection has been confirmed (Cedillos 1987). The cracks of walls or ceilings of shanties in Guayaquil shelter triatomid bugs, which emerge for a blood meal from any available mammal: rat, dog, or person. Chagas' disease is typically asymptomatic, which masks its high incidence, especially among the poor. Years after the initial infection, a frequent outcome is sudden death from cardiopathy.

Rats are implicated in other diseases. Trichinosis can be spread to pigs through trichina-infected rat feces. In Guayaquil, swine are no longer commonly fed municipally hauled garbage, but those raised in shantytowns do come in contact with rats. They may eat those they kill or find dead. However, in the 1950s, more than 3,000 brown rats examined for *Trichinella spiralis* were found free of it (Cárdenas V. 1957). Two other diseases, viral in nature, with proven murine associations elsewhere have not yet been reported in Guayaquil. Hanta virus, which causes hemorrhagic fever and kidney failure in humans, is not yet recorded in Ecuador. Likewise rabies, which is transmitted by rats in Asia, has not been proven to spread this way in Latin America. A disease known as rat-bite fever is caused by *Streptobacillus moniliformis* (Von Lichtenberg 1991:149). The disease has not been reported from Guayaquil, though neither has research on it been conducted yet.

Public health officials in Guayaquil agree that only about a fifth of the actual number of rat bites that occur in the city are recorded. Nevertheless in 1985, 1,622 rat-bite complaints were filed (IHMT, Sección zoonosis, 1986). In the mid-1990s, the five to six rat-bites reported each day to local health authorities suggest that the problem persists. Brown rats inflict the great majority of bites. They are more numerous and also the more aggressive of the two species toward humans. As previously mentioned, Guayaquil housing does not usually have basements, which elsewhere can serve as a buffer between people's main living space and the favorite indoor habitat of the brown rat.

When bitten by a rat, most people do not request medical attention to dress the wound but do express concern about rabies. Anecdotes of jumping rats in living quarters suggest that many of these encounters involve sleeping babies. Gruesome incidents are told of children bitten so badly that death has occurred from infection or loss of blood. If a rat-bite map of the city could be drafted, the concentrations would predictably correspond to city slums and shantytowns. The poor being the principal victims of these attacks has its parallel in North America. In Chicago, for example, rat-bite occurrences are strongly correlated with slum neighborhoods on the South Side (Dzik and Wallen 1987).

Bubonic plague has been absent from Guayaquil now for more than three decades, but it could reenter the complacent city. Gendarmes and market sellers make routine trips to the highlands and could carry the disease or an infected rat flea back to their rat-infested lodgings in the city. Threat of a bubonic return is not monitored. Health units have dismantled their plague-control procedures. Rats are no longer captured to determine if their fleas carry *Yersinia pestis.* A neighborhood rat epizootic would not necessarily come to official attention, because a new generation of inhabitants is without plague experience.

Plague has occurred in different parts of the world through the twentieth century, and the potential for a pandemic has not disappeared. More than twice as many cases were reported in the world in the 1990s than in the 1980s. In Guayaquil plague could easily enter the city as it did before on rats disembarking ships. Vessels using the port are not routinely inspected for rodents; onboard fumigations, though technically obligatory, are bypassed. Hawsers may even lack rat guards. Stocks of antiplague vaccine are unavailable in the city. Recent medical literature published around the world on plague is not locally accessible. Archival records of Guayaquil's experience of plague cases, rat control, and urban sanitation earlier in this century have been destroyed. A large rat population, fitful and desultory inspection procedures, and periodic offloading of live foreign rats at the port make Guayaquil vulnerable to a plague crisis.

Rat Damage

This frayed city has never calculated the extent of rodent-instigated property damage. Its most sensitive aspect is the contamination of food with rat hairs, urine, or feces. Evidence of rats in unprocessed food and food products implies potential dangers to human health, but also reflects negatively on government authorities who are entrusted to enforce sanitation laws. Normally the municipality neglects to deal with noncompliant food enterprises. Periodic health crises like the 1991 cholera epidemic can mobilize efforts. When normalcy returns, personnel shortages cannot sustain these labor-intensive procedures.

Rats also damage or destroy property through gnawing, a physiological necessity that occupies half their waking hours. Especially in old buildings, rats fashion holes in foundations, floors, and walls to gain entry. Their powerful incisors can cut through most materials, including lead pipes and cement. By biting through wire insulation, rats are believed to cause short circuits that result in fires. Fires are usually impossible to trace directly to rodents, but at least some of them listed as undetermined in origin could be blamed on rats.

The dread of plague has had an impact on the urban landscape of Guayaquil. The municipal government changed building codes during the bubonic outbreak. If rats had not been an issue, wooden houses and curved tiles would be more numerous today. Rats were responsible for both replacing the main market building and also displacing its site. Severe rat infestation of several other city-owned buildings led to their destruction and replacement. Most notable was the case of the municipal palace; put up in 1924 near the river, this building was constructed of stone specifically to discourage rat harborage. Rat control formed one of the many reasons for moving the city's port facilities away from the old market. Ironically, rats later gained access to parts of this structure. It took the crisis of bubonic plague to mobilize the national government to establish the Institute of Hygiene and Tropical Medicine in 1941. This institution, named in honor of Dr. Leopoldo Izquieta Pérez, the physician who first identified the presence of *bubónica* in the city, occupies a complex of buildings near the general cemetery, where so many plague victims were buried. The institute has been charged with monitoring health conditions in Ecuador, conducting research on infectious diseases, and distributing certain public health services such as vaccines.

Rat City as Metaphor

If rats were seen in the colonial period as a spontaneous and uncontrollable outgrowth of a murky environmental setting, in modern Guayaquil they are accepted as unwelcome but inevitable urban inhabitants. In four centuries of their habitation there, decisive action controlled them only during the bubonic plague outbreak. The scientific understanding that rats transmit plague emerged scarcely a decade before the disease invaded the city. No durable change of attitude or behavior came out of that zoonosis. The fundamental ecological understanding—that if people deny them food and shelter, commensal rat populations cannot be sustained—has not been generally grasped by guayaquileños. Rat proliferation in Guayaquil is the result of a combination of favorable environmental conditions, human indiscipline, and official laxity. The rodents' adaptation to this city has been made easier by the lack of serious human effort to control them.

Beyond causation is significance. Guayaquil as Rat City is a metaphor for its vocation, situation, and character. First, rats symbolize the port function and external connections that are the city's traditional *raison d'être.* Foreign demands, not local needs, prompted Guayaquil's growth and may even be responsible for its survival. Sea travel, focused directly or indirectly on Europe, brought rats as just part of a long list of foreign introductions that came to dominate this city more than any other in the country.

Second, rats reflect Guayaquil's failures as a civic entity. For a variety of reasons, the city's role in setting standards of elemental hygiene for the urban habitat has not been matched by its enforcement of those standards. By selling or leasing land and not providing basic services to masses of poor people, the municipality itself contributed to rat proliferation. At every level, the rats' ability to pullulate has exceeded the human will to control them.

Third, the poverty that has dominated Guayaquil has the rat as a constant reminder. The poor are most affected by them and least able to control them. Their material circumstances and resignation contribute much to rodent proliferation. Rats for most guayaquileños are a banal accompaniment of daily life, which pales before the struggle to make ends meet. Constraints of culture, economy, and polity make it likely that Guayaquil will continue unabated as Rat City, at least until another serious rat-borne epidemic provokes organized action.

9

Carl Sauer and the Andean Nexus in New World Crop Diversity

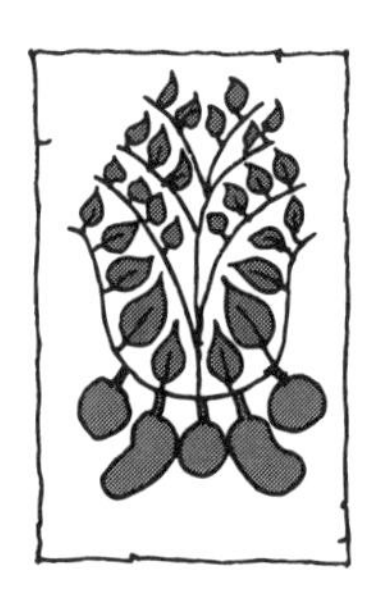

The interrelationships of plants and people were the major thrust of Carl Ortwin Sauer's long and productive life of scholarship. Within that theme his most passionate research focused on crops, especially on those of Latin America. When he began his studies, relatively little was known about most native crop species of the New World. Sauer's writings on cultivated plants of this realm demonstrated most of all his talent for geographical synthesis. He explored the broader topic of biotic domestication as a process. His studies also led him to ponder the value of the old and traditional at a time when the modernist impulse controlled the social and political elite and much of the American middle class. The dangers of moving headlong into modern technology and the biological simplification that implies are now well understood. Over the past half century general interest in and knowledge of New World crops have increased greatly, and it is instructive to probe Sauer's contributions to them. In that effort, assuming the role of a supine disciple is avoided; to Goethe, there should be no veneration without critique.

Background to Sauer's Focus

Sauer's interest in New World crops grew out of his research on aboriginal and colonial cultures, which he started after going to California in 1923. While traveling in Mexico during the 1920s, he observed peasant farming, but not until 1935 did he make crop plants the focus of his investigations. Sauer had become skilled in using the participant-observation methodology, and he profitably applied it to his study of crops as the central livelihood concern of peasant existence. He was equally intrigued with traditional rural life in general; his larger goal was to understand how the contemporary peasant can help us to understand the past. Sauer sought to add time depth to the culture-area concept, an approach he pursued even before most archaeologists did.

In examining peasant life, Sauer truly sought to understand nothing less than the origins and diffusion of agriculture. Assistance came from specialists trained in genetics, especially from Nikolai Vavilov (1887–1943), who more than anyone demonstrated the idea that modern crop production is dependent on traditional agriculture. At various times between 1921 and 1933, Vavilov traveled and collected in the Western Hemisphere. His empirical work on crop origins, first presented in English in 1926, asserted that Mexico was a major agricultural hearth. This excited Sauer, who perceived it as providing evidence of the depth and complexity of Mexico's culture history. With cores and peripheries of its crop regions identified, Vavilov's work appealed to the geographical imagination. Unable to date plant remains, he imputed antiquity from those zones having rich crop diversity. To Vavilov the story of agricultural beginnings was not that of a random spatial process; the earliest distribution of domesticated plants was tied in with that of wild relatives. Vavilov's associates, especially Bukasov (1930), published other monographs which reinforced this approach to understanding crop domestication.

Sauer's interest in Vavilov's ideas may have been clinched when the two met in Berkeley in 1930 (Vavilov 1992:209; Sauer 1952:21). That encounter seems to have stimulated Sauer henceforth to mesh his own field-based knowledge with his understanding of the conditions and environments in which people started to manipulate plants in Mesoamerica: mountainous topography, subtropical temperatures, and a wet-dry seasonal regime. Deserts were excluded; their flora had minimal domesticable potential, and the irrigation works necessary for their cultivation could have emerged only later.

Sauer understood the New World crops north of the Rio Grande to have originally come from southern Mexico. Agricultural invention in what is now the United States and Canada was perceived to be negligible. Of native North American plants, only the sunflower (*Helianthus annuus*) was domesticated. Sauer excluded the Jerusalem artichoke (*H. tuberosus*), which he saw as a weed whose ennoblement occurred after it had been taken to Europe.

Subsequent archaeological techniques have revised our knowledge of agricultural inventiveness in North America. Flotation analysis since the 1960s has revealed that a domesticated complex of plants in eastern North America predated the maize and beans that diffused eastward from the Southwest. Sumpweed (*Iva annua*), knotweed (*Polygonum erectum*), amaranth (*Amaranthus hypochondriacus*), maygrass (*Phalaris caroliniana*), and little barley (*Hordeum pusillum*) were morphologically differentiated through selection to become crops (Smith 1989). The pumpkin (*Cucurbita pepo*), long held to have been introduced from Mexico, now also appears to have been independently domesticated in eastern North America.

South American Horizons

Vavilov designated two major New World centers of crop domestication: southern Mexico–Central America and western South America. Sauer had come to know the former area personally but for the latter had only read the accounts of Andean crop diversity written by the U.S. government agronomist O. F. Cook (1916b, 1925) and the Russian geneticist Bukasov (1930). Sauer wanted to investigate this area for himself. His chance came in 1942 when the Rockefeller Foundation funded him and his son Jonathan to travel in western South America for six months. The official purpose of Sauer's trip was to identify up-and-coming scholars, several of whom could later be encouraged to apply for Rockefeller fellowships to study in the United States. Like most wartime research activities, this trip apparently had an oblique foreign policy objective.

He was able to organize his travel program and, at the same time, satisfy his own curiosity about places, people, and the plants that grew there. The trip itinerary, which stretched 5,500 km from southern Chile to Colombia, included intellectual centers (Santiago, La Paz, Lima, Quito, Medellín) but also small provincial cities, around which peasant farming could be observed on day trips. Sauer believed that educated local people in these regional centers knew more about their surroundings than inhabitants in the big cities knew about rural life outside.

His agricultural objectives sometimes took priority. The island of Chiloé had no scholars to visit, but Sauer nevertheless put this out-of-the-way place on his itinerary. Vavilov had identified Chiloé as the putative homeland of the potato, but Sauer had also learned that that cool rainy land was one of the few places left in Chile where traditional agriculture was still strong. Sauer hoped to find certain crop survivals that Latcham (1936) had recorded but did not see himself. Unsuccessful in that search, Sauer moved northward from Chiloé by land through central and northern Chile, Bolivia, and Peru. The windows of public transportation allowed for only casual observations, but once he was settled in a city, excursions into the surrounding countryside enabled Sauer to discover personally much of what he had read about and more (West 1982). In Peru, Sauer acquired an appreciation of the formidable Andean diversity of maize on a trip through the upper Vilcanota Valley with César Vargas, a young botany professor at the University of Cusco. Accustomed from his Mexico experiences to palavering with farmers in Spanish, Sauer had a strong sense of having descended into a pre-Hispanic world in southern highland Peru. Most peasants still spoke only Quechua and preserved many traits passed on from the Incas.

Sauer's program did not allow him the time to get into more remote parts of the Andes. If he had, new insights on Andean agriculture might have followed, given the regional diversity and his ability to decipher meaning from

Figure 9.1. Carl Sauer in an Indian field on the plateau between Riobamba and Guano in 1942. Dwarf maize, lupine (*Lupinus mutabilis*), and *Fourcroya* are interplanted. Coat and tie was standard field garb of both foreign and local *científicos* until the 1950s. (Photograph courtesy of Jonathan D. Sauer, reprinted with permission)

what he saw. Nor did he make systematic collections of plant material, as the Russian geneticists had. He focused less on the plants themselves than on their role as cultural tracers. Few crops posed problems of identification at the species level. For some verifications he consulted specialists.[1] He sent back seed samples of maize, beans, and cucurbits to Berkeley for study. Peasant folk in the field or market identified cultivars for him. Sauer's empiricism had a scientific foundation, but his larger intentions explain what he did and did not do in working with plants. Above all, he saw them as concrete objects, each with its own history and ecology, which bolstered his speculations about the elaboration and diffusion of culture.

Sauer observed South American crop plants in fields, gardens, markets, and kitchens (figures 9.1 and 9.2). He also pumped local people for information. From these facts he constructed generalizations. Ironically, Sauer was not particularly interested in spatial arrangements of crops. He never mapped a crop field or a garden. It was Edgar Anderson (1952:138–39), a botanist and geneticist, who first showed the spatial pattern of a Latin American garden. Mapping

Figure 9.2. Carl Sauer examining a manioc plant at Hacienda Ocucaje on the Peruvian coast near Ica in 1942. The main crop there was wine grapes, but Sauer did not give his attention to introduced species on that trip. (Photograph courtesy of Jonathan D. Sauer, reprinted with permission)

crops and varieties of those crops in their growing enclosures provides insight into diversity and how it is spatially arranged (Zimmerer 1998:451). Periodic markets in the Andes display to the public a fair share of the crop diversity of a local area as well as its main agricultural emphasis. Unless inventoried through the whole year cycle, market displays cannot, however, provide an accurate view of the range of crops grown in an area. Sauer gathered information in these markets, but even on his one-time visits he never systematically studied the crops' spatial structure. Sauer considered these crops from a culinary perspective: which plant parts were used, how they were cooked, and what their apparent nutritional value was. His reticence to characterize flavor or texture personally may have been simply out of deference to peasant perceptions, which were, in these matters, more relevant than his own.

Sauer's approach to New World agricultural diversity was essentially to report on crop species and varieties as individual organisms. Yet other aspects of agricultural diversity besides richness of species may be as important (Visvanathan 1996:317–18). The "multidimensionality" of crops growing together gives New World agricultural diversity an ecological perspective that is one of its most valuable evolutionary outcomes. Another perspective on diversity is to understand crops in more than economic terms, for to unlettered people, mythological values of biological organisms can be as important as the economic.

Sauer and the *Handbook*

Observations from Sauer's 1942 trip to South America and the wide reading he had done in preparation for it were incorporated into two long chapters he wrote for the *Handbook of South American Indians* (*HSAI*). This six-volume work was conceived and edited by Julian Steward (1902–1972) and was published between 1944 and 1950 by the Bureau of American Ethnology at the Smithsonian Institution.[2] A large number of paid experts wrote separate chapters on the prehistory, history, and contemporary aboriginal life of the continent. Originally Steward had engaged Sauer to prepare the chapter on the physical geography of South America. The editor then asked Sauer to write a chapter on subsistence types, but Sauer desisted, claiming that he knew "too little" about nonagricultural economies (NAA, COS to JS, 6-11-1943). Instead, Sauer suggested a chapter that covered cultivated plants. Initially cool to the idea, Steward relented as the *Handbook* contents jelled in his mind and Sauer's enthusiasm for and knowledge of the subject became apparent. At first the outline called for only the major crops to be considered, but Sauer chose instead to prepare a complete canvas. The ambitiousness of his intent was suggested by his comment in a letter stating that he wanted "to make a survey of a scope not done since Alphonse de Candolle in the 1880s" (NAA, COS to JS, 10-13-1943).

The finished manuscript was titled "Cultivated Plants of South and Central America," thus surpassing the bounds of the series title, which was confined to just South America. Submitted to Steward in 1943, the manuscript presented a state-of-the-art assessment of virtually the entire range of crops grown between the Rio Grande and southern Chile. To Sauer that geographical scope represented a cultural unity on which to build his case for interhemispheric diffusion. He had previously stated to Steward his objection to limiting the *Handbook* to South America. For each crop plant in the Latin American inventory, Sauer sought to puzzle out its origins, diffusion, distribution, ecology, main varieties, and local and regional names and uses. Little known, forgotten, or extinct domesticates at the time got more than passing mention. His own first-hand information was bolstered by a careful sifting of the scattered literature found in the Berkeley libraries and beyond. He enlisted taxonomists to review certain plant groups. Some months before submission, botanist Edgar Anderson critiqued Sauer's sections on cotton, beans, and maize and excised many small and some major errors from the manuscript (SP, EA to COS, 9-1-1943). Three other scientists did the same for other crops.

Sauer's manuscript was originally scheduled for publication in 1945, but the aftermath of the war and high printing costs delayed the appearance of volume 6 until 1950. In those intervening seven years between submission and appearance of the printed text, several important research studies were published that should have changed some of Sauer's assertions. The Bat Cave maize finds hit the archaeological world like a bombshell. If maize was 6,600 years old in the American Southwest, how much older was it likely to be in southern Mexico? On the coast of Peru, Junius Bird (1948) had excavated sites in the Chicama Valley, which provided the earliest evidence then known of agriculture in western South America. Hugh Cutler (1946) published important work on the classification of maize. La Barre (1947) published his emic study of potato taxonomy among the Aymara. Salaman (1949) produced a major monograph on the potato that delved into its pre-Columbian presence in the New World. Hodge (1946) also scooped Sauer on the minor Andean tuber crops by providing the first collective treatment of these plants in English. Less accessible but relevant were Cárdenas' (1948–1950) studies on Andean crops in Bolivia, published in Cochabamba.

None of the above findings got into Sauer's plant chapter. Only the *Gossypium* story was apparently updated and expanded to 2,400 words to incorporate the findings of Hutchinson, Silow, and Stephens (1947) that the New World tetraploid cotton possessed two genomes from a diploid Old World cultivated species as well as two other genomes from a diploid New World wild cotton with which it hybridized. How could Old World cottons have reached the Americas to have effected that cross? To Sauer it gave yet more evidence of a pre-Columbian movement of people across the Pacific.

Sauer developed his ideas about diffusion at a time when archaeology was known as culture history. Diffusion was used to explain historical continuity and change. In the 1960s a paradigm shift in archaeology replaced this approach with a functionalist perspective in which cultures living today are viewed as proxies of the past (Lyman, O'Brien, and Dunnell 1997).

Outcomes of Reflection on Diversity

Sauer's sustained concentration on New World crops led him to several insights. One was the cultural-historical significance of using vegetative parts and seeds as the two modes of planting. Ancient agriculture in South America was seen as being founded on vegetative reproduction ("vegeculture"), whereas that of Mexico–northern Central America strongly emphasized seed farming. To Sauer vegeculture seemed to be the more elemental procedure of planting and therefore of greater antiquity.[3] This idea got a boost when Rouse and Cruxent (1963:5–6, 53–54) found in Venezuela fragments of pottery griddles dated 3000 B.C. They were stratigraphically associated with manioc below the grindstones used for grinding maize. Sauer also thought that root crops originated from wild plants where a wet-dry regime favored the formation of underground food-storage organs. The Caribbean lowlands of Colombia experienced that seasonality and were well-positioned for diffusion northward into Mesoamerica. Archaeology has since largely confirmed Sauer's hunch that root crops came later to Mesoamerica.

As humans moved into the Andean highlands, they encountered an array of different plant families offering raw material for domestication. Lowland-growing manioc (Euphorbiaceae) and *uncucha/yautia* (Marantaceae) were replaced by ajipa (Leguminosae), achira (Cannaceae), *arracacha* (Umbelliferae), and yacon (Compositae) at higher elevations. The potato (Solanaceae) plausibly began in the temperate zone and evolved into different species and varieties at higher altitudes. Oca (Oxalidaceae), *mashua*/añu (Tropaeolaceae), ullucu (Basellaceae), and maca (Cruciferae) shared high Andean field space.

Crops as Relicts

In his *Handbook* plant chapter Sauer sought to establish a rough chronology of the domestication of specific crops in the absence of a sequential archaeological record. He reasoned that cultivated plants with apparent localized distributions or those presumed to be extinct were old domesticates that were subsequently replaced by plants providing better foods. Plants superior in yield, nutrients, environmental adaptability, or positive culinary qualities would displace those inferior in one or several of these factors.

That intuitive logic led him to propose that oca, añu/mashua, ullucu, and maca were crops before the potato was domesticated. Sauer (1950:519) considered them to be "remnants of the oldest Highland agriculture" with food value and productivity lower than the potato. He reasoned that one would not bother to improve a plant that was less productive than one already grown. The state of archaeological knowledge does not permit affirmation of this postulate, but one can wonder if that reasoning would necessarily explain agronomic patterns of the past. Even today, Andean farming places importance on subsistence security, and this largely accounts for the persistence of the minor crops.

Central Andean highland agriculture is now believed to have become established between 4,000 and 5,000 years ago. The temporal sequence of its individual components, however, is not yet clear, and some crops may be much older than that. New techniques have dated the presence of domesticated potato by 6000 B.C. (Browman 1986:149); previously the earliest evidence was dated 2000–1200 B.C. (Ugent, Pozorski, and Pozorski 1982). Ullucu was dated 2300 B.C., and oca 2000 B.C. Maca is found to have been on the coast at least by 1600 B.C. (Pearsall 1992:192). Yet añu has curiously not been identified among any archaeological plant remains. Not enough is known at this point about the prehistory of these high-altitude tubers to arrange them chronologically in Andean agriculture. Even the presumed decline of the lesser tubers within historic times is based on conjecture.

Although he did not see it under cultivation, Sauer placed maca ecologically apart from the other three root crops as one whose unusual resistance to cold enables it to grow where others cannot. Subsequent ethnobotanical research supports that hypothesis (Leon 1964). Maca had fit into a pre-Hispanic agricultural system above 4,000-m elevation; relict field patterns are still visible in that zone (Bonnier 1986). All evidence for contemporary maca cultivation has come from the central highlands of Peru, though the plant has also historically been reported from Bolivia (Vellard 1963:83). Contraction to its Junin core area probably occurred in the early colonial period. People who know this crop appreciate its spicy flavor, but by now it would probably have died out if its medicinal applications had not superseded its use as food (figure 9.3A). Maca root has gained some notoriety as a fertility agent, and there is a market for it outside the region and now even in Japan.

In a later article Sauer (1959) invoked a connection between age and area to explain how other crops fit into an early pattern of domestication. Yautia/uncucha (*Xanthosoma saggitifolium*), achira (*Canna edulis*), ajipa (*Pachyrhizus ahipa*), arracacha (*Arracacia xanthorrhiza*), and yacon (*Polymnia sonchifolia*) were thought to retreat with the spread of efficient competitors. As Sauer conceptualized it, potato usurped them at the temperate limits of their vertical distribution, and manioc overwhelmed those old plants at the macrothermal end of the climatic spectrum. Archaeological evidence is still too fragmentary to validate this displacement hypothesis. Achira, cultivated as early as 3000 B.C.,

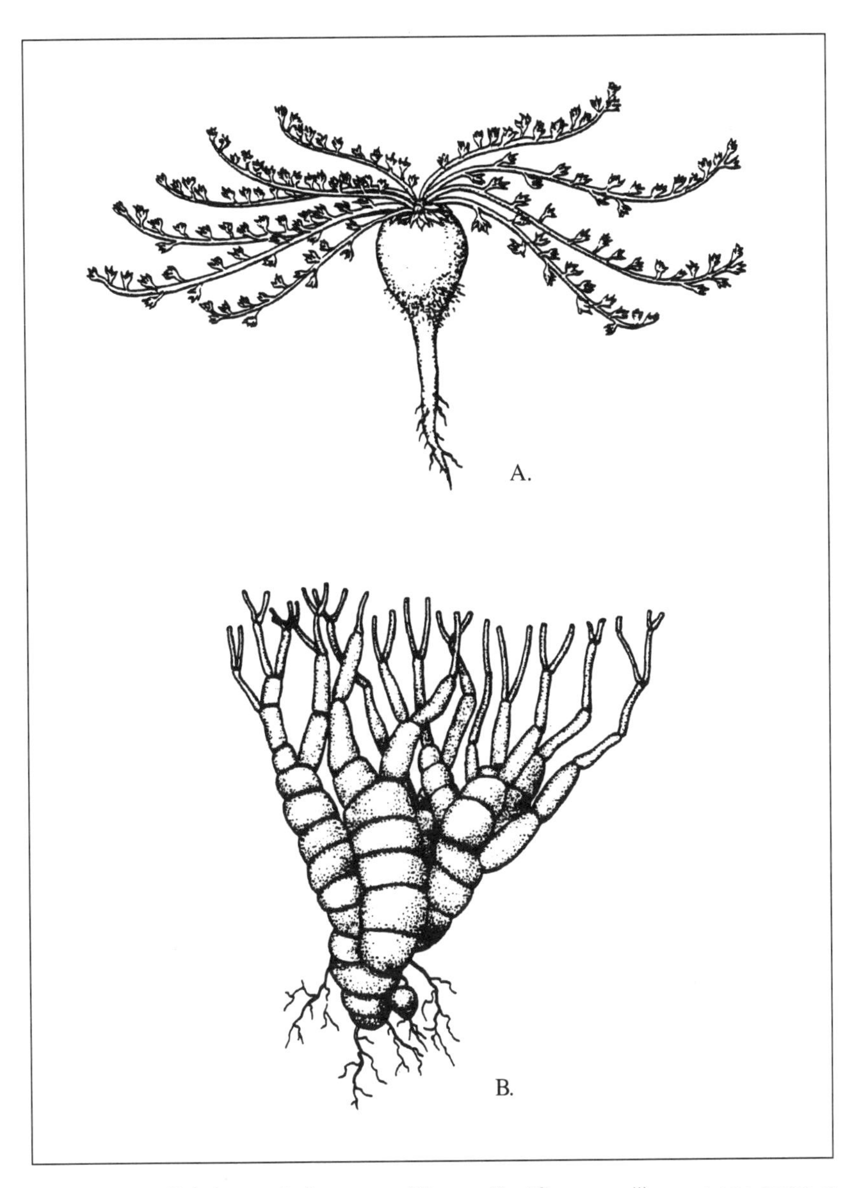

Figure 9.3. Two little-known Andean crops: (A) *maca* (*Lepidium meyenii*), a root crop grown at very high elevations in the central Peruvian highlands; and (B) *mauka* (*Mirabilis expansa*), which was first reported to science in 1968. (Source: Hernández Bermejo and Leon 1992, reprinted with permission of the Food and Agriculture Organization of the United Nations)

and ajipa, which goes back to ca. 2500 B.C., are ancient crops on the irrigated coast. Yacon appears around 1200 B.C., but neither arracacha nor yautia has as yet been dated from any site. Uncertainty of time of domestication is compounded by the lack of early highland dates for any of these above-mentioned crops.

One component in the inventory of Andean domesticated plants is *Mirabilis expansa,* whose place in the time-and-distribution idea is totally unknown (figure 9.3B). First found growing in Bolivia in the 1960s (Rea and Leon 1965), this root crop was subsequently found in other Andean valleys (Seminario Cunya 1993). *Mauka* is one of nine names now associated with it (Hernández Bermejo and Leon 1992:170). In cultivation, this plant of temperate climate is mixed with other crops such as maize in small plots. Roots, subterranean stems, and occasionally the leaves are boiled and used in stews. Its temporal place among Andean domesticates is unknown. Archaeological evidence of mauka is lacking, although name diversity and latitudinal spread suggest a crop of some antiquity. Except for Velasco (1977, 1:155), colonial descriptions of Andean farming do not mention it. That mauka escaped general notice for so long suggests that the Andes may hold other agronomic secrets of pre-Columbian origin.

Seed Plants and Chronology

Sauer's hypothetical crop sequence also included native seed plants. He viewed amaranth and quinoa as representing an indigenous "substrate" of cultivation before maize. He called amaranth an "ancient, possibly very ancient, culture in Mexico and the Andes" (1950:497). On this point, Sauer may have been influenced by Safford (1917), who thought amaranth had been cultivated before maize. In Peru prehistoric evidence of cultivated amaranth is sparse, and it is not known if the amaranth found near Lake Titicaca was in fact a cultivated species. In Mexico domesticated amaranth (*huautli*) is old (6500 B.C.), but still not older than maize there. Wild amaranth seeds have been retrieved in Mexico from as far back as 10,000 B.C. (McClung de Tapia 1992:155).[4] In South America, quinoa, the main cultivated chenopodium species, has been dated from at least 2200 B.C. (Browman 1986:143).

New World leguminous species are less temporally staggered than they are ecologically differentiated. Lima bean prefers wet hot zones; peanut, hot and dry areas; common bean, temperate valleys; and lupine, the cold country. *Canavalia ensiformis,* once grown at low elevations, had poor nutritional quality and high toxicity unless processed. Domesticated *Canavalia* has been dated as occurring before 3000 B.C. in coastal Ecuador, and over the following two millennia was found frequently as charred remains in sites farther south in coastal Peru. It has not been found in the archaeological record of Mesoamerica. As for tarwi, or cultivated lupine, almost nothing is known. Its innate toxicity suggests a pre-*Phaseolus* domestication, but its late appearance in archaeological sites suggests otherwise. The oldest legume in either Mexico or the Andes is the common bean, which appeared ca. 5000 B.C.

A localized crop distribution may denote antiquity, but it may just as well suggest a recent domestication. Sauer thought that disappearing crops were necessarily ancient if they had a nutritional role of long standing. Three native and now ostensibly extinct crops of Chile can be interpreted as either relicts or neocrops. Largely through the comments of Latcham (1936), Sauer learned about *teca,* a plant without even a binomial but reported in historical accounts. Teca was reported to have been toasted, pulverized and mixed with water, and drunk. *Mango* (*Bromus mango*), identified as an edible seed of a grass in the early part of the nineteenth century, was used in the way described for teca. The third crop was *madi* (*Madia sativa*), grown for its oily seeds. Sauer's (1950:495) initial inference was that these crops formed an older substrate. He reasoned that when maize and quinoa were introduced from the north, teca, madi, and mango were unable to compete with them and faded away at some point before the Spanish conquest.

Sauer (1959) later reflected on his early assumption of antiquity for these three Chilean plants and arrived at a diametrically opposite view. Rather than relicts from a pre-maize period, *Bromus* and madi (teca is not mentioned) perhaps were weeds in maize fields and subsequently were selected to become crops themselves. Sauer reasoned that these plants could provide crop insurance in the event that cold or drought destroyed the growing maize. The absence of any archaeological evidence suggests a lack of antiquity, though such an assumption may not be justified.

However interpreted, an age-and-area approach to New World agriculture remains little more than an intriguing idea. Sauer constructed a chronological sequence partly by making idealist assumptions about productive efficiency. However, agriculturalists also may have sought to increase diversity for reasons that have nothing to do with utility. Eventually usefulness became recognized, but even then subsistence security, not productivity, could have been the prime motive for coaxing wild plants into crops. That could explain why inefficient cultigens continue to be grown in spite of their low yields and the time required to process them.

Sauer's Chapter Content

A view from hindsight of Sauer's published work on New World crop plants reveals its shortcomings. His failure to incorporate the scholarship that appeared in the period between the submission of his manuscript and its appearance has already been discussed. He embraced Vavilov's proposition of centers of origin, which is based upon a contestable foundation. The assumption that the place which has diversity of landraces and wild relatives is also the place where the domestication of those derivative crops occurred implies a causal relationship that is not necessarily true. The tomato and cacao first appeared under cultivation in Mesoamerica rather than in South America, where the wild relatives are found. Although Vavilov later amended his statement by referring to "secondary centers," the geographical interlacing of diversity and origin was implicit in Sauer's remarks.

Vavilov and Sauer both worked under the assumption that the rich species and cultivar diversity of the Andes was implicitly responsible for the early beginning of agriculture there. However, though maize diversity in the Andes is as great as it is in Mexico, it is in the latter location that one must look for maize domestication. That a crop's subsequent development of cultivation may influence genetic variations was not taken into account in Vavilov's scheme. Hearths are not necessarily centers of diversity; Harris (1990:15) has earlier made the point that the two must be decoupled as ideas.

Undermining the whole idea of Vavilovian centers is the question of whether such centers are really definable entities. Throughout much of his research career, Harlan (1986) has disputed the proposition that the world's domestications can be spatially clustered into a handful of discrete zones. He pointed out that many crops evolved from weeds, which by definition are "noncentric." Brücher (1969) was another field scientist who disputed the idea of agricultural centers of origin. Near the end of a long career of studying crop plants, Jonathan Sauer (1993) also rejected centers of domestication. Until one knows with some precision where specific plants were actually domesticated, the subject remains open. Yet the center-of-origin concept refuses to disappear, perhaps because of its heuristic and pedagogical values in making sense of the world crop diversity. The identification of zones, however imprecise, helps the human mind to impose order on the welter of facts and to make geographical sense of diversity.

For those who accept them, the definition of centers has not been static. The identification of the Andes as a major world center of crop diversity and plant domestication has gone through various spatial permutations. Vavilov's final map, published in 1940, showed the Andean center to be composed of three discrete subregions: the Colombian highlands; the Central Andes (Ecuador-Peru-Bolivia); and Chiloé and the adjacent mainland of southern Chile. Sauer's (1952) map showed the Andean center as extending from the Cordillera Oriental of Colombia to the highlands of central Bolivia. Subsequently, others defined an Andean center differently. Smith (1995:12–13) placed his center in the "south central Andes," which included only southern Peru and northern Bolivia. He discussed only quinoa and potato, thus ignoring the vertical richness of Andean domestications. At the other extreme is the map of Zhukovsky, who had worked with Vavilov. His center included all of western South America as far south as Chiloé and extended across the Andes to the La Plata estuary (Leppik 1969:130).

Individual Crop Origins

Coconut. Sauer had an accurate take on most crop beginnings, but his judgment was far from infallible. He especially followed the wrong path on the coconut. Sauer leaned toward tropical America as being the place of *Cocos* domestication, even when culture history militated against that view. He seemed to have been persuaded by the "strong arguments" of its propounder, O. F. Cook (1910), and thus turned his back on the findings of Henry Bruman (1944), his own student who had made an intensive study of coconut history in the New World.[5] Bruman, who had read the *HSAI* manuscript after it had been sent to the Bureau of American Ethnology, wrote to Sauer in 1944:

> . . . I am frankly astonished that you still accept O. F. Cook's idea that the coconut is native to this hemisphere and that it was fashioned into a cultigen in the New World. . . . My position is, as it was when I submitted my thesis, that the coconut was native to Indonesia, that it had been brought by ocean currents to the west coast of Central America shortly before the Conquest, and that it was independently introduced by the Spaniards into Puerto Rico and by the Portuguese into Brazil early in the colonial period. (SP, HB to COS, 5-6-1944)

Throughout his life Bruman (1996), who had the support of the eminent botanist Liberty Hyde Bailey, never wavered from his point of view. No one today argues for anything but an Indo-Pacific origin of this pantropical plant (Maloney 1993; Sauer 1993:186–89).

Maize. In the 1940s the origin of maize was a tough call. The debate concerned whether maize had evolved from a wild corn of South America or from the wild grass called *teosinte,* known from Mexico. Sauer was torn between accepting an authoritative opinion that prevailed at the time and accepting his own intuition based on years of studying Mexico. Already before World War II, Paul Mangelsdorf asserted that maize evolved from a still-unidentified pod corn, of which each kernel is covered with an individual sheath. His experimental work in Texas that purported to show that teosinte is a hybrid between maize and the grass *Tripsacum* was seen as providing proof that one could not look to teosinte as the ancestor but as an outcome of crossing two different grass species (Mangelsdorf and Reeves 1939). Intrigued with this evidence and Mangelsdorf's assurance of a primitive pod corn as the ancestor, Sauer on his South American trip had hoped to find it growing there, wild or cultivated (SP, COS to EA, 8-31-1942). Mangelsdorf believed that pod corn would be found in the Paraná-Paraguay Basin, whereas Sauer figured that Colombia would be

Figure 9.4. Flour maize (*maiz blanco,* or *parracay sara* in Quechua), the highest evolutionary achievement in corn agriculture in the Andes. Hacienda Huayoccari near Huayllabamba (Cusco), Peru, had a large commercial production of this maize with the world's largest kernels.

the most plausible place, for he conceptualized northwestern South America as being an environmentally diverse pivot of American agricultural origins, out of which diffusion occurred to the north, south, and east. But Sauer found no primitive maize of that sort in Colombia, nor has it been found anywhere in South America. In fact, the Andes have some of the most highly evolved maize in the New World (figure 9.4).

After return from his South American trip in 1942, Sauer wrote to Edgar Anderson:

> Mangelsdorf seems to consider that the South American origin of maize can be accepted as a fact and to include a late origin of the dents. I don't think we are likely to know that much for a long time. The pre-Inca Paracas is a good dent. . . . Moreover I am impressed with the fact that maize in so many parts of South America is a vegetable rather than a grain. Its domestic uses are much more fundamental and more varied in Middle America than in South America, and its association with religious rites much greater. I don't think we are ready to fit the pieces together as yet, and I fear commitment as to origins. . . . (SP, COS to EA, 10-8-1942)[6]

In spite of these reservations and his better instincts, Sauer accepted, in his *HSAI* chapter, Mangelsdorf's opinion that corn was a South American domesticate that later spread to Mesoamerica. By the time volume 6 of the *Handbook* had been published, the Mangelsdorf hypothesis had lost its luster and was being displaced by Beadle's (1939) concurrent and opposed idea that teosinte is not of hybrid origin. By siding with Mangelsdorf, Sauer was left newly in print promoting a dubious hypothesis. In his later publications Sauer moved toward a Mesoamerican origin of maize, but favored Mangelsdorf's idea that it had come from wild maize.[7] In an attempt to reinforce his claim, Mangelsdorf asserted that he had found wild maize among the lower levels of MacNeish's cob retrieval from the Tehuacan caves (Mangelsdorf, MacNeish, and Galinat 1964). These earliest archaeological specimens of maize are, however, no longer thought to be wild. Mangelsdorf (1974) persisted in his belief in a wild maize until his death in 1989, long after geneticists showed that teosinte is not of hybrid origin and more logically the ancestor.

Molecular evidence has established that an annual form of teosinte (*Zea mays parviglumis*) is, in fact, the ancestor of maize. With that determined, the place of origin of maize can now be argued with greater precision. The zone where the ancestor can still be found as a wild plant lies in the northern Río Balsas Basin between 600 m and 1,600 m on soils derived from limestone (Iltis 1987:213).

Potato. Sauer (1950:513–17) took issue in the *Handbook* with Vavilov about the origin of the potato. In 1926 Vavilov had declared Chiloé to be the cradle of this crop, in spite of never having made his own *Solanum* collection on the island. Previously, Candolle (1967:50–53) had identified Chile as the homeland of the potato on the basis of several reports of wild tuber-bearing solanums there. Sauer's opinion came only after J. G. Hawkes (1944), a *Solanum* specialist, had formulated his disagreement with Vavilov about Chiloé. Brücher (1960, 1979) provided a sounder basis for rejection, since his own explorations of the island turned up no wild or weedy ancestors from which potatoes could have been derived.[8]

Phaseolus *beans.* The story of *Phaseolus* beans is more complicated. For the common bean, *P. vulgaris,* Sauer (1950:503) followed Vavilov's assertion that it originated in Mesoamerica. Vavilov had arrived at that determination on the basis of the great variety of primitive characteristics exhibited by beans growing there (Sauer 1950:503). Gentry's (1969) collection of wild beans from 25 locations in Mexico and Central America bolstered that view. However, the findings of Brücher (1969) and Kaplan (1981) point to the domestication of this plant also in South America, where it appears in the archaeological record even earlier than it does in Mexico. Gepts (1990) has shown that the wild and cultivated populations in Mesoamerica are biochemically different from the wild and cultivated beans in the Andes. That evidence suggests two separate cradles of domestication rather than diffusion from one hearth to the other. As for lima beans (*P. lunatus*), Sauer (1950:501–2) followed the leading expert of the time in accepting a primary center in Guatemala. Subsequently, archaeologic material dated 2500 B.C. was found in coastal Peru. Its domestication in Ecuador–northern Peru yielded a large-seeded bean that contrasts with a smaller-seeded legume from Mesoamerica (Gutierrez Salgado, Gepts, and Debouck 1995).

Other crops. Subsequent investigations and discoveries have caused scientists to reject ideas about geographical origins of three other crops. Using native names and historical references, Sauer took Central America to be the homeland of the papaya (*Carica papaya*), whereas Anderson (1952:181) put it in the Amazon Basin. Prance (1984), studying the distribution of wild *Carica* species, which have the capacity to hybridize, showed that papaya domestication occurred in the northern Andes. Sauer also favored a Mesoamerican origin for the Malabar gourd (*Cucurbita ficifolia*), which had a wide distribution from Mexico to the Andes. In this he seems to have followed Vavilov (1992:347). Without discovering any wild plant from which it might be derived, Sauer fell back on a linguistic explanation. Its Peruvian name of *lacayote* suggested derivation from the Nahuatl name of *chilacayote,* which to Sauer hinted at a spread southward of both the plant and its Mesoamerican name. But using philology to suggest origins was flimsy compared with using information based on archaeological data. Evidence available already before 1950 points to a diffusion of *C. ficifolia* out of South America, where it is identified ca. 3000 B.C.

Time and Space

Sauer's presentation of crop origins and diffusions suffered from the lack of a prehistoric chronology. Archaeological plant materials had been abundantly recovered from the arid coast of Peru and modeled in pre-Inca ceramics, but no real dates could be attached to them. The technique of radiocarbon dating appeared a few years after he submitted his manuscript. In 1948, when Junius Bird was able to date the evidence for agriculture in the Chicama Valley of north Peru as being 5,000 years old, he revolutionized thinking about the antiquity of New World cultivated plants and made them much older than anyone had imagined. Bird's findings implied that irrigated desert valleys were the zones of the earliest plant domestication. That assertion undermined Sauer's hypothesis that the earliest evidence of New World agriculture would be found in the wet or dry tropics. Tello's (1930) idea that the hot lands east of the mountains had played an important part in the culture history of the Andes stimulated Sauer in his elaboration of a putative early farming in the jungle. Although Julian Steward was much opposed to this notion (NAA, JS to COS, 9-2-1943), a circle of archaeologists centered around Donald Lathrap were stimulated by Sauer's claim about the antiquity of jungle cultivation (Oliver 1992:286, 326).

The pre-European intercontinental movement of crops was an idea that may have been suggested to Sauer by Vavilov, who was puzzled by the early presence of *Lagenaria* gourd in the New World. Sauer's cogitation about crop distributions led him to list other possible pre-Columbian diffusions: sweet potato, amaranth, *Canavalia* bean, coconut, jicama (*Pachyrhizus erosus*) cotton, and plantain. Although the editor of the *Handbook* regarded these assertions as tenuous, he was not at the time hostile to diffusionism and allowed them to remain in the published text. The diffusionist angle clearly excited Sauer, for he wrote one of his Ph.D. students: "The fun of doing detective work is one of the main reasons why there are scholars" (SP, COS to JES, 11-4-1950). But another reason was his sense of intellectual contribution. Whereas he was forced to concede to botanists the place of specific crop origins and to archaeologists the time they were probably domesticated, the case for early crop movements offered him more originality.[9] However, the quality of the evidence for early diffusion often did not persuade scientists who sought strong empirical foundations on which to base their judgments. The fascinating possibility of early transfer stimulated further research, but the nature of the resulting data did not lend itself to clinching arguments. Pickersgill and Bunting (1969) assessed the status of knowledge about prehistoric trans-Pacific carriage of plants and declared the evidence for early west-east contacts to be still speculative.

Diffusion by human agency is an idea that has not stood up well over the past half century. Harlan (1986) argued that diffusion has been overrated as a process to explain world agriculture. He insists that most crops have been domesticated more than once. Sauer's notion of a single point of origin in time and space for New World plant domestication assumes that innovation is extremely rare. Where diffusion is indicated, as in cotton and cacao, birds and ocean currents are seen as much more plausible agents of transfer. J. Sauer (1950, 1967, 1993:9–14) repudiated his own earlier ideas about the pre-Columbian spread of cultivated amaranth between the Old and New Worlds in favor of prehistoric domestications in both Asia and the Americas. With regard to *Lagenaria* gourd, independent domestications in both the Western and Eastern Hemispheres seem most plausible on the basis of an ancient pantropical distribution.

J. Sauer (1964) also disproved, on the basis of an exhaustive revision of the genus *Canavalia,* his father's suppositions that humans had probably carried that legume across oceans. Using borrowed museum specimens from around the world he organized the 51 species of *Canavalia* into an evolutionary framework and on that basis reconstructed their histories and distributions. *C. maritima* was the seafaring progenitor that volunteered on ocean beaches all over the world; from it other *Canavalias* evolved, including weedy *C. brasiliensis* (progenitor of the New World domesticate *C. ensiformis*), *C. piperi* (progenitor of the other New World domesticate *C. plagiosperma*), *C. gladiolata* (ancestor of the crop plant *C. gladiata* in Southeast Asia), and *C. virosa* in Africa, which gave rise to the crop *C. regalis. Canavalia maritima* and all cultivated species have the same chromosome number of 2n = 22. Thus no human involvement in transoceanic transfer need be invoked. J. Sauer's work on *Canavalia* revealed the importance of taxonomic studies in crop origins and also demonstrated that essentialist perspectives in which plants form bounded entities cannot be sustained.

Distributions

In the 1940s knowledge of crop distributions in Latin America was in its infancy. Half a century later they still are not well known, for agricultural statistics do not yield even a reasonably accurate assessment of subsistence farm production. By Sauer's own admission, his data base was too incomplete to make useful small-scale maps of any crops,[10] and he resorted to verbal generalizations. Several of Sauer's distributions were off-target. He misconstrued the range of chayote (*Sechium edule*). Sauer had a Central American bias for this cucurbit, but, in fact, it was also important in Brazil.

Coca was grown much more widely than Sauer apparently realized; he wrongly confined it to the yungas valleys of the eastern Andean front. This shrub, with its alkaloid-rich leaf, grew also under irrigation in desert valleys of the west coast as far south as northern Chile (Ríos Bordones and Pizarro Pizarro 1988–1989). Sauer's assertion that coca leaf found in pre-Columbian graves on the coast came from a trans-Andean trade did not take into account the presence of these coastal coca plantations (Rostworowski de Diez Canseco 1973; Netherly 1988). Apparently also unknown to Sauer was that tribal peoples in the lowland tropical forest as far east as the Brazilian Amazon grew coca. Northward, aboriginals still raise coca in the Sierra Nevada de Santa Marta.

Although Sauer's chapter revealed a wide range of agricultural crop diversity, it soon became an unsatisfactory source of information. New data rapidly accumulated as the collection of more plant material led to new taxonomic distinctions. For example, the domesticated capsicum peppers, with their bewildering range of fruit colors, sizes, shapes, and pungency, were divided into five species, not the three that Sauer knew about at the time (Andrews 1995). The archaeological record about prehistoric crops has grown enormously in quantity and relative precision. An example of that is what we now know about cherimoya. Sauer determined this New World fruit to have been pre-Inca on the Peruvian coast on the basis of Safford's erroneous identification of seeds recovered from graves. Pozorski and Pozorski (1997) reevaluated that seed material and other evidence for this plant and have made a persuasive case that cherimoya, in fact, was not introduced to the coast of Peru until the seventeenth century.

Omissions

Sauer intended to make a complete survey, but he fell short of that. A number of crops got no mention, among them nuñas, a distinct group of *Phaseolus* cultivars grown in the Andes between Cajamarca (Peru) and Chuquisaca (Bolivia) (Zimmerer 1992). When heated, these beans pop, a characteristic that may have been selected for fuel-poor high altitudes (Tohme et al. 1995:92). Another plant not mentioned was the Chilean strawberry (*Fragaria chiloensis*), which has been the *frutilla* historically cultivated in the Andes. Originally propagated by Araucanian Indians in Chile, the Spaniards carried it northward into the Central Andes (Poponoe 1921). It is now better known as a germplasm contributor to the cultivated octoploid strawberry (*Fragaria* × *ananassa*) of world importance.

All blossoming New World ornamentals, of which more than three dozen have long been under cultivation in Latin America (Parsons 1992), were ignored in the *Handbook* chapter. Sauer was silent on cacti, even on the prickly pear (*Opuntia ficus-indica*), whose fleshy pads are also a host of the cochineal insect, and the Mesoamerican *xoconochtli* (*Stenocereus stellatus*) (Casas et al. 1997). Also unmentioned was *Leucaena,* a leguminous tree genus which had undergone selection in Mesoamerican culture history (Zárate 1997).

Sauer's discussion left out more than 75 tropical perennials. At the time he was writing, so little was known about the perennials *guaraná* (*Paullinia cupana*), *feijoa* (*Feijoa sellowiana*), and *cupuaçu* (*Theobroma grandiflorum*) that one had to go to the Amazon Basin to see them. Several perennial fruit plants were mentioned because Sauer knew them from Central America. Johannessen (1966) and Balick (1985) have described how the peach palm (*Bactris gaspiaes*) has undergone aboriginal selection for its clustered fruits.

Organization

Any discussion of a wide range of plants must somehow be made intelligible by a system that lays out the connections among them. Sauer's organization followed several schemes. Some plants were grouped according to parts used. Seed versus root was a fundamental distinction to Sauer, who was much concerned with mode of reproduction and its cultural-historical significance. Other crops were categorized by taxonomic affinity: "garden plants of the nightshade family." Still others were placed into an environmentally defined category. Sauer defended the inconsistency of his approach by evoking the targeted audience; he called his chapter a "primer" for anthropologists (NAA, COS to JS, 10-13-1943). However, such an organization results in many overlaps, because in Latin American peasant economies the same plant often has several different uses. Quinoa, for example, was listed under the rubric of "lesser seed crop," yet in some places quinoa is valued just as much for the

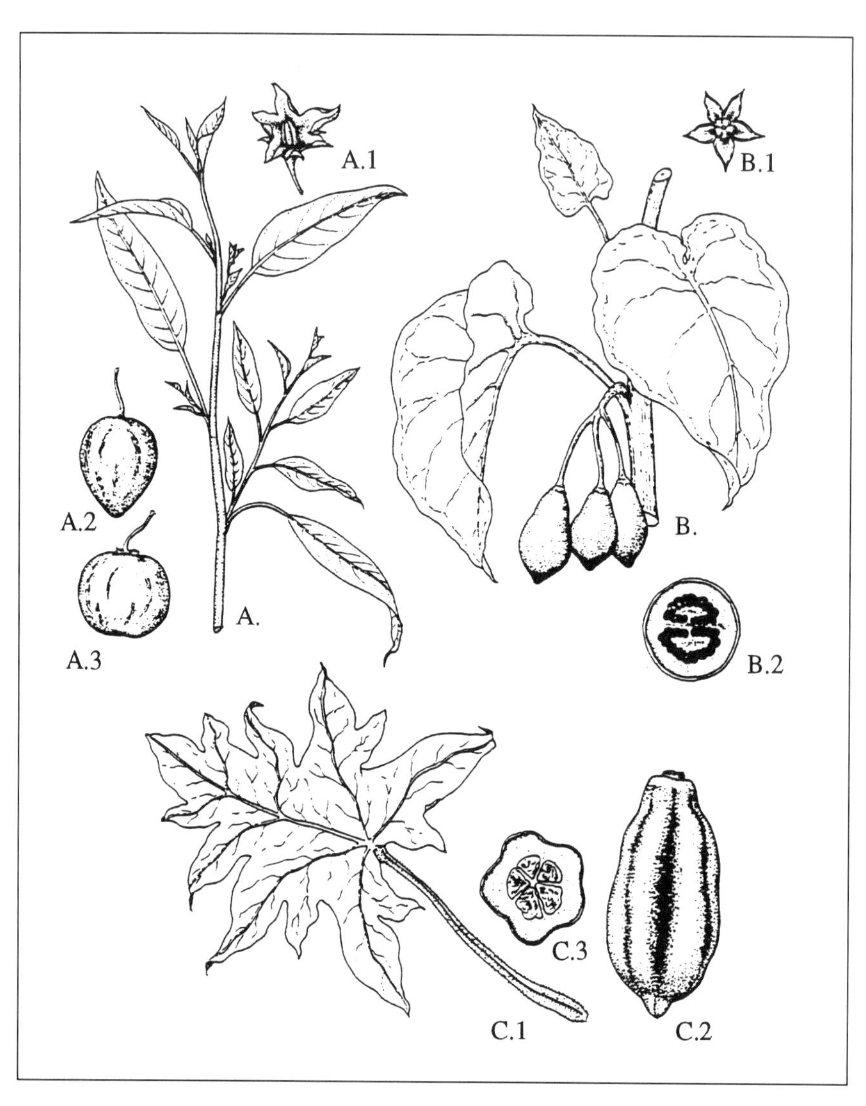

Figure 9.5. Three Andean fruits discussed by Sauer in his *HSAI* plant chapter are still today not well known outside the region: (A) *pepino, Solanum muricatum* ([A.1] flower; [A.2] and [A.3] fruits); (B) tree tomato, *Cyphomandra betacea* ([B.1] flower; [B.2] section of fruit); (C) highland papaya, *Carica pubescens* ([C.1] leaf; [C.2] fruit; [C.3] section of fruit). (Source: Hernández Bermejo and Leon 1992, reprinted with permission of the Food and Agriculture Organization of the United Nations)

ash that can be produced from it. Mixed with water to form a paste, then dried, a bit of this high-calcium material is put in the mouth with coca leaves—not "coca beans," as Sauer (1950:496) stated—to extract alkaloids.

Sauer noted that introduced broad beans (*Vicia faba*) had largely replaced tarwi (*Lupinus mutabilis*) in the Andean ecological zone above maize cultivation, but otherwise he did not consider the implications of native crop diversity in the face of competing Old World crops and practices that diffused rapidly though unevenly after the conquest. The strengths and weaknesses of native agriculture in its confrontation with introductions from Europe are a theme that gives larger meanings to diffusion (Gade 1992) (figure 9.5). Nor did Sauer reflect on Inca agriculture, about which a good deal can be reconstructed from the Spanish colonial chroniclers. He had gone to Latin America seeking specific crops but imposed no overall vision on the agricultural biodiversity of the Andes. He hinted at the ecological and social benefits of diversity, but never developed a strong statement of the value of peasant polyculture. Distinctions need to be made between peasant subsistence on one hand and redistributive economic mechanisms of food surplus instigated by the Inca imperial political apparatus on the other. Zimmerer (1993), by exploring those distinctions, has shown how that Andean diversity resulted from decisions that go beyond the political to the cultural.

Sauer and the Building Block of Crop Knowledge

Carl Sauer was a brick in the edifice of knowledge about New World plants. The *HSAI* discussion of plant diversity far surpassed that presented by Candolle (1967). It communicated the rich array of Latin American crops as known in 1943, and it enhanced awareness that New World agriculture was an artifactual complex from the past. Equally important, Sauer's chapter effectively interpreted the nature of these crops to an audience beyond the plant sciences. It stimulated archaeologists, who had started to pay more attention to the cultivated-plant remains in their middens and trenches. Agriculture came to be seen as key to the development of civilization. By the 1960s, radiocarbon dating had established a broad chronology of domestication. Though much remains to be discovered, enormous amounts have been learned since the 1930s about the origins and diffusion of agriculture.

Retrieval of the Archaic

Sauer brought the content of New World crop diversity to a wide audience of educated readers. He discovered none of these plants for science but synthesized the essentials about them more fully than anyone else had up to that time. Knowledge gaps or uncertainties about a crop launched specialized research. In the 1960s and 1970s, concern was expressed that many native Andean crops

Figure 9.6. Solidly planted *tarwi* field near Chiara, Peru; in many cases the broad bean (*Vicia faba*) has replaced this native crop.

would disappear. Simmonds (1976) predicted the decline of quinoa as a cultivated plant. I thought tarwi a crop on the road to extinction (Gade 1969b), but that prediction did not envision the changes that emerged in the 1980s (figure 9.6). New World crops, especially those of the Andes, became objects of considerable interest. Concern for biotic diversity in general, the search for new specialty crops, and maintenance of indigenous tradition were the driving forces behind arresting the decline.

Amaranth and quinoa are the two seed plants which offer the greatest possibilities for cultivation in the midlatitudes (Risi and Galway 1984; Wood 1988; Johnson 1990). Although both are still specialty foods outside the Andes, they have found a health food niche as an alternative to better-known grains. Harlan (1953) recognized the potential of quinoa as a breakfast food when he wrote that "it is starch, with just enough tang from its pigweed blood to produce a good flavor." Experiments are underway to eliminate the toxicity in the seeds of tarwi, a cultivated lupine, and to develop strains with higher oil content.

The National Research Council (1984, 1989) brought together knowledge about these Andean plants. High Andean tuber crops in addition to potatoes could grow in places with cool growing seasons (Sperling and King 1990). Within the Andes, agronomists have turned their attention to research on native species (Sumar et al. 1992). At an earlier time, these plants were frequently discredited as "Indian food" (*comida de indios*). Now even the most obscure plants are seen to have potential. The Chilean domesticate *Madia sativa* contains high percentages of unsaturated fatty acids, and its seed contains 26 percent oil (Schmeda-Hirschmann 1995). It has not been cultivated anywhere since the 1920s, but it could be again.

For many readers, Sauer's 1950 chapter was the initial source of information about Andean fruit-bearing species. International interest in them has spawned a substantial taxonomic and horticultural literature about them outside the Andean region (National Research Council 1989). Two of them, cherimoya (*Annona cherimolia*) and pepino (*Solanum muricatum*), are especially delicious, though difficulties with fruit set have so far hindered their wider commercialization beyond being exotic novelties (Prohens, Ruiz, and Nuez 1996).

Background to the Masterpiece

Sauer's considerable knowledge of crop biodiversity, combined with his keen sense of long-term processes and perspective on plants as cultural tracers, made possible the grand vision contained in his slim but potent book *Agricultural Origins and Dispersals.* Published in 1952 as an American Geographical Society imprint, it was derived from a series of lectures Sauer had delivered in New York in 1950. After the AGS title fell out of print, rights were sold to the Massachusetts Institute of Technology Press, which republished it and later reprinted it under a different title (Sauer 1969, 1972).[11] Sauer's determination in the late 1940s to learn as much as there was to know about early farming and herding in the Old World led to his developing a grand perspective on plant and animal domestication from around the globe. Although he never traveled to Asia, the Pacific Islands, the Near East, or Ethiopia, more than two decades of observation among peasant folk in Latin America enabled him to construct plausible scenarios of human intervention in biological phenomena even for places he had not personally visited. He determined that most distributions could be explained by the diffusion of ideas and organisms associated with them. Many of Sauer's ideas about the beginnings of domestication crystallized over the decade of the 1940s. For example, in 1946 he wrote in a letter to one of his "Ph.D. boys" that "people who followed stream fishing I more than suspect to be the ancestors of the agriculturists [and] to have been the first planters" (SP, COS to FWM, 2-14-1946). The idea, refined and expanded in *Agricultural Origins and Dispersals,* remains provocative. Sedentary fisherfolk, motivated by curiosity more than by necessity, experimented with plants and later started to cultivate them.[12] Subsequently other ideas, one using a population pressure model (Cohen 1977) or one derived from evolutionary ecology (Rindos 1984; Gremillion 1996), have each gained attention. But consensus about the mechanism by which cultivation first occurred has yet to be reached.

Sauer understood domestication to be a process; this understanding emerged from his field observations, but even more from his discussions with Edgar

Anderson. On this point he acquired from Anderson a way to handle variability: crops are plants along a continuum from being wild and unmodified to being totally dependent on humans for their survival. In the *Handbook* chapter Sauer (1950:488–89) used the word *cultigen* to describe the crops such as maize and manioc that require human intervention for their dispersal and whose ancestors are unknown. From Anderson, Sauer came to appreciate gradation between types—a materialist metaphysic—which countered his natural tendency toward typological thinking. The diffusionist stance in *Agricultural Origins and Dispersals* generated the most controversy. Two botanists, Merrill (1954) and Mangelsdorf (1953), both criticized the book mainly in this regard, though neither was trained in ethnology or culture history. Reconstruction of the early spread of plants over long distances is necessarily speculative. Admittedly diffusion has had an enormous effect on world agriculture. Yet the evidence for pre-Columbian trans-Pacific carriage of maize and plantains is slender.

With those few exceptions, *AOD* received acclaim as a landmark of cultural-historical erudition.[13] Sauer provided a nonpareil synthesis based on an integration of cultural and physical factors. Much of its impact stemmed from the literary effectiveness of the message. Sauer's extraordinary talent for written expression deserves its own study. His spare, elegant style remains a model for anyone wishing to communicate complex ideas. But it may also have compensated for the monograph's scientific weaknesses. Harris (1990:11) suggested that "the book's persuasive influence owed something to the fact that it was beguilingly well written, but factually unsupported. . . ." Some of the factual content of the monograph has now also been superseded, and in time most of it may become obsolete. Yet *AOD* must be recognized as a masterful stepping-stone in the stream of enlightenment about domestication. It has stimulated many people to think about one of the most far-reaching transformations in the history of the earth. Sauer informed a whole generation of scholars in several disciplines, who pondered and debated his ideas.

Sauer and the Moral Value of Biodiversity

Latin American peasant agriculture, in Sauer's view, was inherited from the past. It was well adapted to local conditions because generations of farmers had selected plants to achieve that. He was interested in crops as human artifacts and sources of subsistence, and he recognized already in the 1930s the benefit of agricultural diversity. He became aware of the erosion of traditional crop varieties and species. In that same period Harlan and Martini (1936) noted the replacement of old barley varieties with modern ones. Sauer (1938:770) expressed a broad concern that North American concepts of modern farming had "drastically impoverished the results of biological evolution."

To his way of thinking, monoculture was a form of "destructive exploitation," which Speth (1977:145) defined as the "process of economic devastation and its aftermath associated with occidental commercial culture." [14]

What was to happen later in South America had occurred earlier in Mexico. Carl Sauer contested the biological simplification he saw in Mexican farming. Beginning in 1940, Mexico's agriculture began to change toward the United States' model of fewer crop varieties and the use of fertilizers, pesticides, and machines. In his study of Rockefeller Foundation support of agricultural development, Wright (1984:136) documents the "self-confident production-minded approach of American agribusiness" and moral innocence of its innovators from north of the border. Sauer, in his consultant capacity to the foundation, protested the technical proposals of its research program as deleterious to Mexican peasant economy. Sauer wrote in a letter to the Rockefeller Foundation that "an aggressive bunch of American agronomists and plant breeders could ruin the natural resources for good and all by pushing their American commercial stocks" (Wright 1984:140).

Sauer's voice was not heard, largely because the startling gains in crop yield seduced the professional agricultural scientists. Jennings (1988:191) viewed it as the victory of one style of thought ("the production paradigm") over another ("the social relations paradigm"), the result of the social interests of scientific-based institutions, not impartial discussion. By the 1970s, however, the realization set in that this agribusiness approach, officially licensed by the Mexican government, had not only marginalized millions of Mexican farmers, but had also turned the country into a net food importer. Population increase had outdistanced further yields: the "miracle seeds" had reached their limit. The export of high-energy consumption and capital-intensive farming techniques as models to other countries creates more problems than it solves. Wright suggests that failure to adopt Sauer's kind of humility about American agriculture led American technical experts abroad to act with a combination of arrogance and naive innocence. Marglin (1996:211–17) has documented Sauer's role as the primary antimodern force against the Rockefeller Foundation's involvement in Mexican agriculture and his opposition to the USDA and its policy of encouraging the spread of hybrid corn beyond U.S. borders. It was no coincidence that Henry Wallace, secretary of agriculture, also had a major investment in the largest hybrid seed establishment in the United States.

These attitudes must be placed in the context of the 1930s, when the United States saw itself emerging as the beacon of agricultural enlightenment to the world. High yields and labor efficiency of production were viewed as achievements worthy of imitation everywhere. After World War II, agricultural programs became a centerpiece of American foreign aid to underdeveloped countries. Foreigners bearing tempting ideas of progress were followed by Latin agronomists, who, having studied in the United States, viewed themselves as the vanguard of the agricultural development of their countries' hinterlands. They promised peasants that adoption of modern varieties would lead to higher production and, by extension, to a better life.

Missouri and Mexico were double frames of reference to Sauer, who noted the modernizing trend in both places. In Sauer's lifetime the introduction of machines and a few hybrid varieties squeezed out the kinds of corn that he had known in his midwestern youth. But homogenization of the agronomic base of the U.S. Corn Belt could not compare with the negative consequences of development for the extraordinary maize diversity of Mexico.

Modern assumptions about agriculture reached beyond Mexico into all of Latin America in the 1950s. In more than half of that area, high-yielding cultivars of maize, potatoes, and *Phaseolus* beans have subsequently displaced the old landraces. Genetic loss has occurred in tandem with other aspects of agricultural modernization. In the Andes, however, new crop varieties have often been adopted without corresponding technological inputs. Land abandonment is another cause of germplasm loss. As young people move to cities and their elders die, farming stops.

Beginning in the 1960s, the dangers of biological simplification in world agriculture were increasingly recognized. Van der Plank (1968:140) wrote that "the conservation of nature's vast inheritance of genes is probably this shortsighted world's most urgent single genetic problem." A maize blight in 1970 overran the Corn Belt with lightning speed and revealed how vulnerable the narrow genetic base of this crop made midwestern agriculture. Questions about where more maize germplasm could be found led to the realization that the custodians of this diversity were in Latin America. Decades before, Sauer (1938:771) had pointed out that Acalá cotton from southern Mexico was what made cotton growing in California's San Joaquin Valley possible.

Shifts in Thinking

North American agriculture's heavy investment in powerful chemicals, high use of energy, and vulnerability to crop epidemics received criticism through the decade of the 1980s. "Ecological agriculture" began to enter the mainstream of North American agricultural thinking. Often touted as a "new" approach, it was, in fact, ancient. Fitting crops to particular environments also came under serious discussion. Jojoba (*Simmondsia chinensis*), a high-quality source of oil, has become a desert crop that needs no irrigation. Howard Gentry (1903–1993) was the key figure in acquiring jojoba material from Mexico for eventual cultivation in the United States. His article about it (Gentry 1958) summarized its possibilities. Since the 1970s the plant began to fulfill its promise. Gentry made other contributions to the study of plants and plant use (Cunningham 1994). For several years Gentry was in close contact with Carl Sauer, his undergraduate professor at the University of California, Berkeley, about Indian plant use in northern Mexico.

Other changes have occurred. Perennial polyculture is now discussed as an alternative to annual monoculture in the Midwest (Jackson, Berry, and Colman 1984). Weeds, although they will never be totally redeemed, are increasingly viewed in more complicated ways (Gade 1993). Some of them have their own values; for example, milkweed is grown commercially for its fibers (Knudsen and Zeller 1993). The underlying logic of North American food and fiber production has undergone reevaluation.

The assumptions of the 1950s and 1960s have also been reexamined in parts of Latin America. In the Andes a new-found respect for traditional regional practices has replaced a thirst for foreign models of agriculture. The new ideal is to build on aboriginal achievements of the past, including ridged fields, bank terrace systems, irrigation works, and the inventory of native crops and animals. Considered recuperable elements of a cultural heritage that once sustained empires, they also are viewed as part of the solution for a sustained food production in a difficult environment. International recognition of the Andean achievement bolsters confidence in the worth of its contribution to world agriculture. Old cultivars continue to be lost, but drastic change is not envisioned over the next two decades. The core idea of the Green Revolution is unlikely to be accepted as it was in Mexico or India. Yet the biodiversity issue implicit in this transformation must be seen not as traditional versus modern, for the stability implied in the former has never existed and does not account for peasant ingenuity and resilience (Brush 1992:164).

Attitudes change. Traditional adaptations receive more respect now. Industrial nations appreciate that genetic erosion endangers their own food security. How germplasm should be stored and who has rights to it are matters of debate. *Ex situ* preservation involves the collection of landraces (crop seeds or tissue cultures) for storage in gene banks. A global network of these centers emerged in the 1970s, coordinated later by the Food and Agricultural Organization of the United Nations, to collect, conserve, document, and use germplasm for more than 50 crop species (Williams 1988).

Sauer wrote primarily of a conservation alternative: preserve old crops and varieties by allowing peasants to continue their ways of life without interference from development personnel. Traditional farmers would not only maintain diversity but also bear witness to the domestication process still going on in their midst. However, Sauer would not have favored "freezing the genetic landscape" by artificially providing incentives to keep people prisoners in a way of life that is not of their choosing (Iltis 1974; Altieri and Merrick 1988; Wilkes 1991). He had too much respect for people to engineer them socially into museum pieces for the benefit of industrial societies. Integrating both perspectives negates selecting one over the other. Folk varieties can be used by local farmers but also placed in *ex situ* community seedbanks (Cleveland, Soleri, and Smith 1994).

If still alive today, Carl Sauer would certainly have found the debate over proprietary rights to primitive crop varieties to be a distasteful intrusion of politics or corporate greed. Who owns the intellectual property rights to biological materials has become an international policy issue pitting agribusiness of the industrialized countries against Third World governments. That any entity could "own" the biological diversity in primitive crops and wild plants would have been unthinkable half a century ago. A "farmers' rights" movement seeks compensation for the primitive germplasm used in modern plant breeding. How that compensation would be managed complicates its execution. Given their treatment of peasant agriculturalists in the past, governments could not be counted on to turn over these proceeds to the guardians of this diversity.

The issue is not likely to fade away. Genetic resources are no longer local in their impact. The debate is now framed in terms of "rights" and is ontologically clouded by several conflicting definitions: farmers' rights, cultural rights, human rights, and environmental rights (Cleveland and Murray 1997:489). Competing perspectives will make it difficult, perhaps impossible, to achieve any global consensus on these matters.

Sauerian Thoughtways

Carl Sauer's work on New World crop diversity and its larger meanings came from a conviction that crops are a key to understanding mankind on the planet. Sauer gained extensive knowledge about the corpus of New World domestication, mostly after he had reached the age of 50. He realized that peasant farmers of the Andes have formed a keystone culture which has greatly enhanced crop biodiversity. How long this will continue depends on several factors. Pauperization, rather than reinforcing diversity, could destroy it. Acculturation is often assumed as having a deleterious effect on that diversity, but in fact Zimmerer (1996b) has shown that richer farmers, being part of larger trade networks, may be more inclined to preserve cultivars than those who are poorer.

Sauer's expertise came from observations and a wider reading of relevant studies than had been carried out up to that time. He manifested an intuitive sense about the objects of his study. Some of that came from his respect for peasants' knowledge and an appreciation of their *genre de vie.* At the same time he ceded to the views of scientists he considered to be authorities.

Sauer was somewhat like Goethe, who called himself a *Naturschauer,* an observer using intuitive logic, rather than a *Naturforscher,* an investigator of natural history who follows a set of procedures. An account of how Sauer fashioned knowledge makes clear that his idea of scientific objectivity was a concern for accurate understanding, not the result of controlled procedures. Sauer's metaphysic was sometimes essentialist, sometimes materialist. He fit phenomena into types, but at the same time he realized that the prehistory of crops could not be understood in terms of concrete typologies. Beyond that Sauer's mind can be characterized as biocentric, not spatial-mathematical; in biocentrism the temporal factor animates spatial phenomena with which it is inseparably connected (Gode-von Aesch 1966:80).

Sauer's own empirical observations on crops were not detailed or structured enough to have much lasting value. By favoring imagination over procedure, he provided intriguing suppositions to ponder rather than a body of basic information. His *HSAI* plant chapter was quite heavily cited for slightly more than a decade after it was published. During that time it was seen as the authoritative presentation of the subject (e.g., Horkheimer 1960). By the 1970s the chapter was rarely referenced, a measure of the rapid advances in knowledge about native crops of the Americas.

During the decade of the 1950s this survey made Latin American crop diversity accessible to an English-reading public. In that sense Sauer educated people about the range of this diversity and about the questions related to it. The bold speculations that grew out of this work on the origins of agriculture will continue to be sources of reflection for those interested in origins. His perspective, which overlapped time and space, first came into full focus with his observations in the Andes. Sauer was most comfortable in reconstructing the past, but the meaning of diversity has had its greatest impact on thinking about the future. The understated moral component of his ideas reflected a vision far in advance of his time.

10
Conclusion

This set of biocentric topics on South America has sought to combine the empirical, holistic, and reflexive. Each chapter is a separate constellation with its own interlocking elements, but all are bound in a universalizing theme and embedded in a temporal and spatial context. The configurations that have emerged explicate culture/nature as a totality that incorporates detachment and reflection.

Embrace of the Empirical

My knowing the Andean world in its diversity began with a stroke of curiosity about what I observed. From loose bricks of basic and sometimes naive observation developed an edifice that offered context and interconnectedness. My living in and traveling about the Andes raised endless questions about landscapes and the organisms and people that were part of them. I realized how much of what I learn is based upon what I see. This led me later to appreciate Tyler's (1987:150) insight that the "hegemony of the visual" controls the focus on things. It may be a conceit of geographers as a group that when we think about knowing something, that knowledge is coextensive with what we see.

I noted the forlorn shells of haciendas in Bolivian valleys, a lone tree on an otherwise barren hillside, the nocturnal scurrying of rats in my rudimentary lodgings. Overgrown bank terraces found in the eyebrow of the jungle were a puzzle, just as the strange contents of folk-medicine markets were and the utilization of llamas for everything but their milk. In 1968, Machiguenga Indians living near the Spanish Dominican mission at Koribeni in the Urubamba Valley killed a tapir when I was there, which allowed me to scrutinize the body before they dismembered it for meat. Once these things were examined, there was no authentic way to sustain the classic opposition of nature and culture. The dichotomy best suits armchair theoreticians in the absence of their own data.

Sustained thinking about these observations involved a "venatic methodology" of the sort followed by a hunter who uses clues to locate his prey. In much the same way, an empirical research strategy garners a succession of soft conjectures and hard facts, which initially coexist in a skein of suppositions. With more evidence the conjectures are either discarded, or confirmed, or, if particularly interesting, remain as speculations to challenge the next researcher. This kind of intellectual bricolage combines science with the power of intuition to patch together pragmatically a larger configuration. In our accommodation of ambiguity, subjective aspects do not deny presumed reality, nor does our commending the imagination force us to disregard reason.

No scholar ever starts from a tabula rasa, for everyone operates from certain assumptions, however objective one seeks to be. At the same time, I have consciously resisted theoretical frameworks or hypotheses to be proven. Several chapters in this volume indicate how preconceptions of a given period have blocked critical thinking about the phenomena studied. Spontaneous generation of vermin had Western intelligentsia in its Aristotelian grip long after it was rationally understood that rats reproduce by sexual means. In the eighteenth century the dissonance between the two views became unbearable, and an ancient notion was finally discarded. Causal correlation of environmental variables is an even more persistent idea. Climate has long been attributed to be the key determinant of vegetation landscapes. In the past most trained scientists sought to interpret the absence of trees in the Andes as a response to particular combinations of temperature and rainfall. However untenable that simplistic formula seems today, it nevertheless persisted as a widely accepted deduction in biogeographical circles until after the mid-twentieth century.

Another broad assertion that challenged me to think placed all domesticated ungulates as sources of milk for humans. Llamas and alpacas have never been in the milch category, yet presumptions that they are persist in the popular imagination. Yet another generalization held the Inca as highland dwellers incapable of functioning in the jungle. In fact they not only grew coca in certain warm valleys, but also carefully cultivated it on stone terraces. On still another front, disease etiology since the ancient Greeks had "explained" malaria as being caused by putrescent air emanating from swamps. Not until the late nineteenth century did researchers with microscopes ascertain the link of a swamp-breeding insect carrying blood parasites in its gut. That a magic cure for epilepsy is contained in the tapir foot was a widespread idea of remarkable persistence in South America. It maintained its force in the folk imagination long after scientific understanding of epilepsy emerged in the eighteenth century.

The Goethean strategy of starting with the thing and not the grand idea removes the temptation to highlight voguish notions and to concoct deterministic frameworks. Beyond that, attention to nonrecurring events and to patterns of the past corrects facile explanations of the moment. Pushing back our knowledge of a phenomenon to uncover possible origins enhances understanding even when "causation" cannot be ascertained. A backward gaze also forces one to recognize that others have shared one's curiosity about places and practices.

Sense of the Whole

Each of these chapters integrates ecology, space, and time into an intuitively derived unity that I call the culture/nature gestalt. Like insects in amber, humankind, plants, and animals are imbedded in a matrix of the organic and the inorganic and in relationships that may be parasitic or mutualistic. Two different, but correct, perspectives can be derived from these chapters. Humans are ultimately only fragments of a complex world; yet, at the same time, their vast technological power has enabled them to transform plants, other animals, and whole landscapes. Their control was greatly enhanced with the emergence of agriculture in western South America more than 5,000 years ago. Innumerable interconnections from this rich Andean past cannot yet be clearly reconstructed. Even contemporary examples have a diachronic dimension that elucidates interrelationships better than a functionalist explanation taken from the contemporary moment can. Ecology, time, and place offer a trio of pegs on which to hang the nature/culture gestalt as an interlocking complex that constantly requires contextual clarification.

Ecology

An ecological relationship on a continuum between stability and instability is evident in every chapter in this volume. George Perkins Marsh (1965) in the nineteenth century felicitously saw human uses as "harmonies" and "disturbed harmonies." Holism characteristically evokes harmony but in fact can even more effectively demonstrate the lack of it. Disharmony is salient to this work, perhaps because dysfunctional patterns so compellingly evoke the sense of stewardship lost. Humans have a special responsibility for rectification of the problems they create. Deforestation in the Andes has created wood shortages, high cost for timber and fuel wood, and soil erosion. Introduced organisms, rats and plasmodia, have had negative impacts on human settlement, demography, and health. Disharmonies have always captured the most attention, even when viable solutions to the problems exist but remain unimplemented.

Other ecological processes can be distinguished as adaptation and transformation. Both require attention to long-term changes, and they are not mutually exclusive. After its sixteenth-century introduction to the Andes, malaria took a terrible toll on human life and health. At the same time, a *modus vivendi* emerged by which people adjusted to the presence of the disease through temporary migration and racial displacement. In the Guayaquil rat story, the bubonic crisis did not transform rodents' and people's general accommodation to one another. Without the milking of camelids, pastoral nomadism did not emerge in the Andes, and the subsistence basis remained one of mixed farming and herding.

Regional plant domestications formed the basis of cultural elaboration in Andean civilization. But agriculture, once established, also had great value in creating a stable adaptation to a rugged environment. Deforestation has metamorphosed the landscape into the present; the counterprocess of reforestation has not significantly arrested the trend toward treelessness. A change of more limited scope was the Inca incursion into the jungle fringe to create agricultural space for growing coca, which continued as the motive for settlement in this particular location only about a century after the conquest. Less clear is the role that the tapir cult and the relentless search for tapir parts might have had on the population decline of the mountain tapir and its extermination from the southern part of its original range.

Time

To understand nature/culture as a gestalt is also to conceptualize its permutations through time. Such a perspective opens the door to evaluating the relative stability of linkages between humans and nonhuman phenomena. Pivotal debates concerning Andean culture history involve discussions of continuities and discontinuities from the past. With the arrival of Europeans, a major break in cultural patterns occurred—the *conjoncture,* in Fernand Braudel's historical imagination—because of multiple introductions that affected land use, health, and mortality. Plants and animals new to the Andes expanded land-use possibilities, especially at low elevations. Europeans' introduction of the idea of milking and of Old World animals traditionally milked highlights the absence of independent invention to initiate a practice that would have provided an additional source of human nutrition. A comparison of the Andes with the part of the world that did milk animals offers insight into why such a practice did not emerge in this region. Pest introductions, accidental though seemingly ineluctable, are other manifestations of transoceanic movement from Europe. Rats, in directly competing with humans over food and in carrying diseases to humankind, have had negative effects on the Andean settlements to which they have spread in their relentless movement around the world. Malaria is more complicated in this regard; it was humans, Europeans and Africans, who originally carried the plasmodia in their bloodstreams, but the mosquito vectors are indigenous. The conceit of the upper selva as untouched cannot be reconciled with the cycles of forest removal and regeneration that have prevailed for millennia. Unraveling the last seven centuries of coca cultivation in the eastern Andean valleys belies the myth of the primeval Amazon forest, which the international media and even the scientific literature have fostered as part of an Edenic narrative.

The case for continuity in Andean life comes from a *longue durée* time scale that incorporates development of civilization long before the conquest. Clearing and burning ate away at Andean woodlands for more than 2,000 years, yet opprobrium for much of that practice is characteristically placed on events after the European arrival. Cultivation of the great majority of New World crops is as old a process as deforestation; indeed, the former depended on the latter. Many plants are inefficient producers of food, but they have persisted as part of an Andean polycultural tradition that ensures peasant subsistence. Andean food crops aside from the potato have been of primarily regional importance. Genetic improvement may give them more importance in the future than they have had so far. Andean survivals in the realm of folk medicine have also manifested continuity into the present. The continued use of tapir parts, which involves sympathetic magic rather than any biochemical therapy, testifies to the power of Andean folk tradition.

Space

The nature/culture gestalt focuses on space as concrete places, not as geometric abstractions. A focus on regions, subregions, and localities provides a multiscalar cascade of generalizations that elucidate the tension between linkages. Some geographical levels are transparent; others, complex. At a regional scale, the Central Andes provide an appropriate geographical unit in which to consider agricultural beginnings, exploitation of the camelid, use of wood resources, and integration of the tapir into Andean folklore. At the scale of the subregion, the coca-producing valleys can be examined in the context of the specifics of elevation, climate, vegetation, and transportation routes. At this scale, we can also see that in the valley of Mizque, a few kilometers of vertical distance entirely changed the geography of disease. The local scale of the city permits a focus on the rat problem as one of nodes (e.g., markets) and zones (poor neighborhoods and certain industrial areas). Whether at a small or a large scale, concreteness of place is critical in explicating interconnectedness, for a tenet of holism is that it is impossible to appraise an entity in abstract space. The landscape forms the concrete spatial manifestation of the nature/culture gestalt.

Viewing ecology, time, and space as a complex can open the way for new glimmers of understanding that might escape a reductionist approach. Species, populations, and communities—the three different levels of biological organization dealt with in these chapters—can be further understood when contextuality elicits fresh interpretation. Change in the distribution of the mountain tapir, the realization that *Cedrela* was once a highland tree species, and the spontaneous reappearance of forest in valleys that had been cleared before the European conquest all flow out of that larger context. Seeing these three levels of biological organization from another dimension—that is, through their incorporation into a biotic pattern—reveals additional layers of cultural-historical significance. As the predominant rodent population shifted from the black rat to the brown rat, bubonic plague decreased, then disappeared. The devastation of malaria owed much to the population increase of the mosquitoes living in the stagnant water resulting from irrigated agriculture. Compositions of adaptations and/or transformations derive their validity from the interconnectedness of elements.

Holistic thinking has potential for abuse, however, when the contextual whole becomes a vehicle for deducing organismic function or ideological principle. Without those distortions the concept heightens appreciation of the gestaltic character of life on earth.

Reflexivity and Authorship

A third dimension to these antipodal explorations is the realization that knowledge is affected by both the conscious and the subconscious perspectives brought to it. Our interpretations of phenomena, pursuit of certain topics to the exclusion of others, and essentialist or materialist construction of the world come from the self. Without obtruding myself into the heart of the narrative, I have sought to come to terms with the power of memory, personal identity, and experience, which lie behind these topics and their elaboration. All of what I have learned is conditioned by my own perspectives. However, only a balanced reflexivity is productive; otherwise one risks lapsing into the infinite regress of reflecting on one's reflections to the point of nihilistically debunking everything.

One's early years, although rarely evoked to explain chosen themes, can affect the paths taken. In my case, seeing a friend in epileptic seizure, rats in the backyard, and a live tapir at the zoo and the exhilaration of ascending through the ecological zones of the central Rockies left indelible impressions. Childhood memories played a part, but more pivotal were scholarly influences from certain individuals as transmitters of specific ideas and models of research achievement. Creative scholars trained or influenced by Sauer opened ethnographic lines of thinking about material and nonmaterial cultural patterns; provided a Darwinian focus on plant domestication as a genetic process, but also one influenced by cultural factors; and astutely reconstructed prehistoric geographies that encouraged examination of the Andes as a pre-Columbian achievement rather than as a development failure. Carl Sauer, in control of a vast array of historical knowledge, offered a formidable discourse about human life on earth; his unstated philosophical foundation recalled those of Herder and Goethe. To mention this protagonist of the Berkeley school is to acknowledge the genius of a scholarly point of view whose synergistic creativity has been a bright spot in the history of American geography. It is no accident that none of these chapters promotes the idea of progress, which has not been part of my thinking or that of my professors. But it is from my readings in anthropology that I have learned to appreciate that the enterprise of inquiry is not and should not be disassociated from its results (Rabinow 1977:5).

My Andean quests for knowing have also come from a temperament whose *Ur* lies both in the genes and in psychosocial development. A cardinal element was a Herodotean curiosity and a *feu sacré* to pursue those intuitive flashes of mystery and wonder to the point of organizing them for a larger audience. Those two characteristics have their possible source in a shamaniclike impulse to derive larger meanings from the observation of surrounding phenomena. From the Paleolithic dawn into the present, certain persons have received a call from their inner being to communicate the meaning of the world around them, not through inner contemplation but by looking beyond the self. In decoding the unfamiliar, shamans also intuit wholes in their mind's eye, thereby bunching phenomena into a gestalt. Wonderment about the intricate interrelationships between people and other forms of life and the sense of convergence of body, mind, and spirit flow out of that way of seeing.

Finally, the experience of the Andes as a place, a culture, and an adventure worth sharing with others must be factored into the reflexivity of these accounts. The sheer directness and immediacy of the ties between people and their environment and the rough charm of Andean landscapes, livelihoods, and folklore made a durable impression of life-changing significance. To be sure, my knowledge of the Andes was strongly influenced by perceptions that came from prior conditioning. But as my Andean knowledge expanded over the years, so too did these perceptions change. I may not have always understood the subjectivity of Andean people, but their strong hold on tradition and rootedness in place offered another vision of being.

Living in the Andes formed a core experience which had subsequent transformative value in philosophizing one's own existence. Depending on predisposition, pursuit of knowledge of the outer world can call forth the experience of an inner world. A hardscrabble, mountain peasant setting and culture became for me not only a rendezvous of questions but also a hall of mirrors into the liquescent depths of original being. Working through the multiple contradictions of fieldwork and the *dépaysement* in a peasant milieu provoked cerebration of much broader themes. Out of that germ of reflection, I have been able to contest neat dichotomies inherent in Cartesian rationality, theistic religion, Western superiority, and disciplinary structures.

Scholarly endeavor has long strived to compartmentalize what there is to know and how to study it. This volume has instead celebrated the capacity of overlap and ambiguity to efface the socially constructed borders between things and ideas. In my scheme, culture is naturalized and nature is humanized. A synthesis of scientific discourse and aesthetic sensibility is blended with respect for accurate observation, the accumulation of facts, and humanistic values. Not all readers will necessarily agree with the "correctness" of some of my interpretations. Many of these assertions may fall into the realm of Nietzsche's (1967) "beautiful possibilities" that are meant to provide insights rather than to deliver immutable truths about phenomena. Such a multifaceted approach implies complexity, ambiguity, and mythic dimensions that form an aspect of contextual, integrative, and nonlinear modes of thought. What has been often incommensurable need not be. Adding a brick to knowledge was a motivation, but with the full realization that it is always subject to later revision, for nothing is ever learned once and for all. The conventional disciplinary divisions are sheer academic convenience that hinders the creative formulation of much knowledge. The accounts that I offer as cultural geography may be justifiably considered by others to be anthropology, history, ecology, or perhaps even some offbeat kind of biology. More important than any label of affiliation is what one thinks about it. Subject and object merge, and past and present are part of a continuity. The inextricability of all these phenomena cannot be fully expressed by the categories of language now in use. Blurring traditional boundaries opens unexplored avenues of inquiry to ignite clusters of succulent insights. Reenchanting the pursuit of learning about the world offers in that way one element of an alternative vision for a new millennium.

Notes
References
Index

Notes

Chapter 1. Reflections and Trajectories

1. In 1945 Sauer wrote the botanist Howard Gentry: ". . . I am interested in something less and something more than what is usually called ethnobotany. What I want to see done is a special type of plant geography; a complete record in the first place, of the household plants, meaning those that occupy the fields man has cleared and his backyards, which grow about his habitations, and along his frequented paths. These include the ones he protects and those he tolerates within the space he has modified by clearing and habitation. In all cases there is an interdependence between the plants and man . . ." (SP, COS to HG, 10-19-1945). Geography and ethnobotany have points of contact, especially where observations of the latter move beyond the purely local or it is viewed in landscape terms (Morgan 1995).

2. Carl Sauer's father, William Albert Sauer, was a French and music professor as well as a college botanist at Central Wesleyan College (Kenzer 1987:53). Son Carl dedicated the monograph based on his 1918 Ph.D. dissertation to his father, who ". . . formed his first appreciation of the things that constitute the living world" (Sauer 1920).

3. Anderson was awarded a Guggenheim fellowship from the committee on which Sauer sat; Anderson used it to come to California in 1943 to trade ideas with Sauer. Sauer's contributions on crop geography came partly from what he had learned from Anderson, but both men benefited from the exchange. In a letter, Anderson wrote that coming to wartime Berkeley 16 years earlier was "still the most important event of my professional life" (SP, EA to COS, 2-17-1959).

4. "Unless the scientist reads outside her/his field or takes a sociology class and learns how Kuhnian paradigms as 'cultural baggage' lead humans by the noses in all they think and do, one can go through life ignorantly believing that science is detached, objective, factual, unmythic, and without goal-setting" (Barbour 1995:253, quoting S. Rowe).

5. Beginning in the 1970s, field researchers who have carried out ethnographic studies in the Andes have tended to ennoble collectively, at least in print, the groups and individuals they live among. In fact, however, the researcher-peasant encounter in the field is often one of crippling frustration to the former. In a letter written to Julian Steward, the doctoral candidate Harry Tschopik Jr. wrote, ". . . I'm glad people you have talked with approve of our work here in Chucuito. At times (and right at the moment) I would like to wring the necks of all the bloody Aymara in the countryside, but I suppose you are familiar enough with this reaction!" (NAA, HT to JS, 8-1-1941).

6. Giesecke married into an elite Cusco family and rapidly assimilated to Peruvian society. He later moved to Lima and became a cultural attaché for a time at the U.S. Embassy. On March 13, 1963, Giesecke wrote me a richly detailed letter about multiple contacts that I could make in my field project. Since he was a reluctant author, it took me years to figure out the crucial part he played in fostering regional scholarship in Cusco.

7. In the effusive preface of his major report on Machu Picchu long after he first went there, Bingham (1930) thanked many people, but Giesecke was not among them. Years before, several Cusco professors had confronted Bingham about the disposition of archaeological materials taken from Machu Picchu.

Chapter 2. Andean Definitions and the Meaning of *lo Andino*

1. Physical unity of the Andes is only superficially apparent. This mountain system is now seen as a heterogeneous mass of processes and patterns.

2. Some early definitions of the Andes were broad. For example, the sixteenth-century Spaniard Balthasar Ramírez (1936:19) described the Andes as a "cordillera muy larga . . . que corre desde el brasil o mas adelante, a traves por Venecuela . . . hasta chille y el estrecho de magallanes." Galvano (1969:86), a Portuguese of the sixteenth century, wrote that "certain mountains named the Andes . . . divide Brazil from the empire of the Incas."

3. Humboldt's experience in the mountains did not include southern Peru or Bolivia. He traveled south only as far as Cajamarca, heading from there to Trujillo, where he took the coastal road to Lima.

4. Anthropologists have often constructed a definition of the Andes without elaboration. For example, none of Salomon's (1985) comments on Andean ethnology or any of his 226 references apply to Colombia or Venezuela. Orlove (1985), on the other hand, explicitly limits his definition of "the Andean world" to the Central Andes, which excludes Colombia and Venezuela. McDowell (1992:112), reporting on the Sibundoy region of Colombia, has questioned whether the "Tawantinsuyo-Cuzco nexus . . . should have a preemptory claim to Andeaness."

5. Means (1931) wrote of "Andean textiles" and "Andean social institutions," which at that time was still a novel way to characterize such things. Earlier Cook (1920) had written of the "Andean footplough."

6. In many studies the Central Andes are a region whose boundaries are left undefined (e.g., Ogilvie 1922; Platt 1932; Dresch 1958). Authors who delimit the region do not agree on its boundaries. Caviedes and Knapp (1995:88) place it between the Nudo de Pasto in southern Colombia and northern Chile. Gómez Molina and Little (1981:116–17) put it between Cajamarca, Peru, and Catamarca, Argentina; Brush (1982) positions the Central Andes between northern Ecuador and ca. Antofagasta, Chile; and Cunill (1966) encompasses under "Andes centrales" the zone between Loja, Ecuador, and central Chile. Provincialism reigns in the cases of the Peruvian scholar Millones (1982), who limits the Central Andes to highland Peru, and to the Argentines Llorens and Leiva (1995), whose use of "the Central Andes" refers just to the middle part of western Argentina.

7. *Lo andino* has no one clearly accepted meaning. Salomon (1985:92–93) defines it as the "ideas and institutions whose permutations were felt to pervade Andean culture

history." My broader definition includes the tangible, material achievements that help to characterize the region as one of cultural continuity.

8. Thus Brush (1977) places his research site in an Andean (rather than a Peruvian) valley; Bastien's (1978) subjects live in an Andean (not a Bolivian) *ayllu;* and Mörner (1985) invokes an Andean (not an Ecuadorian-Peruvian-Bolivian) past.

9. *Andeanist* has not yet gained recognition as a dictionary word, but to those who use the term it does not refer to a scaler of high peaks of the Andes, such as is implied by the word *Alpinist.* However, in Spanish an *andinista* is one who climbs, so *andinólogo* has been proposed as the Spanish term equivalent to the English *Andeanist.*

10. Schjellerup (1989) has published an ethnography that covers the kind of empirical detail on sierran adaptations that Andeanists in the next century will mine for its specks of golden insight into peasant material existence.

Chapter 3. Deforestation and Reforestation of the Central Andean Highlands

1. The Spanish crown inspectors recorded these trees in their reports, called *composiciones de tierra,* to transfer lands of a native community considered to be "unused," so they could be awarded to individuals, usually Spaniards. Listing trees implied a state of "noncultivation," when, in fact, these trees themselves were objects of cultivation to be used at periodic if irregular intervals.

2. C. O. Sauer considered Troll to have been wrong about natural vegetation. In a letter he wrote to a colleague, Sauer complained, "He can't be concerned with what 'natural vegetation' means, it's there for him—and I don't even see the concern with 'successions' of the ecologists—much less interest in man as a factor. . . . Troll is a naive environmentalist: climate = vegetation = Wirtschaft" (SP, COS to JJP, 5-27-1959).

Chapter 4. Malaria and Settlement Retrogression in Mizque, Bolivia

1. *Mizque,* meaning "sweet" (*misk'i*) in Quechua, alluded to the abundance of honey once collected from the wild. One kind of native bee (*ilinchupa*) found there deposits its honey inside minuscule balls of wax stored in underground chambers. Collecting this delicious black liquid along with the yellowish wax remains a tradition to some people.

2. Charcas was the first name given to the region which had La Plata (founded 1538) as its settlement focus. In 1559 the *audiencia* of Charcas was formed as an administrative unit within the viceroyalty of Peru, but in 1782 it became part of the viceroyalty of Río de la Plata headquartered in Buenos Aires. Alto Perú was the name used in the colonial period to designate the area that became the republic of Bolivia in 1825. The name of La Plata was changed to Chuquisaca (still the name of the department) in 1825 and to Sucre in 1840.

3. In 1604 Santa Cruz de la Sierra was moved to a new site initially called San Lorenzo de la Barranca, which was 50 leagues (1 league = 5 m) closer to the Andean front, though still only 437 m above sea level. The previous settlement was 130 leagues (ca. 650 km) from Mizque.

4. The place name Salinas del Río Pisuerga represents a toponymic transfer from a town in Palencia Province in Spain, but it was also meant to honor the viceroy of Peru at the time, Luis de Velasco, the marquis of Salinas. Reversion in the eighteenth century to the Quechua name of Mizque evoked the later demise of the town's Spanish

character. The valley in which the town is located has been known as Mizque continuously since the Inca period.

5. The "Libro de Defunciones," in which the name of the deceased and cause of death were recorded, was part of parish rather than governmental records. In these entries, deaths from tercianas were always fewer than those caused by "fiebres" (fevers). While several diseases—most notably typhoid—have a feverish phase, the large number of annual deaths attributed to "fiebre" strongly suggests hyperendemic malaria. Mortality from malaria was probably twice that actually recorded because of periodic absences of the priest and the frequent failure of the next-of-kin outside town to notify him of the deceased.

6. A native name for malaria is in itself not strong evidence for its pre-Columbian presence. The Quechua word *chucchu* could have originally referred to the febrile stage of Carrion's disease (bartonellosis), an indigenous malady of the western Andean slope. After the Spanish conquest the name "chucchu" may have been transferred to the recently introduced malaria.

7. That the mosquito vector might be able to racially discriminate the sources of blood meals was suggested by a controlled experiment in which anophelines preferentially bit people who belonged to blood group O (Wood 1975). American Indians are overwhelmingly of blood group O.

Chapter 5. The Andes as a Dairyless Civilization

1. An earlier version of this chapter was published in Spanish (Gade 1993).

2. The erroneous assertion that llama and sometimes alpaca milk is an item of human consumption in the Andes is widespread and persistent. I have brought together 11 such errors published in encyclopedias, dictionaries, and textbooks which state that llamas were or are milked (Gade 1969a). More recent cases of the same misconception of the llama as a traditional milch animal are those of Carter (1971:13), Coyle (1982:368), Real academia española (1984, 2:848), and Weatherford (1988:94). Cimó (1988:32) and Fitzhugh and Wilhelm (1991:108, 110) were caught in the related trap of supposing that alpacas are milked.

3. In 11 different trips to the Central Andes of Peru and Bolivia between 1963 and 1997, I have never observed any indication, however subtle, of the milking of either llamas or alpacas. Bonavia (1996), who has carefully sifted the literature on South American camelids, is categorical and emphatic about the absence of llama or alpaca milk as food for humans in the past or the present.

4. The insightful contributions of Hahn (1856–1928) are recognized by Plewe (1975–1976) and in three previous biographical sketches translated by West (1990:27–59). West also added an English translation of some of Hahn's work, reviewed the importance of it, and provided his complete bibliography. Hahn had a strong influence on the thought of Carl O. Sauer through his contacts with anthropologist Robert Lowie at Berkeley. Lowie (1937) introduced more German ethnological ideas to North American scholarship than Franz Boas did.

5. Early relations between people and animals include the importance of religious sacrifice in collective human psychology and culture history. Using reasoning derived from cultural materialism, Rodrigue (1992) rejects the religious imperative in the do-

mestication of animals. Yet her statistical sample of early archaeological sites missed some of the places, most notably Catal Hüyük in Turkey, that are most salient in casting light on the ties between religion and cattle as an early phase of their domestication. In contrast, Walsh's (1989) lucid exposition of the long prehistory of female divinities in the Old World and of its progressive transmutation to cow worship makes clear the connection.

Chapter 6. Epilepsy, Magic, and the Tapir in Andean America

1. No taxonomic distinction was made between highland and lowland tapirs until the nineteenth century. Thus La Condamine (1751) made no distinction between the mountain tapir he saw in the woods around Quito and the lowland tapir he noted in the Amazon forest. Following the Spanish example of name transfer, La Condamine called the tapir *élan* (French for the European elk/moose).

2. Tapir enteroliths consist of layers of hydrated phosphates that can be found in the intestine after the animal has been killed (Murphy et al. 1997).

3. The Moors also introduced those same Arabic words, *anta* and *danta,* into Portuguese for what in Africa at an earlier time had referred to the antelope (Real academia española 1992, 2:1172). Neuendorff (1940:219) wrongly placed *anta* and *danta* as words of Guarani origin.

4. Moose parts were by no means the only European folk remedies for epilepsy. Eating cures for the disease included suckling dogmeat, camel's brain, peacock dung, bear testicles, elephant liver, hare's milk, rooster genitalia, and 21 houseflies consumed all at once. Drinking cat's blood or the urine of a black cow also was proclaimed a cure. A talisman of a lynx claw, a walrus tooth, a peacock feather, a live lizard, or a wolfskin was said to prevent the disease.

5. *Elend* was an older name in German applied to the elk/moose. It was also a term for epilepsy. Today *Elend* has only the more general meaning of "misery."

6. Spaniards may have acquired the notion of epilepsy as a disorder of the heart (mal de corazón) from the New World. Medieval Europe explained epilepsy in terms of humors in the brain cavity (Gordonio 1991:113–19). The transfer of an Andean medical idea to Spain would clarify a Spanish name that puzzled Kanner (1930b:125) in his wide-reaching survey of terminology for epilepsy.

Chapter 7. Valleys of Mystery on the Peruvian Jungle Margin and the Inca Coca Connection

1. Renard-Casevitz, Saignes, and Taylor (1984) asserted that the coca-growing region of Avisca along the Río Tono marked the border of Antisuyo. In fact, the Inca concept of Antisuyo included highlands, not just the yungas.

2. Although it is drawn at a small scale, Gagliano's (1994:3) map of coca shows gross inaccuracies in where this plant has been grown. Vast stretches of high puna, where coca cultivation is impossible, are incorrectly shown as part of the "coca production area."

3. For example, Guillaume (1888) placed the Río de Callanga as a tributary flowing into the Río Tono from the southwest.

4. Callanca, Kallanqa, and Kallanqan are alternative spellings of Callanga. Callanga is the Hispanicized derivation from the Quechua Kallanqan.

5. Hassel pursued discovery of Inca roads chiefly out of an antiquarian interest. He apologized to his team members for the deprivations he placed them under in searching for these roads.

6. Alternative spellings for Lacco are Laq'o or Laco. The most recent orthography for Gualla/Hualla is Walla. The names Lacco and Gualla/Hualla have been used to indicate the same place in a broad sense.

7. Especially later in the colonial period, Gualla was often written as Hualla. It is not clear if the place name Gualla/Hualla in the Mapacho drainage was related in some way to the Hualla ethnic group, who lived very near Cusco. Manco Capac, the first Inca, was said to have conquered these people and to have had them killed (Sarmiento de Gamboa 1960). Possibly a group of people near Cusco was moved as a mitimae to the lower Mapacho. Robledo (1899:439) was among the first to have pointed out that Gualla/Hualla and Lacco were part of the same zone.

8. Toledo's *visita* of 1575 mentions Acata as a town in the Gualla Valley (Cook 1975:202). This would seem to be a confusion with Acata, a coca-growing community in the Marcapata Valley, more than 200 km to the south.

9. In 1982, Lacco was detached from Lares District and became part of Yanatile District, with its seat at Quellouno on the lower Yanatile.

Chapter 8. Guayaquil as Rat City

1. Nietschmann (1979:76–78) discusses rats of an unnamed species in a Nicaraguan village in which he lived and offers a local recipe for rat soup. His description of their acrobatics makes it quite clear that he was dealing with *Rattus rattus.* But his tale of a "three-or-four pound (1.37–1.82 kg) rat clenched to his T-shirt" is a gross exaggeration. Neither species of rat can exceed a weight of 1.5 lbs (0.68 kg). In Robinson's (1965:233) detailed study, the biggest brown rat (which as a species is larger than the black rat) weighed in at 1.1 lbs (0.51 kg).

2. Free-living rats engage in marathon sexual activity that seems recreational more than reproductive. I have observed rat orgies in a Pittsburgh backyard, where the same female was mounted repeatedly by different males.

3. Part of a right femur identified as *Rattus* sp. was initially reported from preceramic levels at Huaca Prieta, on the coast of Peru, dated ca. 3000 B.C. That judgment was later revised, for an Old World rat of such antiquity "suggests a disturbance that was undetected during the excavation or an error in the sorting or labeling procedure" (Bird and Hyslop 1985:243).

4. That transatlantic ships have almost invariably carried rats until the twentieth century is suggested by the evidence for rat disembarkation in North America from Florida to Labrador (Armitage 1993).

5. In medieval Spain, cats were not uncommonly eaten by humans. Schwabe (1979:176) offers a recipe of Spanish origin for "Extremaduran cat stew," cooked in olive oil and wine.

6. Most commentators wrote about rats (Cameron 1973:155), but in some sources the rodents eaten on the Magellan journey were "mice" or their French equivalent, *souris* (Pigafetta 1969:24; Pigafetta 1972). That rodential gloss might be explained by the fact that in Pigafetta's native tongue, Italian, *topo* (pl. *topi*) can still refer to either rat or mouse.

7. Both *R. rattus* and *R. norvegicus* have several vernacular names in most European languages, none of which is ideal in differentiating them. In English, black rat is a poor descriptive for *R. rattus* because color mutations of this species may be brown or gray. Also not totally satisfactory for the former are roof rat (in most places also found elsewhere than just roofs), house rat, and ship rat (both ambiguous because both species board ships and live in houses). Especially in the United States, the brown rat is known as the Norway rat, much to the displeasure of Norwegians, who rightly argue that Norway was not its place of origin. Sewer rat, alley rat, dump rat, river rat, water rat, and barn rat are generally useful descriptives for *R. norvegicus;* less accurate are wharf rat, wander rat, field rat, and Hanoverian rat.

8. Patiño (1972:222) placed the invasion of the brown rat in the interior city of Medellín, Colombia, in 1881–1882, which helps to corroborate its arrival at Guayaquil. The speed of its dispersal may owe something to its large litters. The female brown rat typically gives birth to nine young, as compared with six young in a black rat litter.

9. Why, in Castillian Spanish, *ratón* was later limited to the smallest commensal rodent species is a curious reversal of the normal meaning of the augmentative suffix as something large.

10. *Pericote,* a word not used in Spain, has generally referred to the commensal rat. Cobo (1956, 1:351–52) used *pericote* as a synonym for *rat* to distinguish it from the "*pequeñuelo,*" which is probably the mouse. Alcedo (1967, 4:346) wrote that *pericote* was the name given to "large house rats in all of America."

11. Zárate (1947:486) says that the Indian term for rat was *ucucha,* meaning "from the sea," implying that the term followed its introduction from abroad. However, the term probably predates the conquest, for other kinds of small rodents have been on the coast all through the long course of Andean culture history.

12. *R. norvegicus* has a uniform karyotype of 2n = 42, but the chromosome number of *R. rattus* apparently shows geographical variation, for some populations have a diploid number of 38, others 40, and still others 42. The two species have been observed to mate in a few captive situations, but no offspring has ever been recorded (Robinson 1965:6–8).

13. In regard to his travels through Latin America, the writer Paul Theroux's (1979: 279–80) comments about Guayaquil concerned a nest of chirping rats found in a ceiling fixture in his room at the Hotel Humboldt.

14. Hideyo Noguchi, a research bacteriologist at the Rockefeller Institute in New York, confused leptospiral jaundice with yellow fever, even though the latter is caused by a virus. Ironically Noguchi died of yellow fever in Africa in 1928.

Chapter 9. Carl Sauer and the Andean Nexus in New World Crop Diversity

1. In 1945 Sauer carried on extensive and mutually beneficial correspondence with Liberty Hyde Bailey about domesticated cucurbits (BP, LHB and CS).

2. The idea for the *HSAI* started in 1932, but funding for it did not come until it was included in an aggressive appropriations bill under the politically motivated "good neighbor" policy toward Latin America.

3. Harris (1972) subsequently refined the idea that vegeculture was ancient in South America, Africa, and Asia, but because it was nutritionally unbalanced could not expand and was intruded by seed agriculture, which came later. That roots and tubers

serve as storage organs for plants during the dry season may be an oversimplification. Hather (1996) has presented evidence that some crops with useful underground parts originated in the humid tropics as an adaptive strategy for longevity.

4. Dates regarding New World prehistoric agriculture are being revised. AMS (accelerator mass spectrometry) has, in almost all cases, come up with younger dates; for example, maize domestication has now been revised from 5000 B.C. to 2700 B.C., and the age of Bat Cave corn has been cut from 4000 B.C. to 1000 B.C. (Smith 1995). The use of phytoliths and plant pollen to date early crops is still somewhat controversial (Piperno 1994).

5. Sauer appears never to have referenced in print the work of any of his Ph.D. students, perhaps because he thought it was vain to promote in that way those who worked under him. Sauer was critical of some geography professors in Germany who publicly praised the papers of their own students at the Geographentag meeting he attended in Heidelberg in 1957 (SP, COS to JJP, 5-27-1959).

6. Anderson's contribution to the study of maize was great; that of Sauer was minimal. Anderson's identification of key morphological characteristics, notably kernel shape, ear shape, and row number, provided the basis for all subsequent classifications of maize from Latin America (Bohrer 1986:28). The idea of "races of maize" was Anderson's brainchild and was first published in a coauthored paper (Anderson and Carter 1945). Anderson contributed greatly to Sauer's education about crop variation, but his scientific achievements were not limited to crop variation (Stebbins 1978). Yet he also followed a wrong path on some crop origins. From what is known now and was even then, Anderson's (1952) questioning of the New World origins of maize and capsicum seems inexplicable.

7. In *Agricultural Origins and Dispersals,* Sauer was still hedging his bets about the place of origin of maize, but at the 1957 International Congress of Americanists in Costa Rica, he came out in favor of its origin in Mexico (Sauer 1959).

8. Brücher (1915–1991) was an expatriate German geneticist with a keen morphological eye who contributed field-based knowledge to New World crop origins (1977, 1989). He identified the wild relative of ullucu (*Ullucus tuberosus*) and reported on the semidomestication of *Phaseolus* in the Andean foothills of Colombia and Venezuela. He also found that *Pachyrhizus* seeds, normally considered to be poisonous, were in fact eaten (Brücher 1985). In 1991 Brücher was mysteriously murdered at his *finca* near Mendoza, Argentina. Heinz (a.k.a. Enrique) Brücher, a lieutenant colonel in the Gestapo, had in 1943 orchestrated the removal of germplasm material from Soviet plant-breeding stations (Deichmann 1996:258–64). Thus, Brücher's later criticisms of Vavilov's hypotheses about centers of origin may have been motivated by more than scientific differences.

9. Sauer wrote Steward (NAA, COS to JS, 9-1943): "The pre-Columbian trans-Pacific connections can no longer be shushed, but need a thorough and unheated scrutiny whatever Metraux's boiling point may be." Jonathan D. Sauer (1993) evaluated the knowledge about cultivated plant origins. His conclusions regarding crop centers and long-distance diffusion differed sharply from those in C. Sauer's 1952 work, partly the result of some new knowledge but mostly the result of much more caution in interpreting evidence.

10. Sauer defended the absence of maps by writing to Steward: "I think I could

come closer to setting the distributional limits in west Mexico for material culture traits than anyone else. Yet I'd hate to commit myself in a lot of such things in west Mexico, and for the rest of Latin America I am just one of the crowd. I marvel at people who have the courage to draw distributional props for a continental mass and have been scared all along that you might ask me to provide such maps for my crop sections . . ." (NAA, COS to JS, 10-13-1943).

11. In the second edition of this little classic, the maps were in black and white and reduced in scale. Three previously published articles were included. The reprinting of the second edition was retitled, even though none of the content was changed (Sauer 1972). New archaeological information derived from two decades of research would have been in order in these later editions.

12. Dump heaps as microarenas for the earliest cultivation could be inferred in Sauer, but it was Anderson (1952:136–41) who developed the progression involved without realizing that the idea started with Thiess Englebrecht (1916), a scholarly farmer who had an estate in Schleswig-Holstein (Zeven 1973).

13. Edgar Anderson wrote to Sauer, giving his opinion of the book: "This was a true masterpiece with all irrelevance stripped away and partly because the beautifully ordered development of your thought drove me to participate effortlessly in the unrolling as you had planned it. . . . I can't remember when I had my understanding of so many things enlarged in so short a time. . . . I like the style . . . it is closer to Melville than one expects in a scientific essay, but it is readable and effective . . ." (SP, EA to COS, 9-23-1952). That same year Anderson's own book, *Plants, Man and Life,* was published, covering some of the same ground as Sauer had. Visvanathan (1996:315) offered the opinion that "despite its descriptive simplicity, [it] is one of the finest scientific reflections on agriculture." Like Sauer's, Anderson's book had an accessible writing style that attracted a wide audience. Anderson's knowledge of Latin American peasant agriculture was greatly heightened by his many wartime conversations with Sauer, even though they never shared any time together in the field.

14. In February 1942 Sauer wrote about Chiloé in a letter: "Here you are in the ancestral home of all non-Andean potatoes, with an unknown wealth of genes, and the first thing they do is to try to introduce plants from the big commercial regions and to destroy what they have of their own" (quoted in West 1982:53).

References

Manuscript Sources

AGI: Archivo general de Indias, Seville, Spain (All the materials listed for this archive are contained in its Gobierno section.)

Charcas 17. 1603. Carta de la Audiencia de Charcas al rey. La Plata, 12-26-1603.

Charcas 31. 1561. Relación de la ciudad de La Plata a S.M. sobre su sitio y termino. La Plata, 10-8-1561.

Charcas 31. 1611. Relación que haze la ciudad de La Plata de los reinos del Perú a la magestad del rey don Felipe. La Plata.

Charcas 32. 1633. Ciudad de San Lorenço de la Frontera de la Provincia de Santa Cruz de la Sierra.

Charcas 139. 1651. Relación que hace el Obispo de Santa Cruz. Salinas/Mizque, 11-15-1651.

Charcas 140. 1609. Autos de la dibision del Obispado de la ciudad de La Plata. La Plata, 2-16-1609.

Charcas 140. 1612. División del Obispado la ciudad de La Plata.

Charcas 153. 1654. Informe de Obispado de Santa Cruz de la Sierra al rey. Salinas/Mizque, 7-24-1654.

Charcas 388. 1628–1761. Expedientes sobre la translación de la Catedral de Santa Cruz de la Sierra a la Villa de Mizque y declaración de privilegios a los Padres de Comapañía en dipensación de matrimonios. San Lorenzo.

Charcas 410. 1770. Para el consejo por parte del Rev. obispo de S. Cruz de la Sierra. San Lorenzo.

Charcas 464. 1583. Noticia de las provincias del Perú y castas del norte y sur y empleos que en ellas se proveen el en año de 1583.

Contaduria 1786. 1582. El repartimiento de Misque tiene en encomienda doña Izabel Paniagua . . .

Lima 78. 1674. Graves inconvenientes que se sigue a la real hacienda al comercio y vasallos del Perú. (No date).

AGN: Archivo general de Indias, Buenos Aires, Argentina

1771. Colonia Contaduría. Cuerpo 18, A2, No. 1, Legajo 46, 7-29-1771.

1772. Colonia Contaduría. A2, No. 1. Libro dos padrones y revistas de Mizque.

1792. Colonia Contaduría. A2, No. 5, Legajo 50. Padrones de Cochabamba, Lib. 1, Padrones y revistas de indios de Mizque.

1804. Colonia Contaduria. Sala 13, Cuerpo 18, A3, No. 4. Padrones y revistas de indios de Mizque.
ANB: Archivo nacional de Bolivia, Sucre
1692. Padrón de los indios chues de San Sebastián de la Villa de Mizque hecho por el General don José Antonio Ponce de Leon, corregidor y justicia mayor de dicha villa.
1709. Padrón de la ciudad de Mizque levantado por el Corregidor don Juan Baptista de Oliden.
1783, no. 12. Visita y padrón de los indios tributarios de Mizque hecho por don Francisco Robles.
1825, tomo 1, no. 8. Ministerio de Hacienda; Nicholas Jacob Montaño a Cabildo, Cochabamba.
1844. Expedientes sueltos. Padrón de Cliza.
AC (Audiencia de Charcas). Mojos III, 1768. Expediente obrado acerca del remate de la hacienda de Chalguani. La Plata, 9-20-1768.
AM (Archivo de Mizque). 1603. Comunicación mas detallada del cavildo de la Real Audiencia de La Plata de la fundación de la Villa de Salinas del Río Pisuerga.
AM. 1679 (*sic*), No. 1. Padron de los Indios del pueblo de San Sebastian de los Chues levantado por el Maestre Campo Diego de Matos 1669.
E (Encomiendas). 1709, tomo 36.
E. 1714, tomo 34. Hacienda de Perereta en el valle de Omereque.
EC (Expedientes coloniales). 1802, no. 168. Venta de las haciendas Omereque y Viña Perdida.
EC. 1805, no. 67. Autos seguidos entre el convento de San Agustín de Mizque y María Dozal sobre la hacienda de San Juan de Buena Vista.
MI (Ministerio del Interior, República de Bolivia). 1825, tomo 8, no. 63. Antonio de Sucre a cabildo municipal de Mizque. Cochabamba, 7-14-1825.
MI. 1826, tomo 11, no. 13. José María Plaza a Sucre. Cochabamba, 1-4-1826.
MI. 1826, tomo 14, no. 21. Mariano Santibañez a Sucre. 1-13-1826.
MI. 1826, tomo 19, no. 21. Ministerio de Interior al Gobernador de Mizque. Cochabamba, 10-11-1826.
MI. 1827, tomo 16, no. 15. F. Lopez a Facundo Infante. Cochabamba, 10-11-1827.
MI. 1878, tomo 207, no. 15. Al Sr. ministro del estado de Martín Lanza. 3-29-1878.
ASNRA: Archivo del servicio nacional de reforma agraria, La Paz, Bolivia
1957. Expediente 3102. Hacienda Viña Perdida.
1957. Expediente 10191. Hacienda Buena Vista
BP: Bailey, Liberty Hyde, Papers. Cornell University, Ithaca, New York
LHB and CS: Correspondence between Liberty Hyde Bailey and Carl Sauer: 1-13, 1-19, 1-21, 2-23, 3-8, 4-16, 11-10, 11-26, 11-29, 12-4, and 12-6, all in 1945.
IHMT: Instituto de higiene y medicina tropical. Guayaquil, Ecuador
1979–1994. Statistics on leptospirosis.
Sección zoonosis. 1986. Rat bite statistics.
NAA: National Anthropology Archives, National Museum of Natural History, Smithsonian Institution, Washington, D.C.

COS to JS: Letters from Carl O. Sauer to Julian Steward: 6-11-1943, 9-1943, 10-13-1943.
HT to JS: Letter from Harry Tschopik Jr. to Julian Steward: 8-1-1941.
JS to COS: Letter from Julian Steward to Carl O. Sauer: 9-2-1943.
RAH: Real Academia de la historia, Madrid, Spain
Colección Muñoz, legajo 133, 1604. Relación del señor Verrei don Luis de Velasco (Marques de Salinas) al señor (don Gaspar de Acevedo y Zúñiga) Conde de Monterrei sobre el estado del Perú, no. 35, Villa de Salinas i su fundación.
SNM: Servicio nacional de meteorología. La Paz, Bolivia
MSS weather records.
SP: Sauer, Carl, Papers, Bancroft Library, University of California, Berkeley
COS to EA: Letters from Carl O. Sauer to Edgar Andersen: 8-31-1942, 10-8-1942.
COS to FWM: Letter from Carl Sauer to Felix Webster McBryde: 2-14-1946.
COS to HG: Letter from Carl Sauer to Howard Gentry: 10-19-1945.
COS to HSS: Letter from Carl O. Sauer to Henry S. Sterling: 2-2-1939.
COS to JJP: Letter from Carl Sauer to James J. Parsons: 5-27-1959.
COS to JFS: Letter from Carl Sauer to Joseph Spencer: 11-4-1950.
EA to COS: Letters from Edgar Anderson to Carl Sauer: 9-1-1943, 11-1-1943, 9-23-1952, 2-17-1959.
HB to COS: Letters from Henry Bruman to Carl Sauer: 5-6-1944, 6-7-1944.
PLW to COS: Letter from Philip L. Wagner to Carl Sauer: 6-27-1969.

Newspaper Sources

Gazeta municipal (Guayaquil). 1901–1914.
El Heraldo (Cochabamba) 1878.
El País (Cochabamba). 1942.
La Razón (La Paz). 1930–1934.
Revista municipal (Guayaquil). 1926–1930.

Other Sources

Abecía, Valentín. 1939. *Historia de Chuqisaca.* Sucre: Editorial Sucre.
Ackerknecht, Erwin H. 1942. "Problems of primitive medicine." *Bulletin of the History of Medicine* 11:503–21.
Acock, Mary C., John Lydon, Emmanuel Johnson, and Ronald Collins. 1996. "Effects of temperature and light levels on leaf yield and cocaine content in two *Erythroxylon* species." *Annals of Botany* 78:49–53.
Acosta, J. de. 1962. *Historia natural y moral de las Indias.* [1590]. Ed. E. O'Gorman. Mexico City: Fondo de cultura económica.
Aguiló, F. 1982. *Enfermedad y salud según la concepción aymara-quechua.* Sucre: Talleres Gráficos Qorillama.
Alcedo, Antonio. 1967. *Diccionario geográfico de las Indias occidentales o América.* [1786–1789]. 4 vols. Madrid: Ediciones Atlas.
Aldrete Maldonado, F. 1989. "Testimonios." [1590]. Ed. R. Ravines. *Boletín de Lima* 11 (64):3–14.

Allen, Glover Morris. 1942. *Extinct and vanishing mammals of the Western Hemisphere.* Cambridge, Mass.: American Committee for International Wildlife Protection.

Alsedo y Herrera, Dionisio. 1987. *Compendio histórico de la Provincia de Guayaquil.* [1741]. Ed. Pedro Carbo. Guayaquil: Biblioteca de autores ecuatorianos.

Altieri, Miguel A., and Laura C. Merrick. 1988. "Agroecology and in situ conservation of native crop diversity in the Third World." Pp. 361–69 in *Biodiversity,* ed. E. O. Wilson. Washington, D.C.: National Academy Press.

American Geographical Society (AGS). 1938. *Map of Hispanic America: Sheet S.D-18 ("Lima") 1:1,000,000.* New York: AGS.

Amoroso, E. C., and P. A. Jewell. 1963. "The exploitation of the milk-ejection reflex by primitive peoples." Pp. 126–58 in *Man and cattle,* ed. A. E. Mourant and F. E. Zeuner. London: Royal Anthropological Institute.

Anderson, Edgar. 1952. *Plants, man and life.* Boston: Little, Brown.

Anderson, Edgar, and George F. Carter. 1945. "A preliminary survey of maize in the southwestern United States." *Annals, Missouri Botanical Garden* 32:297–322.

Andrews, Jean. 1995. *Peppers: The domesticated capsicums.* 2d ed. Austin: University of Texas Press.

Angles Vargas, Víctor. 1992. *El Paititi no existe.* Cusco: Privately printed.

Ansión, J., and Chris Van Dam. 1986. *El árbol y el bosque en la sociedad andina.* Lima: Proyecto FAO/Holanda/INFOR.

Antúñez de Mayolo R., Santiago E. 1981. *La nutrición en el antiguo Perú.* Lima: Banco central de reserva del Perú.

Aparicio Bueno, Fernando. 1985. *En busca del misterio de Paititi.* Cusco: Editorial Andina.

Apffel-Marglin, Frédérique, ed. 1998. *The spirit of regeneration: Andean culture confronting Western notions of development.* London: Zed Books Ltd.

Araúz, A. 1945. *Informe del prefecto (Cochabamba).* Cochabamba: Editorial América.

Armitage, Philip L. 1993. "Commensal rats in the New World, 1492–1992." *The Biologist* 40(4):174–78.

Arnold, Denise Y., and Juan de Dios Yapita. 1992. "Fox talk: Addressing the wild beasts in the southern Andes." *Latin American Indian Literatures Journal* 8(1):9–37.

Arriaga, Pablo. 1968. "Extirpación de la idolatría en el Perú." [1616]. Pp. 191–278 in *Crónicas peruanas de interés indígena,* ed. F. Esteve Barba. Madrid: Ediciones Atlas.

Arzube-Rodríguez, M. 1968. "Investigación de la fuente alimenticia del *T. dimidiata* Latr. 1811 mediante la reacción de precipitina." *Revista ecuatoriana de higiene y medicina tropical* 23:137–52.

Avery, Kevin J. 1994. *Church's great picture: The heart of the Andes.* New York: Metropolitan Museum of Art.

Baied, C. A. 1991. *Late-Quaternary environments and human occupation of the south-central Andes.* Ann Arbor: University Microfilms International.

Baied, C. A., and Wheeler, J. C. 1993. "Evolution of high Andean puna ecosystems: Environment, climate, and culture change over the last 12,000 years in the Central Andes." *Mountain Research and Development* 13(2):145–56.

Bakewell, Peter. 1984. *Miners of the red mountain: Indian labor in Potosí, 1545–1650.* Albuquerque: University of New Mexico Press.

Balanza, E., and G. Taboada. 1985. "The frequency of lactase phenotypes in Aymara children." *Journal of Medical Genetics* 22(2):128–30.
Baldwin, Martha R. 1993. "Toads and plague: Amulet therapy in seventeenth-century medicine." *Bulletin of the History of Medicine* 67:227–47.
Baleato, Andrés. 1887. *Monografía de Guayaquil el año de 1820.* Guayaquil: Imp. La Nación.
Balée, William, ed. 1998. *Advances in historical ecology.* New York: Columbia University Press.
Balick, M. J. 1985. "Useful plants of Amazonia: A resource of global importance." Pp. 339–68 in *Key environments: Amazonia,* ed. G. T. Prance and T. E. Lovejoy. Oxford: Pergamon Press.
Balslev, H., and T. DeVries. 1982. "Diversidad de la vegetación en cuatro cuadrantes en el páramo arbustivo del Cotopaxi, Ecuador." *Publicaciones del Museo ecuatoriano de ciencias naturales* 3(3):20–32.
Banco de la nación. 1969. *Informe: Zona del Cuzco.* Cusco: División de alcoholes y bebidas.
Barbour, Michael G. 1995. "Ecological fragmentation in the fifties." Pp. 233–55 in *Uncommon ground: Toward reinventing nature,* ed. William Cronon. New York: W. W. Norton.
Barrau, Jacques. 1977. "Histoire naturelle et anthropologie." *L'Espace géographique* 3:203–9.
Bastian, E. 1986. "Grundzüge der Vegetationsdegradation in Südost-Bolivien." Pp. 23–67 in *Bolivien: Beiträge zur physischen Geographie eines Andenstaates,* ed. J. J. Buchholz. Hannover: Geographischen Gesellschaft.
Bastien, Joseph. 1978. *Mountain of the condor: Metaphor and ritual in an Andean ayllu.* St. Paul: West Publishing Co.
Basto Girón, L. J. 1957. *Salud y enfermedad en el campesino peruano del siglo XVII.* Lima: Instituto de etnología y arqueología UNMSM.
Bateson, Gregory. 1979. *Mind over nature: A necessary unity.* New York: Dutton.
Beadle, George W. 1939. "Teosinte and the origin of maize." *Journal of Heredity* 30:245–47.
Beck, Stephen, and Emilia Garcia. 1991. "Flora y vegetación en los diferentes pisos altitudinales." Pp. 65–108 in *Historia natural de un valle en los Andes: La Paz,* ed. Eduardo Forno and Mario Baudoin. La Paz: Instituto de ecología UMSA.
Becker, Barbara. 1988. "Degradation and rehabilitation of Andean ecosystems: An example from Cajamarca." *Angewandte Botanik* 62:147–60.
Bellin, Jacques. 1986. *Descripción geográfica de la Guayana.* Caracas: Ediciones de la presidencia de la república.
Bennett, Wendell C. 1946. "The archeology of the Central Andes." Pp. 61–147 in vol. 2 of *Handbook of South American Indians,* ed. Julian Steward. Bureau of American Ethnology, Bulletin 143. Washington, D.C.: Smithsonian Institution.
Bennett, Wendell C., and Junius B. Bird. 1964. *Andean culture history.* Garden City: Natural History Press.
Berenguer, J., and J. L. Martínez. 1989. "Camelids in the Andes: Rock art environment and myths." Pp. 390–416 in *Animals into art,* ed. H. Morphey. Boston: Unwin Hyman.

Berry, Wendell. 1981. *The gift of good land: Further essays cultural and agricultural.* San Francisco: North Point Press.
Bingham, Hiram. 1930. *Machu Picchu: A citadel of the Incas.* New Haven: Yale University Press and the National Geographic Society.
Bird, Junius B. 1948. "Preceramic cultures in Chicama and Virú." *American Antiquity* 13:21–28.
Bird, Junius B., and John Hyslop. 1985. "The preceramic excavations at the Huaca Prieta, Chicama Valley, Peru." *Anthropological Papers of the American Museum of Natural History* 62(1):1–294.
Blanco, F. 1901. *Diccionario geográfico de la Republica de Bolivia: Departamento de Cochabamba.* La Paz: Oficina nacional de inmigración, estadística y propaganda geográfica.
Blas Valera. 1992. "Relación de las costumbres antiguas de los naturales del Pirú." [1590]. Pp. 43–122 in *Antiguedades del Perú,* ed. H. Urbano and A. Sánchez. Madrid: Historia 16.
Bohrer, Vorsila L. 1986. "Guideposts in ethnobotany." *Journal of Ethnobiology* 6(1):27–43.
Bohs, Lynn. 1989. "Ethnobotany of the genus *Cyphomandra* Solanaceae." *Economic Botany* 43(2):143–63.
Bonavia, Duccio. 1996. *Los camélidos sudamericanos (una introducción a su estudio).* Lima: Instituto francés de estudios andinos.
Bonnier, Elisabeth. 1986. "Utilisation du sol à l'époque préhispanique: Le cas archéologique du Shaka-Palcamayo (Andes centrales)." *Cahiers des sciences humaines* 22(1):97–113.
Borda P., M. R. 1965. "Distribución geográfica de anofelinos en Bolivia." *Revista de la salud pública boliviana* 5(27):19–27.
Borrero Vega, Ana Luz. 1989. *El paisaje rural en el Azuay.* Cuenca: Banco central del Ecuador.
Bortoft, Henri. 1996. *The wholeness of nature: Goethe's way toward a science of conscious participation in nature.* Hudson, N.Y.: Lindisfarne Press.
Bovo de Revello, Julian. 1848. *El brillante porvenir del Cuzco.* Cusco: Privately printed.
Bowman, Isaiah. 1909. "The physiography of the Central Andes. I. The maritime Andes." *American Journal of Science* 28:373–402.
Bowman, Isaiah. 1914. "Results of an expedition to the Central Andes." *Bulletin of the American Geographical Society* 44(3):161–83.
Bowman, Isaiah. 1916. *The Andes of southern Peru.* New York: Henry Holt.
Boyden, S. 1992. *Biohistory: The interplay between human society and the biosphere.* Man and the Biosphere 8. Paris and Park Ridge: UNESCO and Parthenon Publishing Group.
Brandbyge, J. 1992. "Planting of local woody species in the páramo." Pp. 265–75 in *Páramo: An Andean ecosystem under human influence,* ed. H. Balslev and J. L. Luteyn. London: Academic Press.
Brandbyge, J., and L. B. Holm-Nielsen. 1986. *Reforestation of the high Andes with local species.* Risskov, Denmark: The Library, Botanical Institute, University of Aarhus.
Brandin, Abel Victorino. 1955. "Clima, costumbres y enfermedades predominantes en

Guayaquil y Quito." Pp. 1–70 in *Estudios médicos ecuatorianaos raros, importantes y curiosos,* ed. Luis A. León. Quito: Universidad central.
Brotherston, G. 1989. "Andean pastoralism and Inca ideology." Pp. 256–68 in *The walking larder: Patterns of domestication, pastoralism and predation,* ed. J. Clutton-Brock. Boston: Unwin Hyman.
Browman, David L. 1986. "Chenopod cultivation, lacustrine resources, and fuel use at Chiripa, Bolivia." *The Missouri Archaeologist* 47:137–72.
Browman, David L. 1989. "Origins and development of Andean pastoralism: An overview of the past 6000 years." Pp. 269–91 in *The walking larder: Patterns of domestication, pastoralism and predation,* ed. J. Clutton-Brock. Boston: Unwin Hyman.
Browman, David L., and James N. Gundersen. 1993. "Altiplano comestible earths: Prehistoric and historic geophagy of highland Peru and Bolivia." *Geoarchaeology: An International Journal* 8(5):413–25.
Bruce, James. 1913. *South America: Observations and impressions.* New York: Macmillan.
Brücher, Heinz. 1960. "Problematisches zum Ursprung der Kulturkartoffel aus Chiloé." *Zeitschrift für Pflanzenzücht* 43:241–65.
Brücher, Heinz. 1969. "Die Evolution der Gartenbohne (*Phaseolus vulgaris*) aus der südamerikanischen Wildbohne (*Ph. aborigineus*)." *Angewandte Botanik* 42:119–28.
Brücher, Heinz. 1977. *Tropische Nutzpflanzen: Ursprung, Evolution und Domestikation.* Berlin: Springer-Verlag.
Brücher, Heinz. 1979. "Das angebliche 'Genzentrum Chiloé' 50 Jahre nach Vavilov." *Zeitschrift für Pflanzenzücht* 83:133–47.
Brücher, Heinz. 1985. "Der reiche Gen-Fundus vergessener tropischer Nutzpflanzen." *Gewissenschaften in unserer Zeit* 3(1):8–14.
Brücher, Heinz. 1989. *Useful plants of Neotropical origin and their world relatives.* Heidelberg/New York: Springer-Verlag.
Bruhns, Karen Olsen. 1979. "The whole hog (as it were)." *Ñawpa Pacha* 17:87–90.
Bruman, Henry. 1936. "The Russian investigations on plant genetics in Latin America and other bearing on culture history." *Handbook of Latin American Studies* 3:449–58.
Bruman, Henry. 1944. "Some observations on the early history of the coconut in the New World." *Acta Americana* 2:220–43.
Bruman, Henry. 1996. "Recollections of Carl Sauer and research in Latin America." *Geographical Review* 86(3):370–76.
Brush, Stephen B. 1977. *Mountain, farm and family: The economy and human ecology of an Andean valley.* Philadelphia: University of Pennsylvania Press.
Brush, Stephen B. 1982. "The natural and human environment of the Central Andes." *Mountain Research and Development* 2(1):19–38.
Brush, Stephen B. 1992. "Reconsidering the Green Revolution: Diversity and stability in cradle areas of crop domestication." *Human Ecology* 20(2):145–67.
Brush, Stephen B. 1993. "Indigenous knowledge of biological resources and intellectual property rights: The role of anthropology." *American Anthropologist* 95(3):653–86.
Budowski, G. 1968. "La influencia humana en la vegetación natural de montañas tropi-

cales americanas." Pp. 157–62 in *Geo-ecology of the mountainous regions of the tropical Americas,* ed. Carl Troll. Colloquium Geographicum 9. Bonn: F. Dümmlers Verlag.

Bueno, C. 1768. *Descripción de las provincias pertenecientes al Obispado del Cuzco.* Lima.

Buffon, Count de. 1780. *Natural history, general and particular.* 7 vols. Trans. William Smellie. Edinburgh: William Creach.

Bukasov, S. M. 1930. "Vozdelyvaemye rasteniya Meksiki, Gvatemali i Kolumbii" (Cultivated plants of Mexico, Guatemala, and Colombia). *Trudy po Prikladnoi Botanike, Genetike i Selektsii.* Papers on applied botany, genetics, and plant breeding, Supplement 47. Manuscript translation in Library of the Missouri Botanical Garden.

Buksmann, P. E. 1980. "Migration and land as aspects of underdevelopment: An approach to agricultural migration in Bolivia." Pp. 108–28 in *The geographical impact of migration,* ed. P. White and R. Woods. London: Longman.

Bunkse, Edmunds V. 1981. "Humboldt and an aesthetic tradition in geography." *Geographical Review* 71(2):127–46.

Cabello Valboa, Miguel. 1951. *Miscelánea antártica.* Lima: Instituto de etnología y arqueología UNMSM.

Cabrera, Angel L. 1968. "Ecología vegetal de la puna." Pp. 91–116 in *Geo-ecology of the mountainous regions of the tropical Americas,* ed. Carl Troll. Colloquium Gegraphicum 9. Bonn: F. Dümmlers Verlag.

Calancha, Antonio de la. 1939. *Crónica moralizada.* [1638]. La Paz: Ministerio de la educación.

Calderón Viacava, L., A. Cazorla Talleri, and R. León Barúa. 1971. "Incidencia de malabsorción de lactosa en jóvenes peruanos sanos." *Acta Gastroenterologica Latinoamericana* 3:11–16.

Calle Escobar, R. 1984. *Animal breeding and production of American camelids.* Lima: Talleres Gráficos de Abril.

Calvo, Luz, et al. 1994. *Raqaypampa: Los complejos caminos de una comunidad andina.* Cochabamba: CENDA.

Cameron, Ian. 1973. *Magellan and the first circumnavigation of the world.* New York: Saturday Review Press.

Campbell, Joseph. 1991. *The masks of God: Primitive mythology.* New York: Penguin Arkana.

Canby, Thomas Y. 1977. "The rat: Lapdog of the devil." *National Geographic* 152(1):60–87.

Candolle, Alphonse de. 1967. *Origin of cultivated plants.* 2d ed. [1886]. New York: Hafner Publishing Co.

Caravantes, F. Lopez de. 1986. *Noticia general del Perú.* Vol. 2. [1630-1631]. Madrid: Ediciones Atlas.

Cárdenas, Martín. 1948–1950. "Plantas alimenticias nativas de los Andes de Bolivia." *Folia Universitaria* (Cochabamba) 2:36–51 (1948); 3:102–19 (1949); 4:86–108 (1950).

Cárdenas, Martín. 1969. *Manual de plantas económicas de Bolivia.* Cochabamba: Imprenta Ichtus.

Cárdenas V., Bolivar. 1957. "Investigación de *Trichinella spiralis* en las ratas y cerdos de la ciudad de Guayaquil." *Revista ecuatoriana de higiene y medicina tropical* 14(2):49–54.

Cardich, Augusto. 1979. "Cultivation in the Andes." *National Geographic Society Research Reports* 20:85–100.

Cardich, Augusto. 1984–1985. "La agricultura nativa en las tierras altas de los Andes peruanos." *Relaciones de la Sociedad argentina de antropología,* n.s. 16:63–96.

Cardozo, G. A. 1954. *Los auquénidos.* La Paz: Editorial Centenario.

Carmen, M. L. 1984. "Acclimatization of quinoa and cañihua in northern Finland." *Annales Agriculturae Fenniae* 23(3):135–44.

Carter, George F. 1945. *Plant geography and culture history in the American Southwest.* New York: Viking Fund.

Carter, William. 1971. *Bolivia: A Profile.* New York: Praeger.

Casanova Abarca, J. A., et al. 1985. *Evaluación de 21 experimentos de propagación de algunos forestales nativos de la Sierra peruana.* Proyecto de investigacion de los sistemas de cultivos andinos IICA-UNSAAC. Mimeograph.

Casas, Alejandro, Barbara Pickersgill, Javier Caballero, and Alfonso Valiente-Banuet. 1997. "Ethnobotany and domestication in xoconochtli, *Stenocereus stellatus* (Cactaceae) in the Tehuacan Valley and La Mixteca Baja, Mexico." *Economic Botany* 51(3):279–92.

Castillo, M. del 1930a. "Informe sobre las regiones de Mizque." *Boletín de la dirección general de sanidad pública* (La Paz) 2(4):349–67.

Castillo, M. del. 1930b. "Esplendor y decadencia de Mizque." *Boletín de la dirección general de sanidad pública* (La Paz) 2(4):344–47.

Caulin, Antonio. 1958. "Historia corográfica, natural y evangélica de la Nueva Andalucía." [1756]. Pp. 243–567 in vol. 3 (Venezuela) of *Historiadores de Indias,* ed. Guillermo Moron. Madrid: Ediciones Atlas.

Cavero B., Gilberto. 1988. *Supersticiones y medicina quechuas.* 2d ed. Lima: CONCYTEC.

Caviedes, César, and Knapp, Gregory. 1995. *South America.* Englewood Cliffs: Prentice-Hall.

Ceballos Bendezú, Ismael. 1981. "Los mamíferos colectados en el Cusco por Otto Garlepp." *Boletín de Lima* 16-17-18:108–19.

Ceballos Bendezú, Ismael. 1981–1982. "Fauna de Callanga y nueva publicación geográfica de la localidad y del Río Kallanqa en el Cusco." *Cantua* (Cusco) 9-10:31–49.

Cedillos, Rafael A. 1987. "Current knowledge of the epidemiology of Chagas' disease in Central America and Panama." Pp. 41–56 in vol. 1: *Taxonomic, ecological and epidemiological aspects* of *Chagas' disease vectors,* ed. Rodolfo R. Brunner and Angel de la Merced Stoka. Boca Raton: CRC Press.

Central ecuatoriana de servicios agrícolas (CESA). 1992. *El deterioro de los bosques naturales del callejón interandino del Ecuador.* Quito: CESA.

Cerón M., Carlos E., and Elizabeth Pozo. 1994. "El bosque de los arrayanes San Gabriel, Carchi-Ecuador: Importancia botánica." *Hombre ambiente* 31:137–65.

Chepstow-Lusty, A. J., K. D. Bennett, V. R. Switsur, and A. Kendall. 1996. "4000 years of human impact and vegetation change in the central Peruvian Andes—with events paralleling the Maya record?" *Antiquity* 70:824–33.

Chevalier, E., J. Casanova, L. Flores, and G. Mamani. 1990. *Apuntes sobre* Buddleja *spp. en el sur de Puno.* Pomata, Peru: Arbolandino.

Choquehuanca, J. D. 1833. *Ensayo de estadística completa de los ramos económicos políticos de la Provincia de Azangaro en el Departamento de Puno.* Lima: Manuel Cassio.

Cieza de Leon, Pedro. 1986. *Crónica del Perú, primera parte.* [1553]. 2d ed. Lima: Editorial Universidad Católica del Perú.

Cimó, P. 1988. "Alpaca, a symbol and a source of hope in the Andes." *Ceres* 21(1):32–37.

Clément, Jean-Pierre. 1983. "El nacimiento de la higiene urbana en la América española del siglo XVIII." *Revista de Indias* 42(171):77–95.

Cleveland, David A., and Stephen C. Murray. 1997. "The world's crop genetic resources and the right of indigenous farmers." *Current Anthropology* 38(4):477–515.

Cleveland, David A., Daniela Soleri, and Stephen E. Smith. 1994. "Do folk crop varieties have a role in sustainable agriculture?" *BioScience* 44(1):740–51.

Cobo, Bernabé. 1956. *Historia del Nuevo Mundo.* [1653]. 2 vols. Madrid: Ediciones Atlas.

Cohen, Mark N. 1977. *The food crisis in prehistory: Overpopulation and the origins of agriculture.* New Haven: Yale University Press.

Cook, N. D., ed. 1975. *Tasa de la visita general de Francisco de Toledo.* Lima: Universidad nacional mayor de San Marcos.

Cook, N. D. 1981. *Demographic collapse: Indian Peru, 1520–1620.* New York: Cambridge University Press.

Cook, O. F. 1910. "History of the coconut palm in America." *Contributions of the U.S. National Herbarium* 12:271–342.

Cook, O. F. 1916a. "Agriculture and native vegetation in Peru." *Journal of the Washington Academy of Sciences* 6:284–93.

Cook, O. F. 1916b. "Staircase farming of the ancients." *National Geographic Magazine* 29(5):474, 496–534.

Cook, O. F. 1920. "Foot-plow agriculture in Peru." *Annual Report of the Smithsonian Institution for 1918:* 487–92.

Cook, O. F. 1925. "Peru as a center of domestication." *Journal of Heredity* 16:33–46, 95–110.

Córdova Aguilar, Hildegardo. 1993. "Firewood use and the effects on the ecosystem—a case study of the Sierra de Piura, northwestern Peru." *GeoJournal* 26(3):297–309.

Córdova Salinas, D. de. 1957. *Crónica franciscana de las provincias del Perú.* [1651]. Ed. L. G. Canedo. Washington, D.C.: Academy of American Franciscan History.

Corominas, Juan. 1954. *Diccionario crítico etimológico de la lengua castellana.* Madrid: Editorial Gredos.

Corrales Badani, J. 1930. *Profilaxis palúdica y labor de saneamiento en Mizque.* Thesis, Facultad de Medicina, Universidad Mayor San Francisco Xavier de Chuquisaca. Sucre: Imp. Bolivar. 50 pp.

Cowles, Henry C. 1899. "The ecological relations of the vegetation on the sand dunes of Lake Michigan." *Botanical Gazette* 27:95–117, 167–202, 281–308, 361–91.

Coyle, L. P. 1982. *The world encyclopedia of food.* New York: Facts on File, Inc.

Crumley, Caroe L., ed. 1994. *Historical ecology: Cultural knowledge and changing landscapes.* Santa Fe: School of American Research Press.

Cunill, Pedro. 1966. *L'Amérique andine.* Paris: Presses universitaires de France.

Cunningham, Isabel Shipley. 1994. "Howard Scott Gentry 1903–1993, distinguished economic botanist and president, Society for Economic Botany 1974." *Economic Botany* 48(4):359–81.

Cutler, Hugh C. 1946. "Races of maize in South America." *Botanical Museum Leaflets* (Cambridge, Mass.) 12:257–91.

Cutright, Paul Russell. 1940. *The great naturalists explore South America.* New York: Macmillan.

Dalence, J. M. 1851. *Bosquejo estadístico de Bolivia.* Sucre: Imprenta de Sucre.

D'Altroy, T. N., and C. A. Hastorf. 1984. "The distribution and contents of Inca state storehouses in the Xauxa region of Peru." *American Antiquity* 49 (2):334–59.

Davis, J. G. 1955. *A dictionary of dairying.* 2d ed. London: Leonard Hill Ltd.

Deagan, Kathleen A. 1987. "Columbus's lost colony." *National Geographic* 172:672–75.

Debouck, D. G., and D. Libreros Ferla. 1995. "Neotropical montane forests: A fragile home of genetic resources of wild relatives of New World crops." Pp. 561–77 in *Biodiversity and conservation of neotropical montane forests,* ed. Steven P. Churchill et al. New York: New York Botanical Garden.

Deichmann, Ute. 1996. *Biologists under Hitler.* Trans. Thomas Dunlap. Cambridge, Mass.: Harvard University Press.

Demeritt, David. 1994. "Ecology, objectivity and critique in writings on nature and human societies." *Journal of Historical Geography* 20(1):22–37.

Denevan, William M. 1983. "Adaptation, variation and cultural geography." *Professional Geographer* 35(4):399–407.

Denevan, William M. 1992. "The pristine myth." *Annals of the Association of American Geographers* 82(3):369–85.

Descola, Philippe. 1994. *In a society of nature.* New York: Cambridge University Press and Maison des sciences de l'homme.

"Descripción de la Gobernación de Guayaquil." 1973 [1605]. *Revista del Archivo histórico del Guayas* 4:61–93.

"Descripción de la Provincia de Abancay." 1795. *Mercurio peruano* 12:112–63.

"Descripción geográfica del Partido de Piura." 1793. *Mercurio peruano* 8:167–229.

Deyermenjian, Gregory. 1988. "Lost city, sacred mountain." *South American Explorer* 19:18–27.

Deyermenjian, Gregory. 1990. "The 1989 Toporake/Paititi expedition: On the trail of the ultimate refuge of the Incas." *Explorers Journal* 68(2):74–83.

Deyermenjian, Gregory. 1995. "The 1995 Lactapata expedition." Unpublished report, Library of the Explorers Club, New York City.

"Diccionario de algunas voces técnicas de mineralogía y metalurgía." 1791. *Mercurio peruano* 1:73–89.

Dickinson, J. C., III. 1969. "The eucalypt in the Sierra of southern Peru." *Annals of the Association of American Geographers* 59(2):294–307.

Diels, Ludwig. 1937. "Beiträge zur Kenntnis der Vegetation und Flora von Ekuador." *Bibliotheca Botanica* 116:1–190.

Diez de San Miguel, G. 1964. *Visita hecha a la Provincia de Chucuito en el año 1567.* Ed. W. Espinoza Soriano. Lima: Casa de cultura del Perú.

Dilthey, Wilhelm. 1986. *Dilthey: Selected writings.* Ed. H. P. Rickman. New York: Cambridge University Press.

Dirección de sanidad pública. 1923. *La peste bubónica y su prevención.* Cartilla sanitaria No. 1. Guayaquil: Dirección de sanidad pública del Ecuador. 16 pp.

Documentos para la historia argentina. 1929 [1620]. Vol. 20. Buenos Aires: Facultad de filosofía y letra.

Dollfus, Olivier. 1978. "Les Andes intertropicales: Une mosaïque changeante." *Annales: Économies, sociétés, civilisations* 33(5–6):895–903.

Dollfus, Olivier. 1981. *El reto del espacio andino.* Lima: Instituto de estudios peruanos.

Dollfus, Olivier. 1982. "Development of land-use patterns in the Central Andes." *Mountain Research and Development* 2(1):39–48.

Domingo de Santo Tomás, Fray. 1995. *Grammática o parte de la lengua general de los indios de los reynos del Perú.* [1560]. Cusco: Centro de estudios regionales andinos "Bartolomé de las Casas."

Dominguez, B. 1789. "Del mejor medio de curar las calenturas periódicas para precavar sus resultas." *Memorias académicas de la real Sociedad de medicina y demas ciencias de Sevilla* 7:450–69.

Donkin, R. 1977. *Agricultural terracing in the aboriginal New World.* Tucson: University of Arizona Press.

Donkin, R. A. 1997. "A 'servant of two masters'?" *Journal of Historical Geography* 23(3):247–66.

Doyle, Jack. 1985. *Altered harvests: Agriculture, genetics, and the fate of the world's food supply.* New York: Penguin Books.

Dresch, Jean. 1958. "Problèmes morphologiques des Andes centrales." *Annales de géographie* 67:130–51.

Dubos, René J. 1972. *A god within.* New York: Scribner.

Dunn, F. 1965. "On the antiquity of malaria in the Western Hemisphere." *Human Biology* 37:385–93.

Dzik, Anthony J., and Thaila Wallen. 1987. "Geography of rat bites in Chicago, Illinois." *Illinois Geographical Society Bulletin* 29:16–24.

Eamon, William. 1994. *Science and the secrets of nature.* Princeton, N.J.: Princeton University Press.

Eidlitz, K. L. 1969. "Food and emergency food in the circumpolar areas." *Studia Ethnografica Upsaliensia* 32.

Eliade, Mircea. 1964. *Shamanism.* Princeton, N.J.: Princeton University Press.

Ellenberg, Heinz. 1958. "Wald oder Steppe? Die natürliche Pflanzendecke der Anden Perus, I & II." *Die Umschau in Wissenschaft und Technik* 58(20):645–48; (22):679–81.

Ellenberg, Heinz. 1979. "Man's influence on tropical mountain ecosystems in South America." *Journal of Ecology* 67:401–16.

Engelbrecht, T. 1916. "Über die Entstehung einiger feldmässig angebauter Kulturpflanzen." *Geographische Zeitschrift* 22:328–34.

Erickson, Clark L. 1996. *Investigación arqueológica del sistema agrícola de los camellones en la cuenca del Lago Titicaca del Perú.* La Paz: PIWA.

ERTS-GEOBOL. 1978. *Mapa de cobertura y uso actual de la tierra* La Paz: Ministerio de minas.

Eskey, C. R. 1930. "Chief etiological factors of plague in Ecuador and the antiplague campaign." *Public Health Reports* 45(36):2077–115.

Espinosa Soriano, W. 1980. "Los fundamentos linguísticos de la etnohistoria andina y comentarios en torno al anónimo de Charcas de 1604." *Revista española de antropología americana* (Madrid) 10:149–82.

Esquirol, Jean-Etienne Dominique. 1838. *Des maladies mentales.* 2 vols. Paris: J.-B. Baillière.

Estrella, Eduardo. 1977. *Medicina aborigen: La práctica médica aborigen de la Sierra ecuatoriana.* Quito: Editorial Epoca.

Estrella, Eduardo. 1986. *El pan de América: Ethnohistoria de los alimentos aborígenes en el Ecuador.* Madrid: C.S.I.C.

Favre, Henri. 1975. "Le peuplement et la colonisation agricole de la steppe dans le Pérou central." *Annales de géographie* 464:415–41.

Fernández, F. M., and G. Oliver. 1988. "Proteins present in llama milk. I. Quantitative aspects and general characteristics." *Milchwissenschaft* 43(5):299–302.

Fernández de Oviedo y Valdés, Gonzalo. 1959. *Historia general y natural de las Indias.* [1526]. 5 vols. Madrid: Ediciones Atlas.

Fiebrig, Karl. 1911. "Ein Beitrag zur Pflanzengeographie Boliviens." *Botanische Jahrbücher* (Leipzig) 45:1–68.

Figueroa, R. B., E. Melgar, N. O. Jó, and O. L. García. 1971. "Intestinal lactase deficiency in an apparently normal Peruvian population." *American Journal of Digestive Diseases* 16:881–89.

Fitzhugh, H., and A. E. Wilhelm. 1991. "Value and uses of indigenous livestock breeds in developing nations." Pp. 102–16 in *Biodiversity: Culture, conservation, and ecodevelopment,* ed. M. Oldfield and J. Alcorn. Boulder: Westview Press.

Flannery, K. V., J. Marcus, and R. G. Reynolds. 1989. *The flocks of the Wamani: A study of llama herders on the punas of Ayacucho, Peru.* San Diego: Academic Press.

Flores, A. 1929. "El saneamiento de Mizque." *Boletín de la dirección general de sanidad pública* (La Paz, Bolivia) 1(2):189–93.

Flores, Angeles. 1985. "Protagonismo andaluz en la sanidad naval del siglo XVII." Pp. 363–84 in *Andalucía y América en el siglo XVII: Actas de las IV jornadas de Andalucía y América,* ed. Bibiano Torres Ramírez and José Hernandez Palomo. Seville: Escuela de estudios hispano-americanos.

Flores Ochoa, Jorge A. 1968. *Los pastores de Paratía: Una introducción a su estudio.* Mexico City: Instituto indigenista interamericano.

Flores Ochoa, Jorge A. 1977. "Aspectos mágicos del pastoreo: Enqu, enqaychu, illa y khuya rumi." Pp. 211–38 in *Pastores de puna: Uywamichiq punarunakuna,* ed. J. A. Flores Ochoa. Lima: Instituto de estudios peruanos.

Flores Ochoa, Jorge A. 1978. "Classification et denomination des camélidés sud-américains." *Annales: Économies, sociétés, civilisations* 33(5–6):1006–16.

Flores Ochoa, Jorge A. 1984. "Causas que originaron la actual distribución espacial de las alpacas y llamas." *Revista del Museo e instituto de arqueología* (Cusco) 23:223–50.

Flores Ochoa, Jorge. 1985. "Interaction and complementarity in three zones of Cuzco."

Pp. 251–76 in *Andean ecology and civilization,* ed. Shozo Masuda, Izumi Shimada, and Craig Morris. Tokyo: University of Tokyo Press.

Fondahl, Gail. 1989. "Reindeer dairying in the Soviet Union." *Polar Record* 25(155): 285–94.

Foucault, Michel. 1971. *The order of things: An archaeology of the human sciences.* New York: Pantheon.

Fowler, Cary, and Pat Mooney. 1990. *Shattering: Food, politics, and the loss of genetic diversity.* Tucson: University of Arizona Press.

Frank, Erwin H. 1987. "Das Tapirfest der Uni: Eine funktionale Analyse." *Anthropos* 82:151–81.

Friedlander, J. 1977. "Malaria and demography in the Lowlands of Mexico: An ethnohistorical approach." Pp. 113–19 in *Culture, disease, and healing,* ed. D. Landy. New York: Macmillan.

Frontaura A., D. 1935. "Jefatura sanitaria de Mizque." *Boletín de la dirección general de sanidad pública* (La Paz) 5(8):84–93.

Gade, Daniel W. 1969a. "The llama, alpaca and vicuña: Fact vs. fiction." *Journal of Geography* 68(6):339–43.

Gade, Daniel W. 1969b. "Vanishing crops of world agriculture: The case of tarwi (*Lupinus mutabilis*) in the Andes." *Proceedings of the Association of American Geographers* 1:47–51.

Gade, Daniel W. 1970. "The contributions of O. F. Cook to cultural geography." *Professional Geographer* 22(4):206–9.

Gade, Daniel W. 1972. "Comercio y colonización en la zona de contacto entre la Sierra y las tierras bajas del valle de Urubamba, Perú." *Actas y memorias, XXXIX Congreso internacional de Americanistas* 4:207–21.

Gade, Daniel W. 1975. *Plants, man and the land in the Vilcanota Valley of Peru.* Biogeographica no. 6. The Hague: W. Junk.

Gade, Daniel W. 1976. "Naturalization of plant aliens: The volunteer orange in Paraguay." *Journal of Biogeography* 3(3):269–79.

Gade, Daniel W. 1979. "Inca and colonial settlement, coca cultivation and endemic disease in the tropical forest." *Journal of Historical Geography* 5(3):263–79.

Gade, Daniel W. 1991a. "Reflexiones sobre el asentamiento andino de la época toledana hasta el presente." In *Reproducción y transformación de las sociedades andinas: siglos XVI–XX,* ed. F. Salomon and S. Moreno Y. Quito: Ediciones Abya-Yala.

Gade, Daniel W. 1991b. "Weeds in Vermont as tokens of socioeconomic change." *Geographical Review* 81(2):153–69.

Gade, Daniel W. 1992. "Landscape, system and identity in the post-conquest Andes." *Annals of the Association of American Geographers* 82(3):460–77.

Gade, Daniel W. 1993. "Leche y civilización andina: En torno a la ausencia del ordeño de la llama y alpaca." *Yearbook, Conference of Latin Americanist Geographers* 19:3–14.

Gade, Daniel W. 1996. "Carl Troll on nature and culture in the Andes." *Erdkunde* 50(4):301–16.

Gade, Daniel W., and Mario Escobar. 1972. "Canyons of the Apurimac." *Explorers Journal* 51(3):135–40.

Gagliano, Joseph A. 1994. *Coca prohibition in Peru: The historical debates.* Tucson: University of Arizona Press.

Galvano, Antonie. 1969. *The discoveries of the world.* [1555]. Amsterdam/New York: Da Capo Press.

Gandia, E. de. 1939. *Francisco de Alfaro y la condición social de los indios.* Buenos Aires: Librería y Editorial "El Ateneo."

García Quintanilla, J. 1978. "Monografía de la Provincia de Zudañez, Departamento de Chuquisaca." *Boletín de la Sociedad geográfica y histórica "Sucre"* 56(461-462-463):49–84.

García Recio, José Maria. 1988. *Análisis de una sociedad de frontera: Santa Cruz de la Sierra en los siglos XVI y XVII.* Seville: Diputación provincial de Sevilla.

García Rizzo, Carlos E. 1968. "Incidencia de Rattus in Guayaquil." *Revista ecuatoriana de higiene y medicina tropical* 25(3):299–305.

García Rizzo, Carlos E. 1970a. "Síntesis histórica de la peste en Guayaquil." *Revista ecuatoriana de higiene y medicina tropical* 27(1):27–30.

García Rizzo, Carlos E. 1970b. "Contribución al estudio de los roedores (Muridae) de la Provincia de Guayas, Ecuador." *Revista ecuatoriana de higiene y medicina tropical* 27(2):137–60.

Garcilaso de la Vega, El Inca. 1960. *Comentarios reales.* [1609]. Cusco: Editorial universitario.

Gareis, Iris. 1982. *Llama und Alpaca in der Religion der zentralen und südlichen Andengebietes.* Münchner Beitrage zur Amerikanistik, Bd. 6. Höhenshaflorn: Klaus Renner Verlag.

Gentry, Howard S. 1958. "The natural history of jojoba (*Simmondsia chinensis*) and its cultural aspects." *Economic Botany* 12:261–95.

Gentry, Howard S. 1969. "Origin of the common bean, *Phaseolus vulgaris.*" *Economic Botany* 23:55–69.

Gepts, P. 1990. "Biochemical evidence bearing on the domestication of *Phaseolus* (Fabaceae) beans." *Economic Botany* 44(3):28–38.

German, P. 1897. "Voyage d'Asunción (Paraguay) à Mollendo (Pérou)." *Actes de la Société scientifique du Chili* 7:256–96.

Gilt Contreras, Mario A. 1960. "La ciudad de Paititi." *Revista del Instituto americano del arte* (Cusco), no. 10:103–92.

Giménez-Carrazona, Manuel. 1978. "La iglesia de Tomina." *Arte y arqueología* (La Paz) 5-6:23–34.

Girault, Louis. 1984. *Kallawaya: Guérisseurs itinérants des Andes.* Paris: Editions de l'ORSTOM.

Glave, Luis Miguel. 1983. "Trajines: un capitulo en la formacion del mercado interno colonial." *Revista andina* 1:9–76.

Glaser, Gilbert H. 1978. "Epilepsy, hysteria, and 'possession.' " *Journal of Nervous and Mental Disease* 166(4):268–74.

Gode-von Aesch, Alexander. 1966. *Natural science and German romanticism.* New York: AMS Press.

Godoy, Ricardo A. 1984. "Ecological degradation and agricultural intensification in the Andean Highlands." *Human Ecology* 12(4):359–83.

Godoy, Ricardo A. 1990. *Mining and agriculture in highland Bolivia: Ecology, history, and commerce among the Jukumanis.* Tucson: University of Arizona Press.

Göhring, Herman. 1877. *Informe de supremo gobierno del Peru sobre la expedicion a los valles de Paucartambo en 1873 al mando del Coronel D. Baltazar La-Torre.* Lima: Imprenta del Estado.

Göhring, Roberto. 1931. "La región del Cuzco." *Revista universitaria* (Cusco) 20:290–388.

Gómara, F. de Lopez. 1946. "Historia de las Indias." Pp. 155–294 in vol. 1 of *Historiadores primitivos de Indias,* ed. Enrique de Vedia. Madrid: Ediciones Atlas.

Gómez, J. T., E. Arciniegas, and J. Torres. 1978. "Prevalence of epilepsy in Bogotá, Colombia." *Neurology* 28:90–94.

Gómez, Nelson. 1987. *La misión geodésica y la cultura de Quito.* [1745]. Quito: Ediciones Abya-Yala.

Gómez Molina, Eduardo, and Adrienne V. Little. 1981. "Geoecology of the Andes: The natural sciences basis for research planning." *Mountain Research and Development* 1(2):115–44.

González Holguín, F. 1952. *Vocabulario de la lengua de todo el Perú llamado lengua quichua o del Inca.* [1608]. Lima: Universidad nacional mayor de San Marcos.

Gordonio, Bernardo. 1991. *Lilio de medicina.* [1303]. Ed. John Cull and Brian Dutton. Madison: Hispanic Seminary of Medical Studies.

Graf, K. J. 1981. "Palynological investigations of two postglacial peat bogs near the boundary of Bolivia and Peru." *Journal of Biogeography* 8:353–68.

Granada, Daniel. 1896. *Reseña histórico-descriptiva de antiguas y modernas supersticiones del Río de la Plata.* Montevideo: A. Barreiro y Ramos.

Gremillion, Kristen J. 1996. "Diffusion and adaptation of crops in evolutionary perspective." *Journal of Anthropological Archaeology* 15:183–204.

Grieshaber, E. P. 1980. "Survival of Indian communities in nineteenth-century Bolivia: A regional comparison." *Journal of Latin American Studies* 12:223–69.

Grillo Fernández, Eduardo. 1998. "Development or decolonization in the Andes?" Pp. 193–243 in *The spirit of regeneration: Andean culture confronting Western notions of development,* ed. F. Apffel-Marglin. London: Zed Books Ltd.

Guaman Poma de Ayala, Felipe. 1980. *El primer nueva corónica y buen gobierno.* [1615]. 3 vols. Ed. J. V. Murra and R. Adorno. Mexico City: Siglo Veintiuno.

Guevara, José. 1836. "Historia del Paraguay, Río de la Plata y Tucumán." Pp. 491–826 in vol. 1 of *Colección de obras y documentos,* ed. Pedro de Angelis. Buenos Aires: Imprenta del Estado.

Guía de forasteros de la República de Bolivia para 1835. 1835. Sucre.

Guía nacional. 1938. La Paz: Ministerio de interior.

Guillaume, H. 1888. *The Amazon provinces of Peru as a field for European emigration.* London: Wyman and Sons.

Guillet, D. 1973. "Integración sociopolítica de las poblaciones nuevas en Bolivia: Descripción de un caso y discusión." *Estudios andinos* 3(1):111–28.

Guillet, David W. 1985. "Hacía una historia socio-económica de los bosques en los Andes centrales del Perú." *Boletín de Lima* 7(38):79–84.

Guillet, David W. 1992. *Covering ground: Communal water management and the state in the Peruvian highlands.* Ann Arbor: University of Michigan Press.

Gutiérrez Pareja, Salustio. 1984. "Caminos al Antisuyo." *Revista del Museo e instituto de arqueología* (Cusco) 23:63–82.

Gutiérrez Pareja, Salustio. N.d. *Ciudades ocultas del Cusco milenario.* Cusco: Editorial Andina.

Gutiérrez Salgado, A., P. Gepts, and D. G. Debouck. 1995. "Evidence for two gene pools of the Lima bean, *Phaseolus lunatus,* in the Americas." *Genetic Resources and Crop Evolution* 42(1):15–28.

Hackett, L. W. 1945. "The malaria of the Andean region of South America." *Revista del Instituto de salubridad y enfermedades tropicales* (Mexico) 6(4):239–52.

Haenke, T. 1809. "Introduction a l'histoire naturelle de la Province de Cochabamba et des environs et description de ses productions." [1797]. Pp. 389–541 in *Voyage dans l'Amérique méridionale,* ed. F. de Azara. Paris: Dertu.

Hagen, Victor W. von. 1976. *The royal road of the Inca.* London: Gordon and Cremonesi.

Hahn, Eduard. 1896. *Die Haustiere und ihre Beziehungen zur Wirtschaft des Menschen.* Leipzig: Duncker & Humboldt.

Hammerly, Michael T. 1987. *Historia social y económica de la antigua provincia de Guayaquil 1763–1842.* 2d ed. Guayaquil: Banco central del Ecuador.

Hansen, B. C. S., G. O. Seltzer, and H. E. Wright, Jr. 1994. "Late Quaternary vegetational change in the central Peruvian Andes." *Palaeogeography, Palaeoclimatology, Palaeoecology* 109:263–85.

Hansen, B. C. S., H. E. Wright, Jr., and J. P. Bradbury. 1984. "Pollen studies in the Junín area, central Peruvian Andes." *Geological Society of America Bulletin* 95:1454–65.

Hargous, Sabine. 1985. *Les appeleurs d'âme: L'univers chamanique des Indiens des Andes.* Paris: Albin Michel.

Harlan, Harry V. 1953. *One man's life with barley.* New York: Exposition Press.

Harlan, Harry V., and Mary L. Martini. 1936. "Problems and results in barley breeding." *USDA Yearbook of Agriculture 1936:* 303–46.

Harlan, Jack R. 1975. *Crops and man.* Madison: American Society of Agronomy and Crop Science Society of America.

Harlan, Jack R. 1986. "Plant domestication: Diffuse origins and diffusions." Pp. 21–34 in *The origin and domestication of cultivated plants,* ed. C. Barigozzi. New York: Elsevier.

Harris, D. R. 1972. "The origins of agriculture in the tropics." *American Scientist* 60:180–93.

Harris, D. R. 1990. "Vavilov's concept of centres of origin of cultivated plants: Its genesis and its influence on the study of agricultural origins." *Biological Journal of the Linnean Society* 39:7–16.

Hassaurek, Friedrich. 1967. *Four years among the Ecuadorians.* Carbondale: Southern Illinois University Press.

Hassel, Jorge [Georg] M. von. 1898. "Ríos Alto Madre de Dios y Paucartambo." *Boletín de la Sociedad geográfica de Lima* 8(8):288–310.

Hassel, Jorge [Georg] M. von. 1907. "Informe del jefe de la Comisión explorador del Alto Madre de Dios, Paucatambo i Urubamba por la vía del Cusco." Pp. 259–394 in *Ultimas exploraciones ordenadas por la Junta de Vias fluviales.* Lima: Junta de Vias Fluviales.

Hather, John G. 1996. "The origins of tropical vegeculture: Zingiberaceae, Araceae and Dioscoreaceae in Southeast Asia." Pp. 538–51 in *The origins and spread of agriculture and pastoralism in Eurasia,* ed. David R. Harris. Washington, D.C.: Smithsonian Institution Press.

Hawkes, J. G. 1944. *Potato collecting expeditions in Mexico and South America.* Vol. 2: *Systematic classification of the collections.* Cambridge: Imperial Bureau of Plant Breeding and Genetics.

Hawkes, J. G. 1947. "On the origin and meaning of South American potato names." *Journal of the Linnean Society of London* 53:205–50.

Headland, Thomas N. 1997. "Revisionism in ecological anthropology." *Current Anthropology* 38(4):605–30.

Heiser, Charles B., Jr. 1985. *Of plants and people.* Norman: University of Oklahoma Press.

Helms, A. Z. 1807. *Travels from Buenos Ayres by Potosi to Lima.* London: Richard Phillips.

Hendrickson, Robert. 1983. *More cunning than man: A Social history of rats and men.* New York: Stein and Day.

Hernández Bermejo, J. E., and J. Leon, eds. 1992. *Cultivos marginados: Otra perspectiva de 1492.* Rome: Food and Agricultural Organization.

Hernández Morejón, A. 1967. *Historia bibliográfica de la medicina española.* [1845]. Vol. 2. New York: Johnson Reprint Corp.

Herndon, William Lewis, and Lardner Gibbon. 1954. *Exploration of the valley of the Amazon.* 2 vols. Washington, D.C.: A. O. P. Nicholson.

Herrera, F. L. 1933. "El cedro peruano (*Cedrela Herrerae* Harms)." *Boletín de la dirección de agricultura y ganadería* 3(9–10):19–26.

Hershkovitz, Philip. 1954. "Mammals of northern Colombia, preliminary report no. 7: Tapirs (genus *Tapirus*) with a systematic review of American species." *Proceedings of the United States National Museum* 103(3329):465–96.

Herzog, Theodor. 1923. *Die Pflanzenwelt der bolivischen Anden und ihres östlichen Vorlandes.* Leipzig: Englemann.

Hess, Carmen G. 1990. " 'Moving up—moving down': Agro-pastoral land-use patterns in the Ecuadorian paramos." *Mountain Research and Development* 10(4):333–42.

Hildebrandt, Martha. 1969. *Peruanismos.* Lima: Moncloa editores.

Hodge, W. H. 1946. "Three neglected Andean tubers." *Journal of the New York Botanical Garden* 47:214–24.

Hoek, H., and G. Steinmann. 1906. "Erlauterung zur Routenkarte der Expedition Steinmann, Hoek v. Bistram in den Anden von Bolivien 1903–04." *Petermanns Mitteilungen* 52:1–12.

Holden, Clare, and Ruth Mace. 1997. "Phylogenetic analysis of the evolution of lactose digestion in adults." *Human Biology* 69(5):605–28.

Holden, John, James Peacock, and Trevor Williams. 1993. *Genes, crops and the environment.* New York: Cambridge University Press.

Hondermann, Juan. 1988. "El bosque de Zárate (Haurochiri, Lima): La zona 'Gatero,' su composición arborea, distribución diamétrica e intervención antrópica." *Boletín de Lima* 56:67–70.

Horigan, Stephen. 1988. *Nature and culture in Western discourses.* New York: Routledge.

Horkheimer, Hans. 1960. *Nahrung und Nahrungsgewinning im vorspanischen Peru.* Berlin: Colloquium Verlag.

Hornborg, Alf. 1990. "Highland and lowland conceptions of social space in South America: Some ethnoarchaeological affinities." *Folk* 32:61–92.

Hueck, Kurt. 1966. *Die Wälder Südamerikas: Ökologie, Zusammensetzung und wirtschaftliche Bedeutung.* Stuttgart: Gustav Fischer Verlag.

Humboldt, Alexander von. 1850. *Cosmos: A sketch of a physical description of the universe.* 5 vols. in 2. New York: Harper & Brothers.

Humboldt, Alexander von. 1970. *Relation historique du voyage aux régions équinoxiales du nouveau continent.* 3 vols. Ed. H. Beck. Stuttgart: F. A. Brockhaus.

Hutchinson, I. D. 1967. *Informe sobre el estudio preliminar de los bosques de Bolivia.* Rome: Food and Agricultural Organization.

Hutchinson, Joseph B., R. A. Silow, and S. G. Stephens. 1947. *The evolution of Gossypium.* London: Oxford University Press.

Hyslop, John. 1984. *The Inka road system.* Orlando: Academic Press.

Ido, G., R. Hoki, H. Ito, and H. Wani. 1917. "The rat as a carrier of *Spirochaeta icterohaemorrhagia,* the causative agent of Weil's disease." *Journal of Experimental Medicine* 26:341.

Ihering, Rodolpho von. 1968. *Dicionário dos animais do Brasil.* São Paulo: Editora Universidade de Brasilia.

Iltis, Hugh H. 1974. "Freezing the genetic landscape: The preservation of diversity in cultivated plants as an urgent social responsibility of plant geneticist and plant taxonomist." *Maize Genetics Cooperative News Letter* 48:199–200.

Iltis, Hugh H. 1987. "Maize evolution and agricultural origins." Pp. 195–216 in *Grass systematics and evolution,* ed. T. R. Soderstrom et al. Washington, D.C.: Smithsonian Institution Press.

Informe del prefecto de Cochabamba. 1897. Cochabamba: Imprenta de El Heraldo.

Instituto geográfico militar (IGM). 1989. *Atlas del Perú.* Lima: IGM.

Instituto nacional de cultura. 1988. "Evaluación del incendio forestal del parque arqueológico de Machu Picchu." *Boletín de Lima* 11(51):45–50.

Instituto nacional de planificación (INP). 1963–1970. *Atlas histórico-geográfico y de paisajes peruanos.* Lima: INP.

Inventario, evaluación e integración de los recursos naturales de la microregión: Reconocimiento: Departamento de Puno. 1984. Lima: ONERN and CORPUNO.

Inventario y evaluación de los recursos naturales de la zona alto-andina del Perú: Reconocimiento: Departamento del Cuzco. 1985. Lima: ONERN.

Isaac, Erich. 1970. *Geography of domestication.* Englewood Cliffs: Prentice-Hall.

Iwaki Ordóñez, Rubén. 1975. *Operación Paititi.* Cusco: Editorial de la Cultura Andina.

Izquieta Pérez, Leopoldo. 1903. "Profilaxis internacional sobre la peste bubónica." *Boletín de medicina y cirurgía* (Guayaquil) 5(8):113–25, (9):129–32.

Jackson, Wes, Wendell Berry, and Bruce Colman, eds. 1984. *Meeting the expectations of the land: Essays in sustainable agriculture and stewardship.* San Francisco: North Point Press.

Jenness, R. 1974. "The composition of milk." Pp. 3–107 in vol. 3: *Nutrition and biochemistry of milk/maintenance* of *Lactation: A comprehensive treatise,* ed. Bruce L. Larson. New York: Academic Press.

Jennings, Bruce H. 1988. *Foundations of international agricultural research: Science and politics in Mexican agriculture.* Boulder: Westview Press.

Jerez, Francisco de. 1947. "Verdadera relación de la Conquista." [1534]. Pp. 319–46 in *Historiadores primitivos de Indias,* ed. Enrique de Vedia. Madrid: Ediciones Atlas.

Jervis, Oswaldo. 1967. "La peste en el Ecuador de 1908 a 1965." *Boletín de la oficina sanitaria panamericana* 62:418–27.

Jiménez de la Espada, Marcos, ed. 1965. *Relaciones geográficas de Indias.* 3 vols. Madrid: Ediciones Atlas.

Johannessen, Carl L. 1966. "The domestication process in trees reproduced by seed: The pejibaye palm in Costa Rica." *Geographical Review* 46(3):363–76.

Johannessen, S., and C. A. Hastorf. 1990. "A history of fuel management (A.D. 500 to the present) in the Mantaro Valley, Peru." *Journal of Ethnobiology* 10(1):61–90.

Johnson, Duane L. 1990. "New grains and pseudograins." Pp. 122–27 in *Advances in new crops,* ed. Jules Janick and James E. Simon. Portland, Ore.: Timber Press.

Jonstonus, Joanis. 1755. *Theatrum universale: Amnium animalium: Historiae naturales de quadrupedibus.* 5 vols. Heilbronn: C. de Lannoy.

Jordan, E. 1983. "Die Verbreitung von Polylepis-Beständen in der Westkordillere Boliviens." *Tuexenia: Mitteilungen der floristisch-soziologischen Arbeitsgemeinschaft,* n.s. 5:101–16.

Juan, Jorge, and Ulloa, Antonio de. 1978. *Relación histórica del viaje a la América meridional.* [1748]. Vol. 2. Madrid: Fundación universitaria española.

Jung, Carl G. 1964. *Civilization in transition.* New York: Pantheon Books.

Kanner, Leo. 1930a. "The folklore and culture history of epilepsy." *Medical Life* 37(4):185–214.

Kanner, Leo. 1930b. "The names of the falling sickness: An introduction to the study of the folklore and culture history of epilepsy." *Human Biology* 2:109–27.

Kaplan, Lawrence. 1981. "What is the origin of the common bean?" *Economic Botany* 35(2):240–54.

Kautz, R. A. 1980. "Pollen analysis and paleoethnobotany." Pp. 45–64 in *Guitarrero cave: Early man in the Andes,* ed. T. Lynch. New York: Academic Press.

Keating, Philip L. 1997. "Mapping vegetation and anthropogenic disturbances in southern Ecuador with remote sensing techniques: Implications for park management." *Yearbook, Conference of Latin Americanist Geographers* 23:77–90.

Kent, Robert. 1993. "Geographical dimensions of the Shining Path insurgency in Peru." *Geographical Review* 83(4):441–54.

Kenzer, Martin S. 1987. "Like father/like son: William Albert and Carl Ortwin Sauer." Pp. 40–65 in *Carl O. Sauer: A tribute,* ed. M. S. Kenzer. Corvallis: Oregon State University Press.

Kenzer, Martin S. 1988. "Commentary on Carl O. Sauer." *Professional Geographer* 40(3):333–36.

Kessler, Michael. 1995. "Present and potential distribution of *Polylepis* (Rosaceae) forests in Bolivia." Pp. 281–94 in *Biodiversity and conservation of neotropical montane forests,* ed. Steven P. Churchill et al. New York: New York Botanical Garden.

Kinzl, H. 1970. "Bedrohte Natur in den peruanischen Anden." Pp. 253–70 in *Argumenta Geographica,* ed. W. Lauer. Colloquium Geographicum 22. Bonn: F. Dümmlers Verlag.

Kloppenburg, Jack, Jr. 1991. "Social theory and the de/reconstruction of agricultural science: Local knowledge for an alternative agriculture." *Rural Sociology* 56(4):519–48.

Knapp, G. W. 1991. *Andean ecology.* Boulder: Westview Press.

Knudsen, Herbert D., and Richard D. Zeller. 1993. "The milkweed business." Pp. 422–28 in *New crops,* ed. Jules Janick and James E. Simon. New York: John Wiley.

Koepcke, H.-W. 1961. "Synökologische Studien an der Westseite der peruanischen Anden." *Bonner geographische Abhandlungen* 29:1- 320.

Konkola, Karl. 1992. "More than a coincidence? The arrival of arsenic and the disappearance of plague in early modern Europe." *Journal of the History of Medicine and Allied Sciences* 47:186–209.

La Barre, Weston. 1947. "Potato taxonomy among the Aymara Indians of Bolivia." *Acta Americana* 5:83–103.

La Barre, Weston. 1970. *The ghost dance: The origins of religion.* Garden City: Doubleday.

La Condamine, Charles. 1751. *Journal du voyage fait par ordre du roi à l'Equateur.* Paris: Imprimerie royale.

Laegaard, S. 1992. "Influence of fire in the grass páramo vegetation of Ecuador." Pp. 151–70 in *Páramo: An Andean ecosystem under human Influence,* ed. H. Balslev and J. L. Luteyn. London: Academic Press.

La Martinière, Pierre Martin. 1671. *Voyage des pais septentrionaux.* Paris: Chez Louis Vendosme.

Langdon, E. Jean. 1991. "When the tapir is an anaconda: Women and power among the Siona." *Latin American Indian Literatures Journal* 7(1):7–19.

Langer, Erick D. 1989. *Economic change and rural resistance in Southern Bolivia, 1880–1930.* Stanford: Stanford University Press.

Lastres, Juan B. 1951. *Historia de la medicina peruana.* Vol. 1: *La medicina incaica.* Lima: Universidad nacional mayor de San Marcos.

Latcham, Ricardo E. 1936. *La agricultura precolombina en Chile y los países vecinos.* Santiago: Ediciones universidad de Chile.

Lathrop, Donald. 1973. "The antiquity and importance of long-distance trade relationships in the moist tropics of pre-Columbian South America." *World Archaeology* 5:170–86.

Latour, Bruno. 1991. *Nous n'avons jamais été modernes: Essai d'anthropologie symétrique.* Paris: Editions La Découverte.

Lauer, Wilhelm. 1952. "Humide und aride Jahreszeiten in Afrika und Südamerika und ihre Beziehung zu den Vegetationsgürteln." Pp. 15–98 in *Studien zur Klima- und Vegetationskunde der Tropen,* ed. C. Troll. Bonner geographische Abhandlungen 9. Bonn: Geographischen Institut der Universität Bonn.

Lauer, Wilhelm. 1976. "Klimatische Grundzüge der Höhenstufung tropischer Gebirge." Pp. 76–90 in *Tagungsbericht und wissenschaftliche Abhandlung 40 Deutsches Geographentag Innsbruck 1975.* Wiesbaden: Franz Steiner Verlag.

Lauer, Wilhelm. 1981. "Ecoclimatological conditions of the paramo belt in the tropical high mountains." *Mountain Research and Development* 1(3–4):209–21.

Lauer, Wilhelm. 1984. "Nature and man in the ecosystems of tropical high mountains—introductory remarks." Pp. 17–21 in *Natural environment and man in tropical mountain ecosystems,* ed. W. Lauer. Erdwissenschaftliche Forschung 18. Stuttgart: Franz Steiner Verlag.

Lauer, Wilhelm. 1988. "Zum Wandel der Vegetationszonierung in den Lateinamerikanischen Tropen seit dem Höhepunkt der letzten Eiszeit." Pp. 1–45 in *Lateinamerikaforschung: Beiträge zum Gedächtnis-Kolloquium Wolfgang Ericksen.* Jahrbuch der geographischen Gesellschaft zu Hannover. Hannover: Selbstverlag der GGH.

Lauer, Wilhelm. 1993. "Human development and environment in the Andes: A geoecological overview." *Mountain Research and Development* 13(2):157–66.

Lavallée, Danièle. 1970. *Les représentations animales dans la céramique Mochica.* Paris: Institut d'ethnologie.

Lavallée, Danièle, ed. 1985. *Telarmachay: Chasseurs et pasteurs préhistoriques des Andes.* Vol. 1. Paris: Editions recherche sur les civilisations.

Lavallée, Danièle. 1990. "La domestication animale en Amérique du Sud: Le point des connaissances." *Bulletin de l'Institut français d'études andines* 19:25–44.

Laviana Cuetos, Maria Luisa, ed. 1982. "La descripción de Guayaquil por Francisco Requena 1774." *Historiografía y bibliografía americanistas* (Seville) 26:3–134.

LeBaron, A., L. K. Bond, S. P. Aitken, and L. Michaelson. 1979. "An explanation of the Bolivian Highlands grazing-erosion syndrome." *Journal of Range Management* 32(3):201–8.

Leighly, John. 1976. "Carl Ortwin Sauer, 1889-1975." *Annals of the Association of American Geographers* 66(3):337–48.

Leon, Jorge. 1964. *Plantas alimenticias andinas.* Boletín técnico no. 6. Lima: IICA Zona Andina.

León Pinelo, Antonio. 1943. *El paraíso en el Nuevo Mundo.* [1645]. 2 vols. Lima: Imp. Torres Aguirre.

Leppik, E. E. 1969. "The life and work of N. I. Vavilov." *Economic Botany* 23(2):128–32.

Léry, Jean de. 1975. *Histoire d'un voyage fait en la terre du Brésil.* [1578]. Ed. Jean-Claude Moriset. Geneva: Librairie Droz.

Levillier, R., ed. 1918. *La Audiencia de Charcas: Correspondencia de presidentes y oidores.* Madrid: Imprenta de Juan Pueyo.

Levillier, R., ed. 1920. *Gobernación del Tucuman: Papeles de gobernadores en el siglo XVI.* Part 2. Madrid: Juan Pueyo.

Levillier, R., ed. 1925a. *Gobernantes del Perú: Cartas y papeles siglo XVI.* Vol. 9. Madrid: Juan Pueyo.

Levillier, Roberto, ed. 1925b. "Ordenanzas del virrey don Francisco de Toledo." [1575]. *Gobernantes del Perú* 8:304–99.

Levillier, Roberto. 1976. *El Paititi, El Dorado y las Amazonas.* Buenos Aires: Emecé Editores.

Libermann, Máximo. 1986. "Microclima y distribución de *Polylepis tarapacana* en el Parque Nacinal del Nevado Sajama." *Documents Phytosociologiques* (Camerino, Italy) 10(2):235–81.

Libermann Cruz, Máximo, and Seemin Qayum. 1994. *La desertificación en Bolivia.* La Paz: Liga de defensa del medio ambiente.

Lira, Jorge A. 1985. *Medicina andina: Farmacopea y rituales.* Cusco: Centro de estudios regionales andinos "Bartolomé de las Casas."

Livingstone, F. B. 1980. "Natural selection and the origin and maintenance of standard genetic marker systems." *Yearbook of Physical Anthropology* 23:25–42.

Livingstone, F. B. 1984. "The Duffy blood groups, vivax malaria, and malaria selection: A review." *Human Biology* 56:413–25.

Lizarraga, Reginaldo de. 1987. *Descripción del Perú, Tucuman, Río de la Plata y Chile.* [1603]. Ed. Ignacio Ballesteros. Madrid: Historia 16.

Llona, Scipión E., 1904. "Mapa histórico geográfico de los valles de Paucartambo [1:200 000]." *Boletín de la Sociedad geográfica de Lima.*

Llorens, Rosana, and Leiva, Juan C. 1995. "Glaciological studies in the high Central Andes using digital processing of satellite images." *Mountain Research and Development* 15(4):323–30.

Lofstrom, W. 1972. "La administración Sucre y la salud pública en Bolivia." *Illimani: Revista del Instituto de investigaciones historicas y culturales de la H. Municipalidad de La Paz,* no. 4:45–56.

Lofstrom, W. L. 1983. *El Mariscal Sucre en Bolivia.* La Paz: Editorial Alenkar.

Loftus, R. T., et al. 1994. "Evidence for two independent domestications of common cattle." *Proceedings of the National Academy of Sciences of the United States of America* 91:2757–61.

Long, John D. 1930. "La campaña antipestosa en Guayaquil." *Anales de la Sociedad médico-quirúrgica del Guayas* 10(9):614–18.

López de Velasco, Juan. 1971. *Geografía y descripción universal de las Indias.* [1574]. Madrid: Ediciones Atlas.

Lore, Richard, and Kevin Flannelly. 1977. "Rat societies." *Scientific American* 236(5): 106–16.

Lovelock, J. E. 1995. *The ages of Gaia: A biography of the earth.* New York: W. W. Norton.

Lowie, Robert. 1937. *The history of ethnological theory.* New York: Rinehart & Co.

Lozano, Pedro. 1941. *Descripción corográfica del Gran Chaco Gualamba.* Tucumán: Instituto de antropología, Universidad nacional de Tucumán.

Lucena Giraldo, Manuel. 1988. "Entre dos fidelidades: Un catálogo ilustrado de las producciones peruanas." *Revista de Indias* 48(182-183):637–49.

Lyman, R. Lee, Michael J. O'Brien, and Robert C. Dunnell. 1997. *The rise and fall of culture history.* New York: Plenum Press.

Lynch, Thomas. 1983. "Camelid pastoralism and the emergence of Tiawanaku civilization in the south central Andes." *World Archaeology* 15(1):1–14.

Lynch, Thomas F. 1990. "Glacial-age man in South America? A critical review." *American Antiquity* 55(1):12–36.

Lyon, Patricia J. 1981. "An imaginary frontier: Prehistoric highland-lowland interchange in the southern Peruvian Andes." Pp. 3–18 in *Networks of the past: Regional interaction in archaeology,* ed. P. D. Francis, F. J. Kense, and P. G. Duke. Calgary: University of Calgary Archaeology Association.

Lyon, Patricia J. 1994. "El ocaso de los cocales de Paucartambo." *Revista del Instituto americano del arte del Cuzco* 14:1–9.

McClung de Tapia, Emily. 1992. "The origins of agriculture in Mesoamerica and Central America." Pp. 143–71 in *The origins of agriculture: An international perspective,* ed. C. W. Cowan and P. J. Watson. Washington, D.C.: Smithsonian Institution Press.

McDowell, John H. 1992. "Exemplary ancestors and pernicious spirits." Pp. 95–114 in *Andean cosmologies through time,* ed. Robert V. H. Dover, Katharine E. Seibold, and John H. McDowell. Bloomington: Indiana University Press.

Macera, P. 1978. *Bolivia: Tierra y población, 1825–1936.* Lima: Biblioteca Andina.

MacNeish, Richard S. 1992. *The origins of agriculture and settled life.* Norman: University of Oklahoma Press.

Maloney, B. K. 1993. "Palaeoecology and the origins of the coconut." *GeoJournal* 31(4):355–62.

Mangelsdorf, Paul C. 1953. "Review of *Agricultural origins and dispersals* by Carl O. Sauer." *American Antiquity* 19:87–90.

Mangelsdorf, Paul C. 1974. "*Corn: Its origins, evolution and improvement.* Cambridge, Mass.: Belknap Press.

Mangelsdorf, P. C., and R. G. Reeves. 1939. "The origin of Indian corn and its relatives." *Texas Agricultural Experiement Station Bulletin* 574:1–315.

Mangelsdorf, Paul C., Robert S. MacNeish, and W. C. Galinat. 1964. "Domestication of corn." *Science* 143:538–45.

Mann, G. 1968. "Die Ökosysteme Südamerikas." Pp. 171–229 in *Biogeography and ecology in South America,* ed. F. J. Fittkau et al. The Hague: Dr. W. Junk N. V. Publishers.

Manson, Patrick. 1996. *Manson's tropical diseases.* 20th ed. Philadelphia: W. B. Saunders Co.

Manthorne, Katherine E. 1989. *Tropical renaissance: North American artists exploring Latin America, 1839–1879.* Washington, D.C.: Smithsonian Institution Press.

Marcoy, Paul. 1874. *A journey across South America.* London: Blackie & Son.

Marglin, Stephen A. 1996. "Farmers, seedsmen, and scientists: Systems of agriculture and systems of knowledge." Pp. 185–248 in *Decolonizing knowledge: From development to dialogue,* ed. Frédérique Apffel-Marglin and Stephen A. Marglin. Oxford: Clarendon Press.

Marín Moreno, Felipe. 1961. "Panorama fitogeográfico del Perú." *Revista universitaria* (Cusco) 50(120):9–66.

Markgraf, Vera. 1989. "Palaeoclimates in Central and South America since 18,000 B.P. based on pollen and lake-level records." *Quaternary Science Review* 8:1–24.

Markham, Clements. 1991. *Markham in Peru: The travels of Clements R. Markham, 1852–1853.* Ed. Peter Blanchard. Austin: University of Texas Press.

Marsh, George Perkins. 1965. *Man and nature.* [1864]. Ed. David Lowenthal. Cambridge, Mass.: Belknap Press of Harvard University Press.

Martínez, José Luis. 1983. *Pasajeros de Indias: Viajes transatlánticos en el siglo XVI.* Madrid: Alianza editorial.

Matienzo, Juan de. 1875. "Memoria del lic. Matienzo al excel. Sr. don Francisco de Toledo . . . cerca del asiento de la provincia de los Charcas." *Colección de documentos inéditos* (Madrid) 24:149–62.

Matienzo, Juan de. 1967. *Gobierno del Perú.* Ed. G. Lohmann-Villena. Paris/Lima: Institut français d'études andines.

Maúrtua, Victor M., ed. 1906. "Carta del Marques de Monteclaros, Virrey del Perú, año 1616." Pp. 129–69 of *Juicio de límites entre el Perú y Bolivia.* 12 vols. Barcelona: Henrich.

Maúrtua, Victor M., ed. 1907. "Visita a las haciendas de los valles de Tono y Toayma 1658–1691." Pp. 221–30 in vol. 2: *Contestación al alegato de Bolivia: Prueba peruana presentada al gobierno de la Republica Argentina* of *Juicio de límites entre el Perú y Bolivia.* Buenos Aires: G. Kraft.

Means, Philip A. 1931. *Ancient civilizations of the Andes.* New York: Charles Scribner's.

Melean, A., and L. García Mesa. 1905. "El paludismo en Mizque, Bolivia: Estudio de profilaxia." *Anexo de la memoria del Ministerio de gobierno y fomento* (La Paz).

"Memoria del subprefecto de la Provincia del Cercado de Puno, 1874." 1874. *El Peruano* 32(July 23):77–78.

Mendizábel V., Benjamín. 1902. "El valle de Lacco." *Boletín del Centro científico del Cuzco* 1-2:84–89.

Mendoza, Diego. 1976. *Crónica de la Provincia de San Antonio de los Charcas.* [1664]. La Paz: Ministerio de educación.

Mendoza, Jaime. 1925. *El factor geográfico en la nacionalidad boliviana.* Sucre: Privately published.

Mendoza, Jaime. 1931. "El paludismo en Bolivia." Pp. 1–15 in vol. 1 of *Primera conferencia sanitaria boliviana.* Sucre: Instituto médico Sucre.

Merrill, Elmer Drew. 1954. *The botany of Cook's voyages and its unexpected significance in relation to anthropology, biogeography and history.* Waltham: Chronica Botanica.

Meyer, Alfred H. 1936. "The Kankakee 'Marsh' of northern Indiana and Illinois." *Michigan Papers in Geography* 6:359–96.

Middendorf, E. W. 1893–1895. *Peru: Beobachtungen und Studien über das Land und seine Bewohner.* 3 vols. Berlin: Robert Oppenheim.

Mikesell, Marvin W. 1978. "The rise and decline of sequent occupance." Pp. 2–15 in vol. 3: *The nature of change in geographical ideas* of *Perspectives in geography,* ed. Brian J. L. Berry. Dekalb: Northern Illinois University Press.

Miller, G. R. 1977. "Sacrificio y beneficio de camélidos en el sur del Perú." Pp. 193–210 in *Pastores de puna: Uywamichiq punarunakuna,* ed. J. A. Flores Ochoa. Lima: Instituto de estudios peruanos.

Miller, General. 1836. "Notice of a journey to the northward and also to the eastward of Cuzco and among the Chunchos Indians in July 1, 1835." *Journal of the Royal Geographical Society* 6:174–86.

Miller, L. E. 1918. *In the wilds of South America.* New York: Charles Scribner's Sons.

Miller, O. M. 1929. "The 1927–1928 Peruvian expedition of the American Geographical Society." *Geographical Review* 19(1):1–37.

Miller, Robert Ryal. 1968. *For science and national glory: The Spanish scientific expedition to America, 1862–1866.* Norman: University of Oklahoma Press.

Millones O., José. 1982. "Patterns of land use and associated environmental problems of the Central Andes." *Mountain Research and Development* 2(1):49–61.

Mills, C. Wright. 1959. *The sociological imagination.* New York: Oxford University Press.

Mindlin, Betty (and Suruí narrators). 1995. *Unwritten stories of the Suruí Indians of Rondonia.* Austin: Institute of Latin American Studies, University of Texas at Austin.

Ministerio de agricultura y granadería (MAG). 1978. *Mapa ecológico del Ecuador.* Quito: MAG.

Miño, Carlos. 1933. *La peste bubónica en el Ecuador.* Quito: Imprenta nacional.
Möckli-Von Seggern, Marg. 1972. "Die Gemse in der Volksmedizin des Alpenlandes." *Ethnomedizin* 1(3-4):329–51.
Molina, C. 1943. *Relación de la fábulas y ritos de los Inca.* [1575]. Lima: Pequeños grandes libros de historia americana.
Moll, Aristides A., and Shirley Baughman O'Leary. 1945. *Plague in the Americas: An historical and epidemiological survey.* Pub. 225. Washington, D.C.: Pan American Sanitary Bureau.
Monge, Carlos. 1948. *Acclimatization in the Andes.* Trans. Donald F. Brown. Baltimore: Johns Hopkins University Press.
Moreno, J. L. 1879. *Compendio de jeografia de Bolivia.* 2d ed. Santiago de Chile: Sociedad de instrucción primaria.
Morgan, George R. 1995. "Geographic dynamics and ethnobotany." Pp. 250–57 in *Ethnobotany: Evolution of a discipline,* ed. Richard Evans Schultes and Siri von Reis. Portland, Oreg.: Dioscorides Press.
Morison, Samuel E. 1974. *The European discovery of America.* Vol. 2: *The southern voyages A.D. 1492–1616.* New York: Oxford University Press.
Morlon, Pierre, ed. 1992. *Comprendre l'agriculture paysanne dans les Andes centrales.* Paris: Institut National de la Recherche Agronomique.
Mörner, Magnus. 1975. "Continuidad y cambio en una provincia del Cuzco: Calca y Lares desde los años 1680 hasta los 1790." *Historia y cultura* (Lima) 9:79–119.
Mörner, Magnus. 1985. *The Andean past: Land, societies and conflicts.* New York: Columbia University Press.
Mörner, Magnus. 1994. *Local communities and actors in Latin America's past.* Stockholm: Institute of Latin American Studies, Stockholm University.
Moro S., M. 1956. "Contribución al estudio de la leche de las alpacas." *Revista de la facultad de medicina veterinaria* 7-11:117–41.
Moscoso Carrasco, C. 1931–1932. "Resultado práctico del saneamiento antipalúdico en Mizque." *Boletín de la dirección general de sanidad* (La Paz) 3(6):905–15.
Moscoso Carrasco, C. 1963. *Bolivia elimina su malaria.* La Paz: Ministerio de salud pública.
Mujía, J. M., and J. Ondarza. 1901. "Descripciones geográficas y estadísticas del Departamento de Cochabamba." Pp. 160–75 in vol. 1: *Departamento de Cochabamba* of *Diccionario geográfico de la República de Bolivia.* La Paz: Oficina nacional de inmigración, estadistica y propaganda geográfica.
Municipalidad de Guayaquil. 1994. "Mercadeo de productos alimenticios para Guayaquil." 2 vols. Unpublished report.
Municipalidad de Guayaquil. 1995. *Historia de los servicios de recolección y disposición final de las basuras en la ciudad de Guayaquil desde 1988 hasta 1995.* Guayaquil: Dirección de aseo urbano y rural, Municipalidad de Guayaquil. Mimeograph. 65 pp. + anexos.
Murphy, Michael R., et al. 1997. "Tapir (*Tapirus*) enteroliths." *Zoo Biology* 16:427–33.
Murua, Martin de. 1987. *Historia general del Perú.* [1611]. Ed. Manuel Ballesteros. Madrid: Historia 16.
Nabhan, Gary Paul. 1989. *Enduring seeds: Native American agriculture and wild plant conservation.* Tucson: University of Arizona Press.

Nabhan, Gary Paul. 1992. "Native crops of the Americas: Passing novelties or lasting contributions to diversity?" Pp. 143–61 in *Chilies to chocolate: Food the Americas gave the world,* ed. Nelson Foster and Linda S. Cordell. Tucson: University of Arizona Press.

National Research Council. 1984. *Amaranths: Modern prospects for an ancient crop.* Washington, D.C.: National Academy Press.

National Research Council. 1989. *Lost crops of the Incas.* Washington, D.C.: National Academy Press.

Navarete, Carlos. 1987. "El hombre danta en la iconografía del formativo superior de Chiapas y Guatemala." *Anales de la Academia de geografía e historia de Guatemala* 61:219–34.

Netherly, Patricia. 1988. "From event to process: The recovery of late Andean organizational structure by means of Spanish colonial records." Pp. 257–75 in *Peruvian prehistory,* ed. R. Keatinge. New York: Cambridge University Press.

Neuendorff, G. H. 1940. "Säugetiere Spanischamerikas." *Wörter und Sachen,* n.s. 3:200–224.

Neuenschwander, Carlos. 1963. *Pantiacollo.* Lima: Organización peruana del libro.

Nietschmann, Bernard. 1979. *Caribbean edge.* Indianapolis: Bobbs-Merrill Co.

Nietzsche, Friedrich. 1967. *Beyond good and evil.* 4th ed. Trans. Helen Zimmern. London: Allen & Unwin.

Noll, Richard. 1985. "Mental imagery cultivation as a cultural phenomenon: The role of visions in shamanism." *Current Anthropology* 26:443–61.

Nordenskiöld, Erland. 1924. *Forschungen und Abenteuer in Südamerika.* Stuttgart: Strecker and Schröder Verlag.

Novillo Villarroel, A. 1928. *Totora: Notas sobre su pasado.* Cochabamba: Editorial Lopez.

Novoa, C., and J. C. Wheeler. 1984. "Lama [*sic*] and alpaca." Pp. 116–28 in *Evolution of domesticated animals,* ed. I. L. Mason. London and New York: Longman.

Nowak, Ronald M. 1991. *Walker's mammals of the world.* 2 vols. 5th ed. Baltimore: Johns Hopkins University Press.

Nystron, J. 1868. *Informe al supremo gobierno del Perú sobre una expedición al interior de la república.* Lima: Imp. Prugue.

Oberem, Udo. 1974. "Trade and trade goods in the Ecuadorian Montaña." Pp. 346–57 in *Native South Americans,* ed. Patricia J. Lyon. Boston: Little, Brown.

Ocaña, Diego de. 1987. *A traves de la América del Sur.* [1604]. Madrid: Historia 16.

Ogilvie, Alan G. 1922. *Geography of the Central Andes: A handbook to accompany the La Paz sheet of the map of Hispanic America in the millionth scale.* New York: American Geographical Society.

Olarte Estrada, J. 1985. "Agricultura y poblamiento en las vertientes de los Andes centrales." *Revista universitaria* (Cusco): 240–72.

Oldfield, Margery L., and Janis B. Alcorn, eds. 1991. *Biodiversity: Culture, conservation and ecodevelopment.* Boulder: Westview Press.

Oliver, Jose R. 1992. "Donald W. Lathrap: Approaches and contributions in New World archaeology." *Journal of the Steward Anthropological Society* 20(1-2):283–345.

Orbigny, Alcides d'. 1835. *Voyage dans l'Amérique meridionale.* 10 vols. Paris: Pitois Lerrault.

Ordoñez de Ceballos, Pedro. 1614. *Viage del mundo.* Madrid: Luis Sánchez.

Oricaín, Pablo José. 1906. "Compendio breve de discursos varios sobre diferentes materias y noticias geográficas compreensivas a este Obispado del Cuzco que claman remedios spritiuales, año 1790." Pp. 319–77 in vol. 11 of *Juicio de límites entre el Perú y Bolivia,* ed. V. M. Maurtua. Barcelona: Heinrich y Cia.

Orlove, Benjamin S. 1977. *Alpacas, sheep and men.* New York: Academic Press.

Orlove, Benjamin S. 1985. "The history of the Andes: A brief overview." *Mountain Research and Development* 5(1):45–60.

Orlove, Benjamin S., and Ricardo Godoy. 1993. "Putting race in its place: Order in colonial and postcolonial Peruvian geography." *Social Research* 60(2):301–36.

Orton, James. 1870. *The Andes and the Amazon, or, across the continent of South America.* New York: Harper & Bros.

Paige, D. M., E. Leonardo, A. Cordano, J. Nakushima, T. Adrianzen, and G. G. Graham. 1972. "Lactose intolerance in Peruvian children: Effect of age and early nutrition." *American Journal of Clinical Nutrition* 25(3):297–301.

Palacio Pimental, H. Gustavo. 1960. "Relaciones de trabajo entre el patrón y los colonos en los fundos de la provincia de Paucartambo." *Revista universtaria* (Cusco) 49:145–63.

Pareja, Wenceslao. 1916. "Algunos datos sobre la peste bubónica en Guayaquil." Pp. 398–418 in *Actas y trabajos del primer Congreso médico ecuatoriano (9–14 octubre 1915).* Guayaquil: Imprenta municipal.

Pareja, Wenceslao. 1926. *Reglamento especial para combatir la peste bubónica.* Guayaquil: Dirección de sanidad del Distrito Sur. 17 pp.

Parsons, James J. 1977. "Geography as discovery and exploration." *Annals of the Association of American Geographers* 67:1–16.

Parsons, James J. 1982. "The northern Andean environment." *Mountain Research and Development* 2(3):253–62.

Parsons, James J. 1991. "Giant American bamboo in the vernacular architecture of Colombia and Ecuador." *Geographical Review* 81(2):131–52.

Parsons, James J. 1992. "Southern blooms: Latin America and the world of flowers." *Queen's Quarterly* 99(3):542–61.

Parsons, James J. 1996. "Mr. Sauer and the writers." *Geographical Review* 86(1):22–41.

Pärssinen, Martti. 1992. *Tawantinsuyo: The Inca state and the political organization.* Helsinki: SHS.

Patiño, Victor Manuel. 1972. *Factores inhibitorios de la producción agropecuaria.* Cali: Imprenta departamental.

Patzelt, Erwin. 1979. *Fauna del Ecuador.* 2d ed. Quito: Editorial Las Casas.

Paz Soldan, Mariano Felipe. 1865. *Atlas geográfico del Perú.* Paris: Librería Fermin Didot.

Pearsall, Deborah M. 1978. "Paleoethnobotany in western South America: Progress and problems." Pp. 389–416 in *The nature and status of ethnobotany,* ed. Richard I. Ford. Anthropological Papers no. 67. Ann Arbor: Museum of Anthropology, University of Michigan.

Pearsall, Deborah M. 1992. "The origins of plant cultivation in South America." Pp. 173–205 in *The origins of agriculture: An international perspective,* ed. C. W. Cowan and P. J. Watson. Washington, D.C.: Smithsonian Institution Press.

Pease G. Y., Franklin. 1981. "Continuidad y resistencia de lo andino." *Allpanchis* 17-18:105–18.

Peattie, Donald C. 1930. *Flora of the Indiana Dunes.* Chicago: Field Museum of Natural History.

Peñalosa, B. de. 1629. *Libro de las cinco excelencias del español que despueblan a España.* Pamplona.

Pennant, Thomas. 1974. *Arctic zoology.* [1785]. New York: Arno Press.

Pérez, F. L. 1993. "Turf destruction by cattle in the high equatorial Andes." *Mountain Research and Development* 13(1):107–10.

Peterson, Mendel. 1975. *The funnel of gold.* Boston: Little, Brown.

Philo, Chris. 1995. "Animals, geography and the city: Notes on inclusions and exclusions." *Environment and Planning D: Society and Space* 13(6):655–81.

Pickersgill, Barbara, and A. H. Bunting. 1969. "Cultivated plants and the Kon-Tiki theory." *Nature* 222 (5190):225–27.

Pigafetta, Antonio. 1969. *The voyage of Magellan,* ed. S. E. Morison. Englewood Cliffs: Prentice-Hall.

Pigafetta, Antonio. 1972. *The first voyage round the world by Magellan.* Hakluyt Society no. 52. New York: Burt Franklin.

Pineo, Ronn F. 1990. "Misery and death in the Pearl of the Pacific: Health care in Guayaquil, Ecuador, 1870–1925." *Hispanic American Historical Review* 70(4): 609–37.

Pio Aza, José. 1926. "Mapa de las misiones de Santo Domingo de Urubamba y Madre de Dios." *Boletín de la Sociedad geográfica de Lima.*

Piperno, Dolores R. 1994. "On the emergence of agriculture in the New World." *Current Anthropology* 35(5):637–43.

Piso, Guilherme. 1957. *História natural e médica da India ocidental.* Rio de Janeiro: Ministério da educação e cultura.

Platt, Robert S. 1932. "Six farms in the Central Andes." *Geographical Review* 22:245–59.

Plewe, E. 1975–1976. "Lebensbilder: Edward Hahn 1856–1928." Pp. 239–46 in *Geographisches Taschenbuch,* ed. E. Meynen. Wiesbaden: Franz Steiner Verlag.

Plowman, Timothy. 1984. "The origin, evolution and diffusion of coca, *Erythroxylon* spp., in South and Central America." Pp. 125–63 in *Pre-Columbian plant migration,* ed. Doris Stone. Cambridge, Mass.: Peabody Museum of Archaeology and Ethnology, Harvard University.

Plucknett, Donald I., et al. 1987. *Gene banks and the world's food.* Princeton, N.J.: Princeton University Press.

Polentini Wester, Juan Carlos. 1979. *Por las rutas del Paititi.* Lima: Editorial Salesiana.

Polo de Ondegardo, Juan. 1990. *El mundo de los Incas.* [1560]. Ed. Laura Gonzalez and Alicia Alonso. Madrid: Historia 16.

Pomet, Pierre. 1725. *A compleat history of druggs.* 2d ed. London: R. Burwicke.

Poponoe, Wilson. 1921. "The frutilla or Chilean strawberry." *Journal of Heredity* 12: 457–66.

Porta, J. B. della. 1957. *Natural magick.* New York: Basic Books.

Pozorski, Thomas, and Sheila Pozorski. 1997. "Cherimoya and guanabana in the archaeological record of Peru." *Journal of Ethnobiology* 17(2):235–48.

Prance, Ghillean T. 1984. "The pejibaje, *Guilielma gasipaes* (HBK) Bailey, and the papaya, *Carica papaya* L." Pp. 85–104 in *Pre-Columbian plant migration,* ed. Doris Stone. Cambridge, Mass.: Peabody Museum of Archaeology and Ethnology, Harvard University.

Pratt, Harry D., Bayard F. Bjornson, and Kent S. Littig. 1976. *Control of domestic rats and mice.* Publication no. (CDC) 81-8141. Center for Disease Control, Public Health Service, U.S. Department of Health and Human Services. 47 pp.

Presta, A., and M. Del Rio. 1993. "Reflexiones sobre los Churumatas del Sur de Bolivia, siglos XVI–XVIII." *Histórica* (Lima) 17(2):223–37.

Preston, D. A. 1969. "The revolutionary landscape of highland Bolivia." *Geographical Journal* 135(1):1–16.

Preston, David, Mark Macklin, and Jeff Warburton. 1997. "Fewer people, less erosion: The twentieth century in southern Bolivia." *Geographical Journal* 163(2):198–205.

Price, Marie, and Martin Lewis. 1993. "The reinvention of cultural geography." *Annals of the Association of American Geographers* 83(1):1–17.

Prohens, Jaime, Juan J. Ruiz, and Fernando Nuez. 1996. "The pepino (*Solanum muricatum,* Solanaceae): A 'new' crop with a history." *Economic Botany* 50(4):355–68.

Pulgar Vidal, Javier. 1987. *Geografía del Perú.* 9th ed. Lima: Peisa.

Quintanilla P., Víctor G. 1983. "Observaciones fitogeográficas en el páramo de la Cordillera oriental del Ecuador." *Bulletin de l'Institut français d'études andines* 12(1-2):55–74.

Quiroga, R. 1906. "Informe de la Junta agropecuaria de Mizque." *Revista del Ministerio de colonización y agricultura* (La Paz) 2 (13-14-15):250–53.

Quiros, Carlos F., Andrea Epperson, Jinguo Hu, and Miguel Holle. 1996. "Physiological studies and determination of chromosome number in maca, *Lepidium meyenii* (Brassicaceae)." *Economic Botany* 50(2):216–23.

Rabinow, Paul. 1977. *Reflections on fieldwork in Morocco.* Berkeley: University of California Press.

Radin, Paul. 1957. *Primitive religion: Its nature and origin.* New York: Dover Publications.

Ráez, Nemesio A. 1898. "Tayacaja: Monografía de esta provincia, Departamento de Huancavelica." *Boletín de la Sociedad geográfica de Lima* 8(7-8-9):278–320.

Raimondi, Antonio. 1898. "Itinerario de los viajes de Raimondi en el Perú: Cuzco, Quispicanchi, Lucre, Pisac etc. hasta Marcapata." *Boletín de la Sociedad geográfica de Lima* 8(7-8-9):241–77.

Raimondi, Antonio. 1965. *El Perú.* 5 vols. Lima: Editores técnicos asociados.

Ramírez, Balthasar. 1936. "Descripción del reyno del Piru del sitio temple." [1597]. Pp. 1–122 in *Quellen zur Kulturgeschichte des präkolumbischen Amerika,* ed. H. Trimborn. Stuttgart: Strecker und Schröder Verlag.

Ramírez A., M. 1988. *Riqueza forestal de Chuquisaca.* Sucre: Proyecto SURE.

Ramírez del Aguila, Pedro. 1978. *Noticias políticas de Indias.* [1639]. Ed. Jaime Urioste Arana. Sucre: Extensión universitaria.

Ramírez Valverde, M. 1970. "Visita a Pocona." [1557]. *Historia y cultura* (Lima), no. 4:269–308.

Raymond, J. Scott. 1979. "A Huari ceramic tapir foot?" *Ñawpa Pacha* 17:81–86.

Raymond, J. Scott. 1992. "Highland colonization of the Peruvian montana in relation

to the political economy of the Huari empire." *Journal of the Steward Anthropological Society* 20(1-2):17–36.
Rea, J., and J. Leon. 1965. "La mauka (*Mirabilis expansa* Ruiz y Pavon) un aporte de la agricultura prehispánica de Bolivia." *Anales científicos de la Universidad nacional de agronomía* 3(1):38–41.
Real academia española. 1969. *Diccionario de autoridades.* [1737]. 3 vols. Madrid: Editorial Gredos.
Real academia española. 1984. *Diccionario de la lengua española.* 2 vols. 20th ed. Madrid: Espasa-Calpe.
Real academia española. 1992. *Diccionario histórico de la lengua española.* Madrid: Real academia española.
Reichel-Dolmatoff, Gerardo. 1985. "Tapir avoidance in the Colombian northwest Amazon." Pp. 107–44 in *Animal myths and metaphors in South America,* ed. Gary Urton. Salt Lake City: University of Utah Press.
"Relación de Guayaquil." 1994 [1772]. Pp. 495–501 in *Relaciones histórico-geográficas de la Audiencia de Quito (siglo XVI–XIX),* ed. Pilar Ponce Leiva. Quito: MARKA and Instituto de historia y antropología andina.
Renard-Casevitz, Marie-France, Thierry Saignes, and Gerald Taylor. 1984. *L'Inca, l'espagnol et les sauvages.* Paris: Éditions recherche sur les civilisations.
"Reparto de tierras en 1595." 1957. *Revista del Archivo histórico del Cuzco* 8:389–432.
Ressini, Edmundo. 1958. "Notas sobre el 'cebo.' " *Boletín forestal* (La Paz) 1:27–31.
Reumer, Jelle W. F. 1986. "Note on the spread of the black rat, *Rattus rattus.*" *Mammalia* 50(1):118–19.
Reynel, Carlos, and Carmen Felipe-Morales. 1987. *Agroforestería tradicional en los Andes del Perú.* Lima: Ministerio de agricultura/Instituto nacional forestal y de fauna.
Richter, M. 1981. "Klimagegensätze in Südperu und ihre Auswirkungen auf die Vegetation." *Erdkunde* 35:12–30.
Rick, John W. 1980. *Prehistoric hunters of the high Andes.* New York: Academic Press.
Rindos, David. 1984. *The origins of agriculture: An evolutionary approach.* Orlando: Academic Press.
Ríos Bordones, Waldo I., and Elias R. Pizarro Pizarro. 1988–1989. "Cultivos prehispánicos: El caso de la coca en el extremo norte de Chile." *Diálogo andino* (Arica) 7-8:83–99.
Ripinsky-Naxon, Michael. 1993. *The nature of shamanism.* Albany: State University of New York Press.
Risi, C. J., and N. W. Galway. 1984. "The Chenopodium grains of the Andes: Inca crops for modern agriculture." Pp. 145-216 in *Advances in applied botany,* ed. T. H. Coaker. London: Academic Press.
Rivero Luque, Victor. 1990. *La chaquitaqlla en el mundo andino.* Lima: Herrandina.
Robinson, Roy. 1965. *Genetics of the Norway rat.* Oxford: Pergamon Press.
Robledo, Luis. 1899. "La vía fluvial del Urubamba." *Boletín de la Sociedad geográfica de Lima* 8(8):417–49.
Roden, I. Y. 1988. "N. I. Vavilov: Geographer and explorer." *Soviet Geography* 29(7): 658–65.
Rodrigue, Christine M. 1992. "Can religion account for early animal domestication? A

critical assessment of the cultural geographic argument based on Near Eastern archaeological data." *The Professional Geographer* 44(4):417–30.

Rojas, M., and Gaetan Villavicencio. 1988. *El proceso urbano de Guayaquil, 1870–1980.* Guayaquil: CERG.

Rostworowski, María. 1990. "La visita de Urcos de 1562: Un kipu peruano." *Historia y cultura* (Lima) 20:295–317.

Rostworowski de Diez Canseco, María. 1970. "El repartimiento de doña Beatriz Coya en el valle de Yucay." *Historia y cultura* (Lima) 4:153–268.

Rostworowski de Diez Canseco, María. 1973. "Plantaciones prehispánicas de coca en el vertiente del Pacífico." *Revista del Museo nacional* (Lima) 39:193–224.

Rostworowski de Diez Canseco, María. 1981. *Recursos naturales renovables y pesca, siglos XVI y XVII.* Lima: Instituto de estudios peruanos.

Roth, Walter E. 1915. "An inquiry into animism and folk-lore of Guiana Indians." Pp. 103–386 in *Thirteenth annual report of the Bureau of American Ethnology.* Washington, D.C.: Smithsonian Institution.

Rouse, Irving, and José Maria Cruxent. 1963. *Venezuelan archaeology.* New Haven: Yale University Press.

Rundstrom, Robert A., and Martin S. Kenzer. 1989. "The decline of fieldwork in human geography." *Professional Geographer* 41(3):294–303.

Ruthsatz, Barbara. 1983. "Der Einfluss des Menschen auf die Vegetation semiarider bis arider tropischer Hochgebirge am Beispiel der Hochanden." *Berichte der deutschen Botanischen Gesellschaft* 96:535–76.

Safford, William E. 1917. "A forgotten cereal of ancient America." Pp. 12–30 in *Proceedings of the 19th International Congress of Americanists.* Washington, D.C.

Saignes, Thierry. 1980. "Una provincia andina a comienzos del siglo XVII: Pacajes según una relación inédita." *Historiografía y bibliografía americanista* 24:3–21.

Salaman, Redcliffe N. 1949. *The history and social influence of the potato.* Cambridge: Cambridge University Press.

Salazar, R. L. 1918. "El ají: Sus variedades y cultivo." *Boletín de la dirección nacional de estadística y estudios geográficos* (La Paz) 2(3-4-5):13–30.

Salis, Annette. 1985. *Cultivos andinos: Alternativa alimentaria popular?* Cusco: Centro de estudios rurales andinos "Bartolomé de las Casas" and CEDEP-AYLLU.

Salomon, Frank. 1985. "The historical development of Andean ethnology." *Mountain Research and Development* 5(1):79–98.

Sal y Rosas, Federico. 1970. "Indicios de la epilepsia en el Perú antiguo." *Revista de neuro psiquiatría* (Lima) 33:31–44.

Sánchez-Parga, J., and R. Pineda. 1985. "Los yachac de Ilumán." *Cultura* (Quito) 7(21):511–82.

Santillan, Hernando de. 1968. "Relación del origen, descendencia, política y gobierno de los Incas." [1564]. Pp. 97–149 in *Crónicas peruanas de interés indígena,* ed. Francisco Esteve Barba. Madrid: Ediciones Atlas.

Santillana Cantella, Tomás. 1989. *Los viajes de Raimondi.* Lima: Occidental Petroleum Company of Peru.

Sanz, F. P. 1954. "Descripción de las provincias de la Audiencia de Charcas, 1780–1781." *Boletín de la Sociedad geográfica Sucre* 45(441):123–77.

Sarabia Viejo, Ma. Justina, ed. 1989. *Francisco de Toledo: Disposiciones gubernativas para el Virreinato del Peru, 1575–1580.* Sevilla: EEHA/CSIC.

Sarmiento de Gamboa, Pedro. 1960. "Histórica índica." Pp. 193–279 in vol. 4 of *Obras completas del Inca Garcilaso de la Vega,* ed. Carmelo Saenz de Santa María. Madrid: Ediciones Atlas.

Sauer, Carl O. 1920. "The geography of the Ozark highland of Missouri." *The Geographic Society of Chicago, Bulletin No. 7,* 1–245.

Sauer, Carl O. 1936. "American agricultural origins: A consideration of nature and culture." Pp. 279–97 in *Essays in anthropology presented to A. L. Kroeber in celebration of his sixtieth birthday, June 11, 1936.* Berkeley: University of California Press.

Sauer, Carl O. 1938. "Theme of plant and animal destruction in economic history." *Journal of Farm Economics* 20:765–75.

Sauer, Carl O. 1941. "Forward to historical geography." *Annals of the Association of American Geographers* 31:1–24.

Sauer, Carl O. 1950. "Cultivated plants of South and Central America." Pp. 487–543 in vol. 6 of *Handbook of South American Indians,* ed. Julian Steward. Bureau of American Ethnology, Bulletin 143. Washington, D.C.: Smithsonian Institution.

Sauer, Carl O. 1952. *Agricultural origins and dispersals.* New York: American Geographical Society.

Sauer, Carl O. 1956. "The education of a geographer." *Annals of the Association of American Geographers* 46:287–99.

Sauer, Carl O. 1959. "Age and area of American cultivated plants." *Actas del XXIII Congreso internacional de Americanistas* (San José, Costa Rica) 1:215–29.

Sauer, Carl O. 1962. "Seashore—primitive home of man?" *Proceedings of the American Philosophical Society* 106:41–47.

Sauer, Carl O. 1965. "Cultural factors in plant domestication in the New World." *Euphytica* 14:301–6.

Sauer, Carl O. 1969. *Agricultural origins and dispersals: The domestication of animals and foodstuffs.* 2d ed. Cambridge, Mass.: MIT Press.

Sauer, Carl O. 1972. *Seeds, spades, hearths, and herds: The domestication of animals and foodstuffs.* Reprint of 2d ed. of *Agricultural origins and dispersals.* Cambridge, Mass.: MIT Press.

Sauer, Jonathan D. 1950. "The grain amaranths: A survey of their history and cultivation." *Missouri Botanical Garden Annals* 37:561–626.

Sauer, Jonathan D. 1964. "Revision of Canavalia." *Brittonia* 16(2):106–81.

Sauer, Jonathan D. 1967. "The grain amaranths and their relatives: A revised taxonomic and geographic survey." *Missouri Botanical Garden Annals* 54:103–37.

Sauer, Jonathan D. 1993. *Historical geography of crop plants: A select roster.* Boca Raton: CRC Press.

Schauenberg, Paul. 1969. "Contribution à l'étude du tapir pinchaque." *Revue suisse de zoologie* 76:211–56.

Schell, R. 1987. *La flore et la végétation de l'Amérique tropicale.* Vol. 2. Paris: Masson.

Schjellerup, Inge. 1989. *Children of the stones: Hijos de las piedras: A report on the agriculture in Chuquibamba, a district in north-eastern Peru.* Copenhagen: Royal Danish Academy of Sciences and Letters.

Schlaifer, Michel. 1993. "Las especies nativas y la deforestación en los Andes: Una vision histórica, social y cultural en Cochabamba, Bolivia." *Bulletin de l'Institut français d'études andines* 22(2):585–610.

Schmeda-Hirschmann, Guillermo. 1995. "*Madia sativa,* a potential oil crop of central Chile." *Economic Botany* 49(3):257–59.

Schneble, H. 1987. *Krankheit der ungezahlten Namen: Ein Beitrag zur Sozial-, Kultur- und Medizingeschichte der Epilepsie anhand ihrer Benennungen vom Altertum bis zur Gegenwart.* Bern: Huber.

Schneider, Wolfgang. 1968. *Tierische Drogen: Sachwörterbuch zur Geschichte der pharmazeutischen Zoologie.* Frankfurt am Main: Govi-Verlag.

Schulze, J. C., and R. Casanovas S. 1988. *Tierra y campesinado en Potosí y Chuquisaca.* La Paz: CEDLA.

Schurz, W. L. 1921. *Bolivia: A commercial and industrial handbook.* U.S. Department of Commerce. Special Agents Series no. 208, Washington, D.C.: Government Printing Office.

Schwabe, Calvin W. 1979. *Unmentionable cuisine.* Charlottesville: University Press of Virginia.

Schwalm, H. 1927. "Klima, Besiedlung und Landwirtschaft in den Peru- und bolivianische Anden." *Ibero-Americanishes Archiv* 2:17–74, 150–96.

Seamon, David. 1978. "Goethe's approach to the natural world: Implications for environmental theory and education." Pp. 238–50 in *Humanistic geography: Prospects and problems,* ed. David Ley and Marwyn S. Samuels. Chicago: Maaroufa Press.

Seeger, Anthony. 1981. *Nature and society in central Brazil: The Suya Indians of Mato Grosso.* Cambridge, Mass.: Harvard University Press.

Seibert, Paul. 1983. "Human impact on landscape and vegetation in the Central High Andes." Pp. 261–76 in *Man's impact on vegetation,* ed. W. Holzner, M. Werger, and I. Ikusima. The Hague: W. Junk.

Seibert, Paul. 1994. "The vegetation of the settlement area of the Callawaya people and the Ulla-Ulla highlands in the Bolivian Andes." *Mountain Research and Development* 14(3):189–211.

Seminario Cunya, Juan. 1993. "Aspectos etnobotánicos y morfológicos del chago, miso o mauca (*Mirabilis expansa* R. y P.) en el Peru." *Boletín de Lima* 86:71–79.

Senanayake, N., and G. C. Román. 1993. "Epidemiology of epilepsy in developing countries." *Bulletin of the World Health Organization* 71:247–58.

Sherbondy, J. E. 1988. "Mallki: Ancestros y cultivo de árboles en los Andes." Pp. 101–35 in *Sociedad andina: Pasado y presente,* ed. R. Matos Mendieta. Lima: FOMCIENCIAS.

Shweder, Richard A. 1991. *Thinking about cultures: Explorations in cultural psychology.* Cambridge, Mass.: Harvard University Press.

Sick, Wolf-Dieter. 1969. "Geographical substances." Pp. 449–74 in *Biogeography and ecology in South America,* ed. E. J. Fittkau et al. The Hague: W. Junk.

Silistreli, U. 1989. "Köşk Hüyük figürin ve heykelcikleri." *Belleten* 53:497–504.

Silverblatt, Irene. 1987. *Moon, sun and witches: Gender ideologies and class in Inca and colonial Peru.* Princeton, N.J.: Princeton University Press.

Simmonds, N. W. 1976. "Quinoa and relatives." Pp. 29–30 in *Evolution of crop plants,* ed. N. W. Simmonds. London: Longman.

Simoons, Frederick J. 1970. "Primary lactose intolerance and the milking habit: A problem in biological and cultural interrelations, II—A culture historical hypothesis." *American Journal of Digestive Disease* 15(8):695–710.

Simoons, Frederick J. 1971. "The antiquity of dairying in Asia and Africa." *Geographical Review* 61:431–39.

Simoons, Frederick J. 1973. "New light on ethnic differences in adult lactose intolerance." *American Journal of Digestive Diseases* 18(7):595–611.

Simoons, Frederick J., and Elizabeth S. Simoons. 1968. *A ceremonial ox of India.* Madison: University of Wisconsin Press.

Simpson, Beryl B. 1979. *A revision of the genus* Polylepis *(Rosaceae: Sanguisorbeae).* Washington, D.C.: Smithsonian Institution Press.

Smith, Bruce D. 1989. "Origins of agriculture in eastern North America." *Science* 246:1566–71.

Smith, Bruce D. 1995. *The emergence of agriculture.* New York: Scientific American Library.

Smith, C. Earle, Jr. 1980a. "Plant remains from Guitarrero Cave." Pp. 87–120 in *Guitarrero Cave: Early man in the Andes,* ed. T. F. Lynch. New York: Academic Press.

Smith, C. Earle, Jr. 1980b. "Vegetation and land use near Guitarrero Cave." Pp. 65–83 in *Guitarrero Cave: Early man in the Andes,* ed. T. F. Lynch. New York: Academic Press.

Sperling, Calvin R., and Steven R. King. 1990. "Andean tuber crops: Worldwide potential." Pp. 428–35 in *Advances in new crops,* ed. J. Janick and J. E. Simon. Portland, Ore.: Timber Press.

Speth, William W. 1977. "Carl Ortwin Sauer on destructive exploitation." *Biological Conservation* 11:145–60.

Speth, William W. 1987. "Historicism: The disciplinary world view of Carl O. Sauer." Pp. 11–39 in *Carl O. Sauer: A tribute,* ed. Martin S. Kenzer. Corvallis: Oregon State University Press.

Speth, William W. 1993. "Carl O. Sauer's use of geography's past." *Yearbook of the Association of Pacific Coast Geographers* 55:37–65.

Starn, Orin. 1991. "Missing the revolution: Anthropologists and the war in Peru." *Cultural Anthropology* 6:63–91.

Starn, Orin. 1994. "Rethinking the politics of anthropology: The case of the Andes." *Current Anthropology* 35(1):13–38.

Stebbins, G. Ledyard. 1978. "Edgar Anderson: November 9, 1897–June 18, 1969." *Biographical Memoirs: National Academy of Sciences–Washington* 49:2–23.

Stern, Margaret J. 1995. "An inter-Andean forest relict: Vegetation change in Pasochoa Volcano, Ecuador." *Mountain Research and Development* 15(4):339–48.

Stevens, Anthony. 1993. *The two million year old self.* College Station: Texas A & M University Press.

Steward, Julian, ed. 1944–1950. *Handbook of South American Indians.* 6 vols. Bureau of American Ethnology, Bulletin 143. Washington, D.C.: Smithsonian Institution.

Steward, Julian, and Louis C. Faron. 1959. *Native peoples of South America.* New York: McGraw-Hill.

Stierlin, Henri. 1984. *Art of the Incas and its origins.* New York: Rizzoli.

Stoddart, D. R. 1987. "To claim the high ground: Geography for the end of the century." *Transactions of the Institute of British Geographers* 12(3):327–36.
Stone, Peter B., ed. 1992. *The state of the world's mountains: A global report.* London: Zed Books Ltd.
Stouse, P. A. D. 1970. "The distribution of llamas in Bolivia." *Proceedings of the American Association of American Geographers* 2:136–39.
Strube Erdmann, León. 1963. *Vialidad imperial de los Incas.* Serie histórica 33. Córdoba: Instituto de estudios americanistas, Universidad nacional de Córdoba.
Sumar, J. 1988. "Present and potential role of South American camelids in the high Andes." *Outlook on Agriculture* 17(1):23–29.
Sumar, Luis, et al. 1992. "Grain amaranth research in Peru." *Food Reviews International* 8(1):87–124.
Super, John C. 1994. "History, Indians and university reforms in Cuzco." *The Historian* 56(2):325–38.
Sutter, Patrick de. 1985. "Arquitectura andina tradicional y sus problemas." *Cultura* (Quito) 7(21):145–214.
Tamayo Herrera, José. 1978. *Historia social del Cuzco republicano.* Lima: Industrialgrafica.
Tello, Julio C. 1930. "Andean civilization: Some problems of Peruvian archaeology." *Proceedings of the International Congress of Americanists* 23:259–90.
Temkin, O. 1971. *The falling sickness: A history of epilepsy from the Greeks to the beginning of modern neurology.* Baltimore: Johns Hopkins University Press.
Theroux, Paul. 1979. *The old Patagonian express: By train through the Americas.* Boston: Houghton Mifflin.
Thiermann, Alejandro B. 1977. "Incidence of leptospirosis in the Detroit rat population." *American Journal of Tropical Medicine and Hygiene* 26:970–74.
Thomas, Oldfield. 1921. "Report on the mammalia collected by Mr. Edmund Heller." *Proceedings U.S. National Museum* 58(2333):21–249.
Tissot, Samuel. 1770. *Traité de l'épilepsie.* Paris: Didot.
Tohme, Joseph, et al. 1995. "Variability in Andean nuña common beans (*Phaseolus vulgaris,* Fabaceae)." *Economic Botany* 49(1):78–95.
Torrance, Robert M. 1994. *Spiritual quest: Transcendance in myth, religion and science.* Berkeley: University of California Press.
Torres, B. de. 1974. *Crónica agustina del Perú.* 3 vols. Ed. I. Prado Pastor. Lima: Universidad nacional mayor de San Marcos.
Torres, R. 1981. *El control de los vectores.* Sucre: Servicio nacional para la erradicación de malaria. Typescript. 23 pp.
Torres de Mendoza, L., ed. 1868. "Descripción de la villa de Santiago de la Frontera de Tomina y su distrito." *Colección de documentos inéditos* 9:317–46.
Tosi, Joseph A. 1960. *Zonas de vida natural en el Perú: Memoria explicativa sobre el mapa ecológico del Perú.* Washington, D.C.: Instituto interamericano de ciencias agrícolas de la OEA zona andina.
Tovar, Oscar. 1973. "Comunidades vegetales de la reserva nacional de vicuñas de Pampa Galeras, Ayacucho, Perú." *Publicaciones del Museo de historia nacional "Javier Prado,"* Serie B (Botánica) 27:1–32.
Towle, Margaret A. 1961. *The ethnobotany of pre-Columbian Peru.* Chicago: Aldine.

Treacy, John M. 1994. *Las chacras de Coporaque: Andenería y riego en el valle del Colca.* Lima: Instituto de estudios peruanos.
Trimble, Michael R. 1991. *The psychoses of epilepsy.* New York: Raven Press.
Troll, Carl. 1929. "An expedition to the Central Andes, 1926–28." *Geographical Review* 19:234–47.
Troll, Carl. 1931. "Die geographischen Grundlagen der andinen Kulturen und des Inkareiches." *Ibero-Amerikanishces Archiv* 5:258–94.
Troll, Carl. 1935. "Gedanken und Bemerkungen zur ökologischen Pflanzengeographie." *Geographische Zeitschrift* 41:380–88.
Troll, Carl. 1952. "Die Lokalwinde der Tropengebirge und ihr Einfluß auf Niederschlag und Vegetation." Pp. 124–82 in *Studien zur Klima- und Vegetationskunde der Tropen,* ed. C. Troll. Bonner geographische Abhandlungen 9. Bonn: F. Dümmlers Verlag.
Troll, Carl. 1959. *Die tropische Gebirge.* Bonner geographische Abhandlungen 28. Bonn: F. Dümmlers Verlag.
Troll, Carl. 1968. "The cordilleras of the tropical Americas." Pp. 15–55 in *Geoecology of the mountain regions of the tropical Americas,* ed. C. Troll. Colloquium Geographicum 9. Bonn: Geographisches Institut.
Troll, Carl. 1985. *Tagebücher der Reisen in Bolivien 1926/27.* Ed. Felix Monheim. Stuttgart: Franz Steiner Verlag.
Tschudi, J. J. von. 1844. *Untersuchungen über die Fauna peruana.* St. Gallen: Sheitlin v. Zollikofer.
Tschudi, J. J. von. 1849. *Travels in Peru during the years 1838–1842.* New York: Putnam.
Tschudi, J. J. von. 1885. "Das Lama (Auchenia Lama Fisch) in seinen Beziehungen zum altperuanischen Volksleben." *Zeitschrift für Ethnologie* 17:193–209.
Tyler, Stephen A. 1987. *The unspeakable.* Madison: University of Wisconsin Press.
Ugent, Donald, Sheila Pozorski, and Thomas Pozorski. 1982. "Archaeological potato tuber remains form the Casma Valley of Peru." *Economic Botany* 38:417–32.
Uhle, Max. 1909. "La esfera de influencias del país de los Incas." *Revista histórica* (Lima) 4:5–40.
Ulloa, Antonio de. 1990. *Viaje a la América meridional.* [1748]. Madrid: Historia 16.
Ulloa Ulloa, Carmen, and Peter Moller Jorgensen. 1993. *Arboles y arbustos de los Andes del Ecuador.* Aarhus: Botanical Institute, University of Aarhus.
UNESCO. 1975. *Informe final: Reunión regional sobre investigación ecológica integrada y formación de especialistas en el area amdina, La Paz, 10–15 junio de 1974.* Serie Informes del MAB no. 23. Paris: UNESCO.
UNESCO. 1981. *Carte de la végétation d'Amérique du Sud: Notice explicative.* Paris: UNESCO.
Unidad sanitaria de Cochabamba. 1968. *Primer seminario boliviano sobre coordinación e integración de los programas de erradicación de la malaria y viruela en los servicios generales de salud.* June 19–22, 1968. Cochabamba: Unidad sanitaria de Cochabamba. Typescript. 30 pp.
Unzueta Q., O. 1975. *Mapa ecológico de Bolivia: Memoria explicativa.* La Paz: República de Bolivia, Ministerio de asuntos campesinos y agropecuarios.
Urioste, Jaime. 1983. *Hijos de Pariya Qapa: La tradición oral de Waru Chiri.* Syracuse: Syracuse University Press.

Urlegaza Uranga, E. 1934. "Reseña histórica del paludismo." *Trabajo de la cátedra de historia crítica de la medicina* (Madrid), no. 2:183–88.

Urquidi, G. 1954. *Monografía del Departamento de Cochabamba.* Cochabamba: Municipalidad de Cochabamba.

Vaca Guzman, G. 1915. "El agua: Aguas potables de Sucre." *Revista del I. Municipalidad de Sucre* 33:554–88.

Valcárcel, Luis. 1978. *Historia del Perú antiguo.* 3 vols. Lima: Editorial Juan Mejía Baca.

Valdizán, Hermilio, and Angel Maldonado. 1922. *La medicina popular peruana.* 2 vols. Lima.

Valenzuela, A. J. 1932. "Razones histórico-epidemiológicas en contra del *Leptospira icteroide.*" *Anales de la Sociedad médico-quirúrgica del Guayas* 12(1):1–4.

Van der Hammen, T., and G. W. Noldus. 1985. "Pollen analysis of the Telarmachay rockshelter (Peru)." Pp. 379–87 in vol. 1 of *Telarmachay: Chasseurs et pasteurs préhistoriques des Andes,* ed. Danièle Lavallée. Paris: Editions recherche sur les civilisations.

Van der Plank, J. E. 1968. *Disease resistance in plants.* New York: Academic Press.

Vargas C., César. 1938. "El *Podocarpus glomerata* Don (intimpa) y la silvicultura nacional." *Actas de la Academia de ciencias exactas fisicas y naturales de Lima* 1(1):27–31.

Vargas C., César. 1946. *Diez años al servicio de la botánica en la Universidad del Cuzco.* Cuzco: Universidad nacional del Cuzco, Cuzco.

Vargas Machuca, Bernardo de. 1892. *Milicia y descripción de las Indias.* [1599]. Madrid: Pedro Madrigal.

Vavilov, N. I. 1926. "Studies on the origin of cultivated plants." *Bulletin of Applied Botany* (Leningrad) 16(2):1–248.

Vavilov, N. I. 1931. "Mexico and Central America as the principle centre of origin of cultivated plants of the New World." *Bulletin of Applied Botany* (Leningrad) 26:135–99.

Vavilov, N. I. 1992. *Origin and geography of cultivated plants.* Trans. Doris Löve. New York: Cambridge University Press.

Vayda, Andrew. 1986. "Holism and individualism in ecological anthropology." *Reviews in Anthropology* 13:295–313.

Vázquez de Espinosa, Antonio. 1623. *Tratado verdadero del viage y navigación deste año de 1622 que hice la flota de Nueva España y Honduras.* Málaga: n.p.

Vázquez de Espinosa, Antonio. 1948. *Compendio y descripción de las Indias occidentales.* [1620]. Smithsonian Miscellaneous Collections no. 108. Washington, D.C.: Government Printing Office.

Vázquez de Espinosa, Antonio. 1969. *Compendio y descripción de las Indias occidentales.* [1628]. Madrid: Ediciones Atlas.

Velasco, Juan de. 1977. *Historia del reino de Quito en la América meridional.* [1682]. 3 vols. Quito: Editora casas de la cultura ecuatoriana.

Vélez, Francisco. 1613. *Historia de los animales mas recebidos en el uso de medicina.* Madrid: Imprenta Real.

Vellard, Jean. 1963. *Civilisation des Andes.* Paris: Gallimard.

Venero G., J. L., and H. Macedo R. 1983. "Relictos de bosques en la puna del Perú." *Boletín de Lima* 5(30):19–26.

Venero, José Luis, and Rayner Hostnig. 1987. "Un rodal de *Puya raimondi* en Andahuaylas (Apurimac)." *Boletín de Lima* 9(54):5–7.

Venero González, José Luis, and Alfredo Tupayachi Herrera. 1989. "Ampay: Santuario nacional." *Boletín de Lima* 11(61):57–64.

Viedma, F. de. 1969. *Descripción geográfica y estadística de la Provincia de Santa Cruz de la Sierra.* Cochabamba: Editorial "Los Amigos del Libro."

Villagomez, Pedro. 1919. *Exortaciones e instrucción acerca de las idolatrías de los indios del Arzobispado de Lima.* [1589]. Colección de libros y documentos referentes a la historia del Perú 12. Lima: San Marti.

Villanueva Urteaga, Horacio. 1970. "Documentos sobre Yucay en el siglo XVI." *Revista del Archivo histórico del Cuzco* 13:1–148.

Villanueva Urteaga, Horacio, ed. 1982. *Cuzco 1689: Economía y sociedad en le sur andino: Informes de los párracos al obispo Mollinedo.* Cusco: Centro de estudios rurales andinos "Bartolomé de las Casas."

Viscarra, E. 1967. *Apuntes para la historia de Cochabamba: Casos históricos y tradiciones de la ciudad de Mizque.* [1907]. Cochabamba: Editorial "Los Amigos del Libro."

Visvanathan, Shiv. 1996. "Footnotes to Vavilov: An essay on gene diversity." Pp. 306–39 in *Decolonizing knowledge: From development to dialogue,* ed. Frédérique Apffel-Marglin and Stephen A. Marglin. Oxford: Clarendon Press.

Vokral, Edita. 1989. *Küchenorganisation und Agrarzyklos auf dem Altiplano.* Bonn: Holos Verlag.

Vollmer, Günter. 1967. *Bevölkerungspolitik und Bevölkerungsstruktur im vizekonigreich Peru zu Ende der Kolonialzeit (1741–1821).* Bad Homburg: Verlag Gehlen.

Von Lichtenberg, Franz. 1991. *Pathology of infectious diseases.* New York: Raven Press.

Wachtel, N. 1982. "The mitimas of the Cochabamba Valley: The colonization policy of Huayna Capac." Pp. 199–235 in *The Inca and Aztec states, 1400–1800: Anthropology and history,* ed. G. A. Collier, R. I. Rosaldo, and J. D. Worth. New York: Academic Press.

Wachtel, Nathan, 1992. "Note sur le problème des identités collectives dans les Andes meridionales." *L'Homme* 32(2-3-4):39–52.

Wagner, Philip L. 1970. "The hemisphere revisited." *East Lakes Geographer* 6:26–36.

Wallach, Bret. 1998. "In Memoriam: James J. Parsons, 1915–1997." *Annals of the Association of American Geographers* 88(2):316–28.

Walsh, Kathleen D. 1989. "Cult of the moon: Religious and ritual in the domestication and early history of *Bos primigenius* in the Mediterranean region." Ph.D. dissertation, University of California, Berkeley.

Walter, Heinrich, 1971. *Ecology of tropical and subtropical vegetation.* Trans. D. Mueller-Dombois. New York: Van Nostrand Reinhold.

Walter, Heinrich. 1979/1985. *Vegetation of the earth and ecological systems of the geobiosphere.* 2d and 3d eds. Berlin/New York: Springer-Verlag.

Walton, W. 1811. *An historical and descriptive account of the four species of Peruvian sheep called carneros de la tierra.* London: Longman.

Warf, Barney. 1992. "The third plague pandemic." *Ontario Geography* 39:13–22.

Weatherford, Jack. 1988. *Indian givers: How the Indians of the Americas transformed the world.* New York: Crown Publishers.

Weberbauer, A. 1911. *Die Pflanzenwelt der peruanischen Anden.* Leipzig.

Weberbauer, A. 1936. "Phytogeography of the Peruvian Andes." Pp. 13–81 in *Flora of Peru,* ed. J. F. MacBride. Publication 351. Chicago: Field Museum of Natural History.

Webster, Steven. 1973. "Pastoralism in the south Andes." *Ethnology* 2(2):115–34.

Weddell, H. A. 1853. *Voyages dans le nord de la Bolivie.* Paris: Bertrand.

Wernsdorfer, W. H. 1980. "The importance of malaria in the world." Pp. 1–93 in vol. 1: *Epidemiology, chemotherapy, morphology and metabolism* of *Malaria,* ed. Julius P. Kreier. New York: Academic Press.

West, Robert C. 1979. *Carl Sauer's fieldwork in Latin America.* Dellplain Latin American Studies 3. Ann Arbor: University Microfilms International.

West, Robert C., ed. 1982. *Andean reflections.* Dellplain Latin American Studies 11. Boulder: Westview Press.

West, Robert C., ed. 1990. *Pioneers of modern geography: Translations pertaining to German geographers of the late nineteenth and early twentieth centuries.* Geoscience and Man 28. Baton Rouge: Louisiana State University.

West, T. L. 1987. "The burning bush." Pp. 151–69 in *Arid land use strategies and risk management in the Andes,* ed. D. L. Browman. Boulder: Westview Press.

Wheeler, Jane C. 1984. "On the origin and early development of camelid pastoralism in the Andes." Pp. 395–410 in vol. 3: *Early herders and their flocks* of *Animals and archaeology,* ed. J. Clutton-Brock and C. Grigson. Oxford: BAR International Series 202.

Wheeler, Jane C. 1995. "Evolution and present situation of the South American Camelidae." *Biological Journal of the Linnean Society* 54:271–95.

Wheeler, Jane C., A. J. F. Russel, and Hilary Redden. 1995. "Llamas and alpacas: Pre-Conquest breeds and post-Conquest hybrids." *Journal of Archaeological Science* 22:833–40.

White, Richard. 1995. "Are you an environmentalist or do you work for a living? Work and nature." Pp. 177–85 in *Uncommon ground: Toward reinventing nature,* ed. William Cronon. New York: W. W. Norton.

White, Stuart. 1985. "Relations of subsistence to the vegetation mosaic of Vilcabamba, southern Peruvian Andes." *Yearbook, Conference of Latin Americanist Geographers* 11:5–12.

White, Stuart, and Fausto Maldonado. 1991. "The use and conservation of natural resources in the Andes of southern Ecuador." *Mountain Research and Development* 11(1):37–55.

Whitten, Norman E. 1976. *Sacha Runa: Ethnicity and adaptation of Ecuadorian jungle Quichua.* Urbana: University of Illinois Press.

Whymper, Edward. 1892. *Travels amongst the great Andes of the equator.* New York: Charles Scribner's Sons.

Wilkes, Garrison. 1991. "In situ conservation of agricultural systems." Pp. 86–101 in *Biodiversity: Culture, conservation and ecodevelopment,* ed. M. L. Oldfield and J. B. Alcorn. Boulder: Westview Press.

Williams, J. Trevor. 1988. "Identifying and protecting the origins of our food plants." Pp. 240–47 in *Biodiversity,* ed. E. O. Wilson. Washington, D.C.: National Academy Press.

Williams, Michael. 1994. "The relations of environmental history and historical geography." *Journal of Historical Geography* 20(1):3–21.

Wilson, Edward O. 1978. *On human nature.* Cambridge, Mass.: Harvard University Press.

Wilson, Edward O. 1998. *Consilience: The unity of knowledge.* New York: Knopf.

Wing, Elizabeth. 1986. "Domestication of Andean mammals." Pp. 246–64 in *High altitude tropical biogeography,* ed. F. Vuilleumier and M. Monasterios. New York: Oxford University Press.

Winterhalder, Bruce, Robert Larsen, and R. Brooke Thomas. 1974. "Dung as an essential resource in a highland Peruvian community." *Human Ecology* 2(2):89–104.

Winterhalder, Bruce, and R. Brooke Thomas. 1978. *Geoecology of southern highland Peru: A human adaptation perspective.* Occasional paper 27. Boulder: Institute of Arctic and Alpine Research, University of Colorado.

Wood, C. S. 1975. "New evidence for a late introduction of malaria into the New World." *Current Anthropology* 16:93–104.

Wood, R. W. 1988. *Quinoa: The supergrain.* Tokyo: Japan Publications.

Worster, Donald. 1995. "Nature and the disorder of history." Pp. 65–85 in *Reinventing nature? Responses to postmodern deconstruction,* ed. Michael Soulé and Gary Lease. Washington, D.C.: Island Press.

Woytkowski, Felix. 1978. *Peru, my unpromised land.* Trans. Z. Zienkiewicz. Warsaw: National Center for Scientific, Technical and Economic Information.

Wright, Angus. 1984. "Innocents abroad: American agricultural research in Mexico." Pp. 135–51 in *Meeting the expectations of the land: Essays in sustainable agriculture and stewardship,* ed. Wes Jackson, Wendell Berry, Bruce Coleman. San Francisco: North Point Press.

Wright, H. E., Jr. 1983. "Vegetational and climatic changes in the Peruvian Andes." *National Geographic Society Research Reports* 19:735–45.

Wunder, Sven. 1996. "Deforestation and the uses of wood in the Ecuadorian Andes." *Mountain Research and Development* 16(4):367–82.

Young, Kenneth R. 1993: "National park protection in relation to the ecological zonation of a neighboring community: An example form northern Peru." *Mountain Research and Development* 13(3):267–80.

Young, Kenneth R., and Blanca Leon. 1995. "Distribution and conservation of Peru's montane forests: Interactions between the biota and human society." Pp. 363–76 in *Tropical montane cloud forests,* ed. L. S. Hamilton, J. O. Juvik, and F. W. Scatena. New York: Springer-Verlag.

Zalles Flossbach, Teresa. 1984. "El taladro del eucalypto (*Phorancantha semipunctata* Fabr.): Peligro para las plantaciones de este género en el Valle de Cochabamba?" *Ecología en Bolivia* 5:39–51.

Zárate, Agustin de. 1947. "Historia del descubrimiento y conquista de la Provincia del Perú." [1555]. Pp. 459–574 in vol. 2 of *Historiadores primitivos de Indias,* ed. Enrique de Vedia. Madrid: Ediciones Atlas.

Zárate, Sergio. 1997. "Domestication of cultivated *Leucaena* (Leguminosae) in Mexico: The sixteenth century documents." *Economic Botany* 51(3):238–50.

Zardini, Elsa. 1992. "*Madia sativa* Mol. (Asteraceae-Heliantheae-Madinae): An ethnobotanical and geographical disjunct." *Economic Botany* 46(1):34–44.

Zerries, Otto. 1990. "Die Rolle des Tapirs bei außerandinen Indianern." Pp. 589–626 in *Circumpacifica: Festschrift für Thomas S. Barthel,* ed. Bruno Illius and Matthias Laubscher. Frankfurt am Main: Peter Lang.

Zeven, A. C. 1973. "Dr. Th. H. Engelbrecht's views on the origin of cultivated plants." *Euphytica* 22:279–86.

Zimmerer, Karl S. 1992. "Biological diversity and local development: 'Popping beans' in the Central Andes." *Mountain Research and Development* 12(1):47–61.

Zimmerer, Karl S. 1993. "Agricultural biodiversity and peasant rights to subsistence in the Central Andes during Inca rule." *Journal of Historical Geography* 19(1):15–32.

Zimmerer, Karl S. 1996a. "Ecology as cornerstone and chimera in human geography." Pp. 161–88 in *Concepts in human geography,* ed. C. Earle, K. Mathewson, and M. Kenzer. Totowa: Rowman and Littlefield.

Zimmerer, Karl S. 1996b. *Changing fortunes.* Berkeley: University of California Press.

Zimmerer, Karl S. 1998. "The ecogeography of Andean potatoes." *BioScience* 48(6): 445–54.

Zinsser, Hans. 1934. *Rats, lice and history.* Boston: Atlantic Monthly Press.

Index